Theory and Application of Fixed Point

Theory and Application of Fixed Point

Editors
Erdal Karapinar
Juan Martínez-Moreno
Inci M. Erhan

MDPI • Basel • Beijing • Wuhan • Barcelona • Belgrade • Manchester • Tokyo • Cluj • Tianjin

Editors
Erdal Karapinar
China Medical University
Taiwan

Juan Martínez-Moreno
University of Jaén
Spain

Inci M. Erhan
Atılım University
Turkey

Editorial Office
MDPI
St. Alban-Anlage 66
4052 Basel, Switzerland

This is a reprint of articles from the Special Issue published online in the open access journal *Axioms* (ISSN 2075-1680) (available at: https://www.mdpi.com/journal/axioms/special_issues/application_fixed_point).

For citation purposes, cite each article independently as indicated on the article page online and as indicated below:

LastName, A.A.; LastName, B.B.; LastName, C.C. Article Title. *Journal Name* **Year**, *Volume Number*, Page Range.

ISBN 978-3-0365-2071-1 (Hbk)
ISBN 978-3-0365-2072-8 (PDF)

© 2021 by the authors. Articles in this book are Open Access and distributed under the Creative Commons Attribution (CC BY) license, which allows users to download, copy and build upon published articles, as long as the author and publisher are properly credited, which ensures maximum dissemination and a wider impact of our publications.

The book as a whole is distributed by MDPI under the terms and conditions of the Creative Commons license CC BY-NC-ND.

Contents

About the Editors vii

Preface to "Theory and Application of Fixed Point" ix

Reny George, Zoran D. Mitrović, and Stojan Radenović
On Some Coupled Fixed Points of Generalized T-Contraction Mappings in a $b_v(s)$-Metric Space and Its Application
Reprinted from: *Axioms* 2020, 9, 129, doi:10.3390/axioms9040129 1

Yaé Ulrich Gaba, Erdal Karapınar, Adrian Petruşel and Stojan Radenović
New Results on Start-Points for Multi-Valued Maps
Reprinted from: *Axioms* 2020, 9, 141, doi:10.3390/axioms9040141 15

Kazeem Olalekan Aremu, Chinedu Izuchukwu, Hammed Anuolwupo Abass and Oluwatosin Temitope Mewomo
On a Viscosity Iterative Method for Solving Variational Inequality Problems in Hadamard Spaces
Reprinted from: *Axioms* 2020, 9, 143, doi:10.3390/axioms9040143 27

Kengo Kasahara and Yasunori Kimura
Iterative Sequences for a Finite Number of Resolvent Operators on Complete Geodesic Spaces
Reprinted from: *Axioms* 2021, 10, 15, doi:10.3390/axioms10010015 41

Olawale Kazeem Oyewole and Oluwatosin Temitope Mewomo
A Strong Convergence Theorem for Split Null Point Problem and Generalized Mixed Equilibrium Problem in Real Hilbert Spaces
Reprinted from: *Axioms* 2021, 10, 16, doi:10.3390/axioms10010016 53

Hassan Almusawa, Hasanen A. Hammad and Nisha Sharma
Approximation of the Fixed Point for Unified Three-Step Iterative Algorithm with Convergence Analysis in Busemann Spaces
Reprinted from: *Axioms* 2021, 10, 26, doi:10.3390/axioms10010026 73

Anil Kumar and Aysegul Tas
Note on Common Fixed Point Theorems in Convex Metric Spaces
Reprinted from: *Axioms* 2021, 10, 28, doi:10.3390/axioms10010028 85

Binayak S. Choudhury, Nikhilesh Metiya, Debashis Khatua and Manuel de la Sen
Fixed-Point Study of Generalized Rational Type Multivalued Contractive Mappings on Metric Spaces with a Graph
Reprinted from: *Axioms* 2021, 10, 31, doi:10.3390/axioms10010031 93

Miroslav Hristov, Atanas Ilchev, Diana Nedelcheva and Boyan Zlatanov
Existence of Coupled Best Proximity Points of p-Cyclic Contractions
Reprinted from: *Axioms* 2021, 10, 39, doi:10.3390/axioms10010039 101

Gana Gecheva, Miroslav Hristov, Diana Nedelcheva, Margarita Ruseva and Boyan Zlatanov
Applications of Coupled Fixed Points for Multivalued Maps in the Equilibrium in Duopoly Markets and in Aquatic Ecosystems
Reprinted from: *Axioms* 2021, 10, 44, doi:10.3390/axioms10020044 115

Based Ali, Mohammad Imdad and Salvatore Sessa
A Relation-Theoretic Matkowski-Type Theorem in Symmetric Spaces
Reprinted from: *Axioms* **2021**, *10*, 50, doi:10.3390/axioms10020050 . . . . **131**

Kapil Jain and Jatinderdeep Kaur
Some Fixed Point Results in b-Metric Spaces and b-Metric-Like Spaces with New Contractive Mappings
Reprinted from: *Axioms* **2021**, *10*, 55, doi:10.3390/axioms10020055 . . . . **141**

Hsien-Chung Wu
Banach Contraction Principle and Meir–Keeler Type of Fixed Point Theoremsfor Pre-Metric Spaces
Reprinted from: *Axioms* **2021**, *10*, 57, doi:10.3390/axioms10020057 . . . . **157**

Mustapha Sabiri, Abdelhafid Bassou, Jamal Mouline and Taoufik Sabar
Fixed Points Results for Various Types of TricyclicContractions
Reprinted from: *Axioms* **2021**, *10*, 72, doi:10.3390/axioms10020072 . . . . **177**

Samera M. Saleh, Salvatore Sessa, Waleed M. Alfaqih and Fawzia Shaddad
Common Fixed Point Results for Almost $\mathcal{R}_g$-Geraghty Type Contraction Mappings in b_2-Metric Spaces with an Application to Integral Equations
Reprinted from: *Axioms* **2021**, *10*, 101, doi:10.3390/axioms10020101 . . . . **189**

About the Editors

Erdal Karapinar received his Ph.D. on "Isomorphisms of ℓ—Köthe spaces" in 2004 from Graduate School of Natural Applied Sciences, Middle East Technical University (METU), Ankara, Turkey, under the supervision of Professor V. P. ZAKHARYUTA and Professor M.YURDAKUL. He is the author and co-authors of several research papers, and he is also the co-author of the book "Fixed Point Theory in Metric Type Spaces". He is the Editor-in-Chief of "Result in Nonlinear Analysis and Advances in the Theory of Nonlinear Analysis and its Applications". He is also the associate editors of several journals. He was also a Highly Cited Researcher of Web of Science (WoS) of Clarivate Analysis from 2015–2019. He is currently a full professor at Cankaya University, Ankara, Turkey and visiting professor at China Medical University. His current research interests are focused on fixed point theory and its applications.

Juan Martínez-Moreno is currently a Full Professor in the Department of Mathematics, University of Jaen (UJA), Spain. He is the author of more than 50 papers published in ISI journals and is involved in teaching mathematics to students of economics. His current research interests include fixed point theory, fuzzy sets, iterative procedures, involving linear operators and its techniques for image processing. He has supervised 6 doctoral theses. He has participated in the organization of more than 10 national and international conferences.

Inci M. Erhan was born on 9 April 1971, in Russe, Bulgaria. She received her Ph.D. in 2003 from the Middle East Technical University (METU), Ankara, Turkey. Her main areas of research are fixed point theory and its applications, numerical solutions of differential equations and dynamic equations on time scales. Currently, she is a full professor at the Department of Mathematics, Atılım University, Ankara, Turkey

Preface to "Theory and Application of Fixed Point"

This book collects the published manuscripts submitted to a Special Issue of Axioms entitled "Theory and Application of Fixed Point".

Fixed point theory is initiated with the famous Banach contraction mapping principle, and has been a subject of considerable and increasing interest. It has applications in many areas of mathematics, science, engineering, economics and even medicine. The pioneering work of Banach and the huge application potential of fixed point theory have inspired numerous researchers to advance the theoretical studies in different directions related to the conditions on the contraction mappings and the relevant spaces. This has resulted in many important achievements in the field.

This book contains some very recent theoretical results related to some new types of contraction mappings defined in various types of spaces. There are also studies related to applications of the theoretical findings to mathematical models of specific problems, and their approximate computations. In this sense, this book will contribute to the area and provide directions for further developments in fixed point theory and its applications.

Erdal Karapinar, Juan Martínez-Moreno, Inci M. Erhan
Editors

Article

On Some Coupled Fixed Points of Generalized T-Contraction Mappings in a $b_v(s)$-Metric Space and Its Application

Reny George [1,2,*], **Zoran D. Mitrović** [3,*] **and Stojan Radenović** [4]

[1] Department of Mathematics, College of Science and Humanities in Al-Kharj, Prince Sattam bin Abdulaziz University, Al-Kharj 11942, Saudi Arabia
[2] Department of Mathematics and Computer Science, St. Thomas College, Bhilai, Chhattisgarh 491022, India
[3] Faculty of Electrical Emgineering, University of Banja Luka, Patre 5, 78000 Banja Luka, Bosnia and Herzegovina
[4] Faculty of Mechanical Engineering, University of Belgrade, Kraljice Marije 16, 11000 Beograd, Serbia; radens@beotel.net
* Correspondence: r.kunnelchacko@psau.edu.sa (R.G.); zoran.mitrovic@etf.unibl.org (Z.D.M.)

Received: 16 October 2020; Accepted: 7 November 2020; Published: 9 November 2020

Abstract: Common coupled fixed point theorems for generalized T-contractions are proved for a pair of mappings $S : X \times X \to X$ and $g : X \to X$ in a $b_v(s)$-metric space, which generalize, extend, and improve some recent results on coupled fixed points. As an application, we prove an existence and uniqueness theorem for the solution of a system of nonlinear integral equations under some weaker conditions and given a convergence criteria for the unique solution, which has been properly verified by using suitable example.

Keywords: common coupled fixed point; $b_v(s)$-metric space; T-contraction; weakly compatible mapping

1. Introduction

In the last three decades, the definition of a metric space has been altered by many authors to give new and generalized forms of a metric space. In 1989, Bakhtin [1] introduced one such generalization in the form of a b-metric space and in the year 2000 Branciari [2] gave another generalization in the form a rectangular metric space and generalized metric space. Thereafter, using the above two concepts, many generalizations of a metric space appeared in the form of rectangular b-metric space [3], hexagonal b-metric space [4], pentagonal b-metric space [5], etc. The latest such generalization was given by Mitrović and Radenović [6] in which the authors defined a $b_v(s)$-metric space which is a generalization of all the concepts told above. Some recent fixed point theorems in such generalized metric spaces can be found in [6–9]. In [10–12], one can find some interesting coupled fixed point theorems and their applications proved in some generalized forms of a metric space. In the present note, we have given coupled fixed point results for a pair of generalized T-contraction mappings in a $b_v(s)$-metric space. Our results are new and it extends, generalize, and improve some of the coupled fixed point theorems recently dealt with in [10–12].

In recent years, fixed point theory has been successfully applied in establishing the existence of solution of nonlinear integral equations (see [11–15]). We have applied one of our results to prove the existence and convergence of a unique solution of a system of nonlinear integral equations using some weaker conditions as compared to those existing in literature.

2. Preliminaries

Definition 1. [6] *Let X be a nonempty set. Assume that, for all $x, y, \in X$ and distinct $u_1, \cdots, u_v \in X - \{x, y\}$, $d_v : X \times X \to R$ satisfies :*

1. $d_v(x,y) \geq 0$ and $d_v(x,y) = 0$ if and only if $x = y$,
2. $d_v(x,y) = d_v(y,x)$,
3. $d_v(x,y) \leq s[d_v(x,u_1) + d_v(u_1,u_2) + \cdots + d_v(u_{v-1},u_v) + d_v(u_v,y)]$, for some $s \geq 1$.

Then, (X, d_v) is a $b_v(s)$-metric space.

Definition 2. *[6] In the $b_v(s)$-metric space (X, d_v), the sequence $<u_n>$*

(a) *converges to $u \in X$ if $d_v(u_n, u) \to 0$ as $n \to \infty$;*
(b) *is a Cauchy sequence if $d_v(u_n, u_m) \to 0$ as $n, m \to +\infty$.*

Clearly, $b_1(1)$-metric space is the usual metric space, whereas $b_1(s)$, $b_2(1)$, $b_2(s)$, and $b_v(1)$-metric spaces are, respectively, the b-metric space ([1]), rectangular metric space ([2]), rectangular b-metric space ([3]), and v-generalized metric space ([2]).

Lemma 1. *[6] If (X, d_v) is a $b_v(s)$-metric space, then (X, d_v) is a $b_{2v}(s^2)$-metric space.*

Definition 3. *An element $(u,v) \in X \times X$ is called a coupled coincidence point of $S : X \times X \to X$ and $g : X \to X$ if $g(u) = S(u,v)$ and $g(v) = S(v,u)$. In this case, we also say that $(g(u), g(v))$ is the point of coupled coincidence of S and g. If $u = g(u) = S(u,v)$ and $v = g(v) = S(v,u)$, then we say that (u,v) is a common coupled fixed point of S and g.*

We will denote by $COCP\{S, g\}$ and $CCOFP\{S, g\}$ respectively the set of all coupled coincidence points and the set of all common coupled fixed points of S and g.

Definition 4. *$S : X \times X \to X$ and $g : X \to X$ are said to be weakly compatible if and only if $S(g(u), g(v)) = g(S(u,v))$ for all $(u,v) \in COCP\{S, g\}$.*

3. Main Results

We will start this section by proving the following lemma which is an extension of Lemma 1.12 of [6] to two sequences:

Lemma 2. *Let (X, d_v) be a $b_v(s)$-metric space and let $<u_n>$ and $<v_n>$ be two sequences in X such that $u_n \neq u_{n+1}, v_n \neq v_{n+1}$ $(n \geq 0)$. Suppose that $\lambda \in [0,1)$ and c_1, c_2 are real nonnegative numbers such that*

$$K_{m,n} \leq \lambda K_{m-1,n-1} + c_1 \lambda^m + c_2 \lambda^n, \text{ for all } m, n \in \mathbb{N}, \tag{1}$$

where $K_{m,n} = \max\{d_v(u_m, u_n), d_v(v_m, v_n)\}$ or $K_{m,n} = d_v(u_m, u_n) + d_v(v_m, v_n)$. Then, $<u_n>$ and $<v_n>$ are Cauchy sequences.

Proof. From (1), we have

$$\begin{aligned} K_{n,n+1} &\leq \lambda K_{n-1,n} + c_1 \lambda^n + c_2 \lambda^{n+1} \\ &\leq \cdots \\ &\leq \lambda^n K_{0,1} + c_1 n \lambda^n + c_2 n \lambda^{n+1} \\ &\leq \lambda^n K_{0,1} + C_0 n \lambda^n. \end{aligned} \tag{2}$$

For $m, n, k \in N$, by (1), we have

$$\begin{aligned}
K_{m+k,n+k} &\leq \lambda \max\{K_{m+k-1,n+k-1}, c_1\lambda^{m+k-1} + c_2\lambda^{n+k-1})\} \\
&\leq \lambda K_{m+k-1,n+k-1} + c_1\lambda^{m+k} + c_2\lambda^{n+k}) \\
&\cdots \\
&\leq \lambda^k K_{m,n} + kC_1\lambda^k(\lambda^m + \lambda^n).
\end{aligned} \quad (3)$$

Since $0 < \lambda < 1$, we can find a positive integer q_k such that $0 < \lambda^{q_k} < \frac{1}{s}$. Now, suppose $v \geq 2$. Then, by using condition 3. of a $b_v(s)$-metric and inequalities (2) and (3), we have

$$\begin{aligned}
K_{m,n} &\leq s[K_{m,m+1} + K_{m+1,m+2} + \cdots + K_{m+v-3,m+v-2} + K_{m+v-2,m+q_k} + K_{m+q_k,n+q_k} + K_{n+q_k,n}] \\
&\leq s[\lambda^m + \lambda^{m+1} + \cdots + \lambda^{m+v-3}]K_0 + sC_0[m\lambda^m + (m+1)\lambda^{m+1} + \cdots + (m+v-3)\lambda^{m+v-2}] \\
&\quad + s[\lambda^m K_{v-2,q_k} + m\lambda^m(\lambda^{v-2} + \lambda^{q_k})K_0] \\
&\quad + s[\lambda^{q_k} K_{m,n} + q_k\lambda^{q_k}(\lambda^m + \lambda^n)K_0] + s[\lambda^n K_{q_k,0} + n\lambda^n(\lambda^{q_k} + 1)K_0].
\end{aligned}$$

Then,

$$\begin{aligned}
K_{m,n} &\leq \frac{s\lambda^m}{(1-s\lambda^{q_k})(1-\lambda)} K_{0,1} + \frac{s(m+v-3)\lambda^m}{(1-\lambda)(1-s\lambda^{q_k})} \\
&\quad + \frac{s}{1-s\lambda^{q_k}}[\lambda^m K_{v-2,q_k} + m\lambda^m(\lambda^{v-2} + \lambda^{q_k})K_{0,1}] \\
&\quad + \frac{s}{1-s\lambda^{q_k}}[q_k\lambda^{q_k}(\lambda^m + \lambda^n)K_{0,1}] + \frac{s}{1-s\lambda^{q_k}}[\lambda^n K_{q_k,0} + n\lambda^n(\lambda^{q_k} + 1)K_{0,1}].
\end{aligned}$$

Thus, from the definition of $K_{m,n}$, we see that, as $m, n \to +\infty$, $d_v(u_m, u_n) \to 0$ and $d_v(v_m, v_n) \to 0$ and thus $<u_n>$ and $<v_n>$ are Cauchy sequences. $\square$

3.1. Coupled Fixed Point Theorems

We now present our main theorems as follows:

Theorem 1. *Let (X, d_v) be a $b_v(s)$-metric space, $T\colon X \to X$ be a one to one mapping, $S\colon X \times X \to X$ and $g\colon X \to X$ be mappings such that $S(X \times X) \subset g(X)$, $Tg(X)$ is complete. If there exist real numbers λ, μ, ν with $0 \leq \lambda < 1, 0 \leq \mu, \nu \leq 1$, $\min\{\lambda\mu, \lambda\nu\} < \frac{1}{s}$ such that, for all $u, v, w, z \in X$*

$$\begin{aligned}
d_v(TS(u,v), TS(w,z)) &\leq \lambda \max\{d_v(Tgu, Tgw), d_v(Tgv, Tgz), \mu d_v(Tgu, TS(u,v)), \mu d_v(Tgv, TS(v,u), \\
&\quad \nu d_v(Tgw, TS(w,z)), \nu d_v(Tgz, TS(z,w))\}
\end{aligned} \quad (4)$$

then the following holds :

1. *There exist w_{x_0}, w_{y_0} in X, such that sequences $<Tgu_n>$ and $<Tgv_n>$ converge to Tgw_{x_0} and Tgw_{y_0} respectively, where the iterative sequences $<gu_n>$ and $<gv_n>$ are defined by $gu_n = S(u_{n-1}, v_{n-1})$ and $gv_n = S(v_{n-1}, u_{n-1})$ for some arbitrary $(u_0, v_0) \in X \times X$.*
2. $(w_{x_0}, w_{y_0}) \in \text{COCP}\{S, g\}$.
3. *If S and g are weakly compatible, then S and g have a unique common coupled fixed point.*

Proof. 1. We shall start the proof by showing that the sequences $<Tgu_n>$ and $<Tgv_n>$ are Cauchy sequences, where $<gu_n>$ and $<gv_n>$ are as mentioned in the hypothesis.

By (4), we have

$$
\begin{aligned}
d_v(Tgu_n, Tgu_{n+1}) &= d_v(TS(u_{n-1}, v_{n-1}), TS(u_n, v_n)) \\
&\leq \lambda \max\{d_v(Tgu_{n-1}, Tgu_n), d_v(Tgv_{n-1}, Tgv_n), \mu d_v(Tgu_{n-1}, TS(u_{n-1}, v_{n-1})), \\
&\quad \mu d_v(Tgv_{n-1}, TS(v_{n-1}, u_{n-1})), \nu d_v(Tgu_n, TS(u_n, v_n)), \nu d_v(Tgv_n, TS(v_n, u_n))\} \\
&\leq \lambda \max\{d_v(Tgu_{n-1}, Tgu_n), d_v(Tgv_{n-1}, Tgv_n), d_v(Tgu_{n-1}, Tgu_n), \\
&\quad d_v(Tgv_{n-1}, Tgv_n), d_v(Tgu_n, Tgu_{n+1}), d_v(Tgv_n, Tgv_{n+1})\}.
\end{aligned}
\qquad(5)
$$

Similarly, we get

$$
\begin{aligned}
d_v(Tgv_n, Tgv_{n+1}) &\leq \lambda \max\{d_v(Tgv_{n-1}, Tgv_n), d_v(Tgu_{n-1}, Tgu_n), d_v(Tgv_{n-1}, Tgv_n), \\
&\quad d_v(Tgu_{n-1}, Tgu_n), d_v(Tgv_n, Tgv_{n+1}), d_v(Tgu_n, Tgu_{n+1})\}.
\end{aligned}
\qquad(6)
$$

Let $K_n = \max\{d_v(Tgu_n, Tgu_{n+1}), d_v(Tgv_n, Tgv_{n+1})\}$. By (5) and (6), we get

$$
K_n \leq \lambda \max\{d_v(Tgv_{n-1}, Tgv_n), d_v(Tgu_{n-1}, Tgu_n), d_v(Tgv_n, Tgv_{n+1}), d_v(Tgu_n, Tgu_{n+1})\}. \qquad(7)
$$

If

$$
\max\{d_v(Tgv_{n-1}, Tgv_n), d_v(Tgu_{n-1}, Tgu_n), d_v(Tgv_n, Tgv_{n+1}), d_v(Tgu_n, Tgu_{n+1})\} \\
= d_v(Tgv_n, Tgv_{n+1}) \text{ or } d_v(Tgu_n, Tgu_{n+1}),
$$

then (7) will yield a contradiction. Thus, we have

$$
\max\{d_v(Tgv_{n-1}, Tgv_n), d_v(Tgu_{n-1}, Tgu_n), d_v(Tgv_n, Tgv_{n+1}), d_v(Tgu_n, Tgu_{n+1})\} \\
= \max\{d_v(Tgv_{n-1}, Tgv_n), d_v(Tgu_{n-1}, Tgu_n)\},
$$

and then (7) gives

$$
K_n \leq \lambda \max\{d_v(Tgv_{n-1}, Tgv_n), d_v(Tgu_{n-1}, Tgu_n)\} = \lambda K_{n-1} \preceq \lambda^2 K_{n-2} \preceq \cdots \preceq \lambda^n K_0. \qquad(8)
$$

For any $m, n \in N$, we have

$$
\begin{aligned}
d_v(Tgu_m, Tgu_n) &= d_v(TS(u_{m-1}, v_{m-1}), TS(u_{n-1}, v_{n-1})) \\
&\leq \lambda \max\{d_v(Tgu_{m-1}, Tgu_{n-1}), d_v(Tgv_{m-1}, Tgv_{n-1}), \\
&\quad \mu d_v(Tgu_{m-1}, TS(u_{m-1}, v_{m-1})), \mu d_v(Tgv_{m-1}, TS(v_{m-1}, u_{m-1})), \\
&\quad \nu d_v(Tgu_{n-1}, TS(u_{n-1}, v_{n-1})), \nu d_v(Tgv_{n-1}, TS(v_{n-1}, u_{n-1}))\} \\
&\leq \lambda \max\{d_v(Tgu_{m-1}, Tgu_{n-1}), d_v(Tgv_{m-1}, Tgv_{n-1}), d_v(Tgu_{m-1}, Tgu_m), \\
&\quad d_v(Tgv_{m-1}, Tgv_m), d_v(Tgu_{n-1}, Tgu_n), d_v(Tgv_{n-1}, Tgv_n)\}.
\end{aligned}
$$

Then, by using (8), we get

$$
\begin{aligned}
d_v(Tgu_m, Tgu_n) &\leq \lambda \max\{d_v(Tgu_{m-1}, Tgu_{n-1}), d_v(Tgv_{m-1}, Tgv_{n-1})\} \\
&\quad + (\lambda^m + \lambda^n) K_0.
\end{aligned}
\qquad(9)
$$

Similarly, we have

$$
\begin{aligned}
d_v(Tgv_m, Tgv_n) &\leq \lambda \max\{d_v(Tgu_{m-1}, Tgu_{n-1}), d_v(Tgv_{m-1}, Tgv_{n-1})\} \\
&\quad + (\lambda^m + \lambda^n)K_0\}.
\end{aligned}
\tag{10}
$$

Let $K_{m,n} = \max\{d_v(Tgu_m, Tgu_n), d_v(Tgv_m, Tgv_n)\}$. By (9) and (10), we get

$$K_{m,n} \leq \lambda K_{m-1,n-1} + (\lambda^m + \lambda^n)K_0.$$

Thus, we see that inequality (1) is satisfied with $c_1 = c_2 = K_0$. Hence, by Lemma 2, $<Tgu_n>$ and $<Tgv_n>$ are Cauchy sequences. For $v = 1$, the same follows from Lemma 1. Since $(Tg(X), d)$ is complete, we can find $w_{x_0}, w_{y_0} \in X$ such that

$$\lim_{n\to\infty} Tgu_n = Tgw_{x_0} \text{ and } \lim_{n\to\infty} Tgv_n = Tgw_{y_0}.$$

2. Now,

$$
\begin{aligned}
d_v(TS(w_{x_0}, w_{y_0}), Tgw_{x_0}) &\leq s[d_v(TS(w_{x_0}, w_{y_0}), TS(u_n, v_n)) + d_v(TS(u_n, v_n), TS(u_{n+1}, v_{n+1})) \\
&\quad + \cdots + d_v(TS(u_{n+v-2}, v_{n+v-2}), TS(u_{n+v-1}, v_{n+v-1})) + d_v(TS(u_{n+v-1}, v_{n+v-1}), Tgw_{x_0}) \\
&\leq s[\lambda \max\{d_v(Tgw_{x_0}, Tgu_n), d_v(Tgw_{y_0}, Tgv_n), \mu d_v(Tgw_{x_0}, TS(w_{x_0}, w_{y_0})), \\
&\quad \mu d_v(Tgw_{y_0}, TS(w_{y_0}, w_{x_0})), \nu d_v(Tgu_n, TS(u_n, v_n)), \nu d_v(Tgv_n, TS(v_n, u_n))\} \\
&\quad + d_v(Tgu_{n+1}, Tgu_{n+2}) + \cdots + d_v(Tgu_{n+v-1}, Tgu_{n+v}) + d_v(Tgu_{n+v}, Tgw_{x_0}) \\
&\leq s[\lambda \max\{d_v(Tgw_{x_0}, Tgu_n), d_v(Tgw_{y_0}, Tgv_n), \mu d_v(Tgw_{x_0}, TS(w_{x_0}, w_{y_0})), \\
&\quad \mu d_v(Tgw_{y_0}, TS(w_{y_0}, w_{x_0})), \nu d_v(Tgu_n, Tgu_{n+1}), \nu d_v(Tgv_n, Tgv_{n+1})\} \\
&\quad + d_v(Tgu_{n+1}, Tgu_{n+2}) + \cdots + d_v(Tgu_{n+v-1}, Tgu_{n+v}) + d_v(Tgu_{n+v}, Tgw_{x_0}).
\end{aligned}
\tag{11}
$$

Note that, since $<Tgu_n>$ and $<Tgv_n>$ are Cauchy sequences, by definition, $d_v(Tgu_n, Tgu_{n+1}) \to 0$, $d_v(Tgv_n, Tgv_{n+1}) \to 0$ as $n \to \infty$. Thus, from (11), as $n \to \infty$, we get

$$d_v(TS(w_{x_0}, w_{y_0}), Tgw_{x_0}) \leq s\lambda \max\{\mu d_v(Tgw_{x_0}, TS(w_{x_0}, w_{y_0})), \mu d_v(Tgw_{y_0}, TS(w_{y_0}, w_{x_0}))\}.$$

Similarly, we get

$$d_v(TS(w_{y_0}, w_{x_0}), Tgw_{y_0}) \leq s\lambda \max\{\mu d_v(Tgw_{x_0}, TS(w_{x_0}, w_{y_0})), \mu d_v(Tgw_{y_0}, TS(w_{y_0}, w_{x_0}))\}.$$

Thus, we have

$$
\begin{aligned}
\max\{d_v(TS(w_{x_0}, w_{y_0}), Tgw_{x_0}), d_v(TS(w_{y_0}, w_{x_0}), Tgw_{y_0})\} & \\
\leq s\lambda\mu \max\{d_v(Tgw_{x_0}, TS(w_{x_0}, w_{y_0})), d_v(Tgw_{y_0}, TS(w_{y_0}, w_{x_0}))\}.&
\end{aligned}
\tag{12}
$$

Proceeding along the same lines as above, we also have

$$
\begin{aligned}
\max\{d_v(Tgw_{x_0}, TS(w_{x_0}, w_{y_0})), d_v(Tgw_{y_0}, TS(w_{y_0}, w_{x_0}))\} & \\
\leq s\lambda\nu \max\{d_v(Tgw_{x_0}, TS(w_{x_0}, w_{y_0})), d_v(Tgw_{y_0}, TS(w_{y_0}, w_{x_0}))\}.&
\end{aligned}
\tag{13}
$$

Using (12) and (13) along with the condition $\min\{\lambda\mu, \lambda\nu\} < \frac{1}{s}$, we get $TS(w_{x_0}, w_{y_0}) = Tgw_{x_0}$ and $TS(w_{y_0}, w_{x_0}) = Tgw_{y_0}$. As T is one to one, we have $S(w_{x_0}, w_{y_0}) = gw_{x_0}$ and $S(w_{y_0}, w_{x_0}) = gw_{y_0}$. Therefore, $(w_{x_0}, w_{y_0}) \in COCP\{S, g\}$.

3. Suppose S and g are weakly compatible. First, we will show that, if $(w_{x_0}^*, w_{y_0}^*) \in COCP\{S,g\}$, then $gw_{x_0}^* = gw_{x_0}$ and $gw_{y_0}^* = gw_{y_0}$, or in other words the point of coupled coincidence of S and g is unique. By (5), we have

$$
\begin{aligned}
d_v(Tgw_{x_0}^*, Tgw_{x_0}) &= d_v(TS(w_{x_0}^*, w_{y_0}^*), TS(w_{x_0}, w_{y_0})) \\
&\leq \lambda \max\{d_v(Tgw_{x_0}^*, Tgw_{x_0}), d_v(Tgw_{y_0}^*, Tgw_{y_0}), \mu d_v(Tgw_{x_0}^*, TS(w_{x_0}^*, w_{y_0}^*)), \\
&\quad \mu d_v(Tgw_{y_0}^*, TS(w_{y_0}^*, w_{x_0}^*)), \nu d_v(Tgw_{x_0}, TS(w_{x_0}, w_{y_0})), \nu d_v(Tgw_{y_0}, TS(w_{y_0}, w_{x_0}))\} \\
&\leq \lambda \max\{d_v(Tgw_{x_0}^*, Tgw_{x_0}), d_v(Tgw_{y_0}^*, Tgw_{y_0})\}.
\end{aligned}
$$

Similarly, we have

$$d_v(Tgw_{y_0}^*, Tgw_{y_0}) \leq \lambda \max\{d_v(Tgw_{x_0}^*, Tgw_{x_0}), d_v(Tgw_{y_0}^*, Tgw_{y_0})\}.$$

Thus, from the above two inequalities, we get

$$\max\{d_v(Tgw_{x_0}^*, Tgw_{x_0}), d_v(Tgw_{y_0}^*, Tgw_{y_0})\} \leq \lambda \max\{d_v(Tgw_{x_0}^*, Tgw_{x_0}), d_v(Tgw_{y_0}^*, Tgw_{y_0})\}$$

which implies that $Tgw_{x_0}^* = Tgw_{x_0}$ and $Tgw_{y_0}^* = Tgw_{y_0}$. Since T is one to one, we get $gw_{x_0}^* = gw_{x_0}$ and $gw_{y_0}^* = gw_{y_0}$, which is the point of coupled coincidence of S and g is unique. Since S and g are weakly compatible and, since $(w_{x_0}, w_{y_0}) \in COCP\{S,g\}$, we have

$$ggw_{x_0} = gS(w_{x_0}, w_{y_0}) = S(gw_{x_0}, gw_{y_0})$$

and

$$ggw_{y_0} = gS(w_{y_0}, w_{x_0}) = S(gw_{y_0}, gw_{x_0})$$

which shows that $(gw_{x_0}, gw_{y_0}) \in COCP\{S,g\}$. By the uniqueness of the point of coupled coincidence, we get $ggw_{x_0} = gw_{x_0}$ and $ggw_{y_0} = gw_{y_0}$ and thus $(gw_{x_0}, gw_{y_0}) \in CCOFP\{S,g\}$. Uniqueness of the coupled fixed point follows easily from (4). □

Our next result is a generalized version of Theorem 2.1 of Gu [10].

Theorem 2. *Let (X, d_v), T, S and g be as in Theorem 1 and suppose there exist $\beta_1, \beta_2, \beta_3$ in the interval $[0,1)$, such that $\beta_1 + \beta_2 + \beta_3 < 1$, $\min\{\beta_2, \beta_3\} < \frac{1}{5}$ and for all $u, v, w, z \in X$*

$$d_v(TS(u,v), TS(w,z)) + d_v(TS(v,u), TS(z,w)) \leq \beta_1(d_v(Tgu, Tgw) + d_v(Tgv, Tgz)) + \beta_2(d_v(Tgu, TS(u,v)) + d_v(Tgv, TS(v,u))) + \beta_3(d_v(Tgw, TS(w,z)) + d_v(Tgz, TS(z,w))). \quad (14)$$

Then, conclusions 1, 2, and 3 of Theorem 1 are true.

Proof. Let $K_n' = d_v(Tgu_n, Tgu_{n+1}) + d_v(Tgv_n, Tgv_{n+1})$ and $K_{m,n}' = d_v(Tgu_m, Tgu_n) + d_v(Tgv_m, Tgv_n)$. From condition (14), we obtain

$$
\begin{aligned}
d_v(Tgu_n, Tgu_{n+1}) + d_v(Tgv_n, Tgv_{n+1}) &= d_v(TS(u_{n-1}, v_{n-1}), TS(u_n, v_n)) + \\
& \quad d_v(TS(v_{n-1}, u_{n-1}), TS(v_n, u_n)) \\
&\leq \beta_1[d_v(Tgu_{n-1}, Tgu_n) + d_v(Tgv_{n-1}, Tgv_n)] + \beta_2[d_v(Tgu_{n-1}, TS(u_{n-1}, v_{n-1})) \\
&\quad + d_v(Tgv_{n-1}, TS(v_{n-1}, u_{n-1}))] + \beta_3[d_v(Tgu_n, TS(u_n, v_n)) + d_v(Tgv_n, TS(v_n, u_n))] \\
&\leq (\beta_1 + \beta_2)[d_v(Tgu_{n-1}, Tgu_n) + d_v(Tgv_{n-1}, Tgv_n)] \\
&\quad + \beta_3[d_v(Tgu_n, Tgu_{n+1}) + d_v(Tgv_n, Tgv_{n+1})].
\end{aligned}
$$

Therefore,
$$d_v(Tgu_n, Tgu_{n+1}) + d_v(Tgv_n, Tgv_{n+1}) \leq \lambda'[d_v(Tgu_{n-1}, Tgu_n) + d_v(Tgv_{n-1}, Tgv_n)],$$

where $\lambda' = \dfrac{\beta_1 + \beta_2}{1 - \beta_3} < 1$. Thus, we get

$$K'_n \leq \lambda' K'_{n-1} \leq \cdots \leq \lambda'^n K'_0. \tag{15}$$

For any $m, n \in \mathbb{N}$, we have

$$\begin{aligned}
d_v(Tgu_m, Tgu_n) &+ d_v(Tgv_m, Tgv_n) = d_v(TS(u_{m-1}, v_{m-1}), TS(u_{n-1}, v_{n-1}) + \\
&\quad d_v(TS(v_{m-1}, u_{m-1}), TS(v_{n-1}, u_{n-1}) \\
&\leq \beta_1[d_v(Tgu_{m-1}, Tgu_{n-1}) + d_v(Tgv_{m-1}, Tgv_{n-1})] \\
&\quad + \beta_2[d_v(Tgu_{m-1}, TS(u_{m-1}, v_{m-1})) + d_v(Tgv_{m-1}, TS(v_{m-1}, u_{m-1}))] \\
&\quad + \beta_3[d_v(Tgu_{n-1}, TS(u_{n-1}, v_{n-1})) + d_v(Tgv_{n-1}, TS(v_{n-1}, u_{n-1}))] \\
&\leq \beta_1[d_v(Tgu_{m-1}, Tgu_{n-1}) + d_v(Tgv_{m-1}, Tgv_{n-1})] + \beta_2[d_v(Tgu_{m-1}, Tgu_m) \\
&\quad + d_v(Tgv_{m-1}, Tgv_m)] + \beta_3[d_v(Tgu_{n-1}, Tgu_n) + d_v(Tgv_{n-1}, Tgv_n)].
\end{aligned}$$

Then, by using (15), we get

$$\begin{aligned}
d_v(Tgu_m, Tgu_n) + d_v(Tgv_m, Tgv_n) &\leq \beta_1[d_v(Tgu_{m-1}, Tgu_{n-1}) + d_v(Tgv_{m-1}, Tgv_{n-1})] \\
&\quad + (\beta_2 \lambda'^m + \beta_3 \lambda'^n) K'_0\}.
\end{aligned}$$

That is,

$$K'_{m,n} \leq \lambda K'_{m-1, n-1} + (\lambda^m + \lambda^n) K'_0$$

where $\lambda' = \beta_1 + \beta_2 + \beta_3 < 1$. Now for $m, n, r \in N$. Thus, we see that inequality (1) is satisfied with $c_1 = c_2 = K_0$. Hence, by Lemma 2, $<Tgu_n>$ and $<Tgv_n>$ are Cauchy sequences. For $v = 1$, the same follows from Lemma 1.

Since $(Tg(X), d)$ is complete, we can find $w_{x_0}, w_{y_0} \in X$ such that

$$\lim_{n \to \infty} Tgu_n = Tgw_{x_0} \text{ and } \lim_{n \to \infty} Tgv_n = Tgw_{y_0}.$$

Again, from condition 3 in Definition 1, we have

$$\begin{aligned}
d_v(TS(w_{x_0}, w_{y_0}), Tgw_{x_0})) &\leq s[d_v(TS(w_{x_0}, w_{y_0}), TS(u_n, v_n)) + d_v(TS(u_n, v_n), TS(u_{n+1}, v_{n+1})) + \cdots + \\
&\quad + d_v(TS(u_{n+v-2}, v_{n+v-2}), TS(u_{n+v-1}, v_{n+v-1})) + \\
&\quad d_v(TS(u_{n+v-1}, v_{n+v-1}), Tgw_{x_0}))]
\end{aligned}$$

and

$$\begin{aligned}
d_v(TS(w_{y_0}, w_{x_0}), Tgw_{y_0})) &\leq s[d_v(TS(w_{y_0}, w_{x_0}), TS(v_n, u_n)) + d_v(TS(v_n, u_n), TS(v_{n+1}, u_{n+1})) + \cdots + \\
&\quad d_v(TS(v_{n+v-2}, u_{n+v-2}), TS(v_{n+v-1}, u_{n+v-1})) + \\
&\quad d_v(TS(v_{n+v-1}, u_{n+v-1}), Tgw_{x_0}))].
\end{aligned}$$

Therefore,

$$d_v(TS(w_{x_0}, w_{y_0}), Tgw_{x_0}) + d_v(TS(w_{y_0}, w_{x_0}), Tgw_{y_0}) \leq s[d_v(TS(w_{x_0}, w_{y_0}), TS(u_n, v_n))$$
$$+d_v(TS(w_{y_0}, w_{x_0}), TS(v_n, u_n))$$
$$+d_v(TS(u_n, v_n), TS(u_{n+1}, v_{n+1})) + \cdots + d_v(TS(u_{n+v-2}, v_{n+v-2}), TS(u_{n+v-1}, v_{n+v-1}))$$
$$+d_v(TS(v_n, u_n), TS(v_{n+1}, u_{n+1})) + \cdots + d_v(TS(v_{n+v-2}, u_{n+v-2}), TS(v_{n+v-1}, u_{n+v-1}))$$
$$+d_v(TS(u_{n+v-1}, v_{n+v-1}), Tgw_{x_0}) + d_v(TS(v_{n+v-1}, u_{n+v-1}), Tgw_{y_0})]$$
$$\leq s[\beta_1(d_v(Tgw_{x_0}, Tgu_n) + d_v(Tgw_{y_0}, Tgv_n)) + \beta_2(d_v(Tgw_{x_0}, TS(w_{x_0}, w_{y_0})) +$$
$$d_v(Tgw_{y_0}, TS(w_{y_0}, w_{x_0})) + \beta_3(d_v(Tgu_n, TS(u_n, v_n)) + d_v(Tgv_n, TS(v_n, u_n)))\}$$
$$+d_v(Tgu_n, Tgu_{n+1}) + \cdots + d_v(Tgu_{n-1}, Tgu_n) + +d_v(Tgv_n, Tgv_{n+1}) + \cdots + d_v(Tgv_{n-1}, Tgv_n)$$
$$+d_v(Tgu_{n+v-1}, Tgw_{x_0}) + d_v(Tgv_{n+v-1}, Tgw_{y_0})].$$

As $n \to \infty$, we get

$$d_v(TS(w_{x_0}, w_{y_0}), Tgw_{x_0}) + d_v(TS(w_{y_0}, w_{x_0}), Tgw_{y_0})$$
$$\leq s\beta_2[d_v(Tgw_{x_0}, TS(w_{x_0}, w_{y_0})) + d_v(Tgw_{y_0}, TS(w_{y_0}, w_{x_0}))]. \quad (16)$$

Similarly, we can show that

$$d_v(Tgw_{x_0}, TS(w_{x_0}, w_{y_0})) + d_v(Tgw_{y_0}, TS(w_{y_0}, w_{x_0}))$$
$$\leq s\beta_3[d_v(Tgw_{x_0}, TS(w_{x_0}, w_{y_0})) + d_v(Tgw_{y_0}, TS(w_{y_0}, w_{x_0}))] \quad (17)$$

Using (16) and (17) along with the condition $\min\{\beta_2, \beta_3\} < \frac{1}{s}$, we get $d_v(Tgw_{x_0}, TS(w_{x_0}, w_{y_0})) + d_v(Tgw_{y_0}, TS(w_{y_0}, w_{x_0})) = 0$, i.e., $TS(w_{x_0}, w_{y_0}) = Tgw_{x_0}$ and $TS(w_{y_0}, w_{x_0}) = Tgw_{y_0}$. As T is one to one, we have $S(w_{x_0}, w_{y_0}) = gw_{x_0}$ and $S(w_{y_0}, w_{x_0}) = gw_{y_0}$. Therefore, $(w_{x_0}, w_{y_0}) \in COCP\{S, g\}$.

If $(w_{x_0}^*, w_{y_0}^*) \in COCP\{S, g\}$, then, by (14), we have

$$d_v(Tgw_{x_0}^*, Tgw_{x_0}) + d_v(Tgw_{y_0}^*, Tgw_{y_0}) = d_v(TS(w_{x_0}^*, w_{y_0}^*), TS(w_{x_0}, w_{y_0})) + d_v(TS(w_{y_0}^*, w_{x_0}^*), TS(w_{y_0}, w_{x_0}))$$
$$\leq \beta_1[d_v(Tgw_{x_0}^*, Tgw_{x_0}) + d_v(Tgw_{y_0}^*, Tgw_{y_0})] + \beta_2[d_v(Tgw_{x_0}^*, TS(w_{x_0}^*, w_{y_0}^*))$$
$$+d_v(Tgw_{y_0}^*, TS(w_{y_0}^*, w_{x_0}^*))] + \beta_3[d_v(Tgw_{x_0}, TS(w_{x_0}, w_{y_0})) + d_v(Tgw_{y_0}, TS(w_{y_0}, w_{x_0}))]$$
$$\leq \beta_1[d_v(Tgw_{x_0}^*, Tgw_{x_0}) + d_v(Tgw_{y_0}^*, Tgw_{y_0})].$$

Thus, $d_v(Tgw_{x_0}^*, Tgw_{x_0}) + d_v(Tgw_{y_0}^*, Tgw_{y_0}) = 0$, which implies that $Tgw_{x_0}^* = Tgw_{x_0}$ and $Tgw_{y_0}^* = Tgw_{y_0}$. Since T is one to one, we get $gw_{x_0}^* = gw_{x_0}$ and $gw_{y_0}^* = gw_{y_0}$, which is the point of coupled coincidence of S, and g is unique. The remaining part of the proof is the same as in the proof of Theorem 1. □

The next results can be proved as in Theorems 1 and 2 and so we will not give the proof.

Theorem 3. *Theorem 1 holds if we replace condition (4) with the following condition:*
There exist $\beta_i \in [0, 1), i \in \{1, \ldots, 6\}$ such that $\sum_{i=1}^{6} \beta_i < 1$, $\min\{\beta_3 + \beta_4, \beta_5 + \beta_6\} < \frac{1}{s}$ and for all $u, v, w, z \in X$,

$$d_v(TS(u, v), TS(w, z)) \leq \beta_1 d_v(Tgu, Tgw) + \beta_2 d_v(Tgv, Tgz) + \beta_3 d_v(Tgu, TS(u, v))$$
$$+\beta_4 d_v(Tgv, TS(v, u)) + \beta_5 d_v(Tgw, TS(w, z)) + \beta_6 d_v(Tgz, TS(z, w)). \quad (18)$$

Taking T to be the identity mapping in Theorems 1–3, we have the following:

Corollary 1. Let (X, d_v), S, g, λ, μ and ν be as in Theorem 1 such that, for all $u, v, w, z \in X$, the following holds:

$$d_v(S(u,v), S(w,z)) \leq \lambda \max\{d_v(gu, gw), d_v(gv, gz), \mu d_v(gu, S(u,v)), \mu d_v(gv, S(v,u)), \nu d_v(gw, S(w,z)), \nu d_v(gz, S(z,w))\}. \tag{19}$$

Then, $COCP\{S, g\} \neq \phi$. Furthermore, if S and g are weakly compatible, then S and g has a unique common coupled fixed point. Moreover, for some arbitrary $(u_0, v_0) \in X \times X$, the iterative sequences ($<gu_n>$, $<gv_n>$) defined by $gu_n = S(u_{n-1}, v_{n-1})$ and $gv_n = S(v_{n-1}, u_{n-1})$ converge to the unique common coupled fixed point of S and g.

Corollary 2. Corollary 1 holds if the condition (19) is replaced with the following condition:
There exist $\beta_1, \beta_2, \beta_3$ in the interval [0,1), such that $\beta_1 + \beta_2 + \beta_3 < 1$, $\min\{\beta_2, \beta_3\} < \frac{1}{s}$ and for all $u, v, w, z \in X$

$$d_v(S(u,v), S(w,z)) + d_v(S(v,u), S(z,w)) \leq \beta_1(d_v(gu, gw) + d_v(gv, gz)) + \beta_2(d_v(gu, S(u,v)) + d_v(gv, S(v,u))) + \beta_3(d_v(gw, S(w,z)) + d_v(gz, S(z,w))). \tag{20}$$

Corollary 3. Corollary 1 holds if the condition (19) is replaced with the following condition:
There exist $\beta_i \in [0,1), i \in \{1, \ldots 6\}$ such that $\sum_{i=1}^{6} \beta_i < 1$, $\min\{\beta_3 + \beta_4, \beta_5 + \beta_6\} < \frac{1}{s}$ and, for all $u, v, w, z \in X$,

$$d_v(S(u,v), S(w,z)) \leq \beta_1 d_v(gu, gw) + \beta_2 d_v(gv, gz) + \beta_3 d_v(gu, S(u,v)) + \beta_4 d_v(gv, S(v,u)) + \beta_5 d_v(gw, S(w,z)) + \beta_6 d_v(gz, S(z,w)). \tag{21}$$

Remark 1. Since every b-metric space is a $b_1(s)$ metric space, we note that Theorem 1 is a substantial generalization of Theorem 2.2 of Ramesh and Pitchamani [11]. In fact, we do not require continuity and sub sequential convergence of the function T.

Remark 2. Note that condition (2.1) of Gu [10] implies (20) and hence Corollary 2 gives an improved version of Theorem 2.1 of Gu [10].

Remark 3. Condition (3.1) of Hussain et al. [12] implies (18) and hence Theorem 3 is an extended and generalized version of Theorem 3.1 of [12].

3.2. Application to a System of Integral Equations

In this section, we give an application of Theorem 1 to study the existence and uniqueness of solution of a system of nonlinear integral equations.

Let $X = C[0, A]$ be the space of all continuous real valued functions defined on $[0, A]$, $A > 0$. Our problem is to find $(u(t), v(t)) \in X \times X$, $t \in [0, A]$ such that, for $f : [0, A] \times R \times R \to R$ and $G : [0, A] \times [0, A] \to R$ and $K \in C([0, A]$, the following holds:

$$u(t) = \int_0^A G(t,r) f(t, u(r), v(r)) dr + K(t)$$
$$v(t) = \int_0^A G(t,r) f(t, v(r), u(r)) dr + K(t). \tag{22}$$

Now, suppose $F : X \times X \to X$ is given by

$$F(u(t), v(t)) = \int_0^A G(t,r) f(t, u(r), v(r)) dr + K(t).$$

$$F(v(t), u(t)) = \int_0^A G(t,r) f(t, v(r), u(r)) dr + K(t).$$

Then, (22) is equivalent to the coupled fixed point problem $F(u(t), v(t)) = u(t)$, $F(v(t), u(t)) = v(t)$.

Theorem 4. *The system of Equation (22) has a unique solution provided the following holds:*

(i) $G: [0, A] \times [0, A] \to R$ and $f: [0, A] \times R \times R \to R$ are continuous functions.

(ii) $K \in C([0, A])$.

(iii) For all $x, y, u, v \in X$ and $t \in [0, A]$, we can find a function $g: X \to X$ and real numbers $p \geq 1$, λ, μ, ν with $0 \leq \lambda < 1$, $0 \leq \mu, \nu \leq 1$, minimum $\{\lambda\mu, \lambda\nu\} < \frac{1}{3^{s-1}}$ satisfying

$(iii-a):|f(t,u(r),v(r))) - f(t,x(r),y(r)))|^p \leq \lambda^p \max\{|g(u(r)) - g(x(r))|^p, |g(v(r)) - g(y(r))|^p,$
$\mu|g(u(r)) - F(u(r),v(r))|^p, \mu|g(v(r)) - F(v(r),u(r))|^p,$
$\nu|g(x(r)) - F(x(r),y(r))|^p, \nu|g(y(r)) - F(y(r),x(r))|^p\}.$

$(iii-b)$ $F(g(u(t)), g(v(t))) = g(F(u(t), v(t)))$.

(iv) $\sup_{t \in [0,A]} \int_0^A |G(t,r)|^p dr \leq \frac{1}{\lambda^{p-1}}$.

Moreover, for some arbitrary $u_0(t), v_0(t)$ *in X, the sequence* $(< gu_n(t) >, < gv_n(t) >)$ *defined by*

$$gu_n(t) = \int_0^A G(t,r) f(t, u_{n-1}(r), v_{n-1}(r)) dr + K(t)$$

$$gv_n(t) = \int_0^A G(t,r) f(t, v_{n-1}(r), u_{n-1}(r)) dr + K(t) \qquad (23)$$

converges to the unique solution.

Proof. Define $d_v: X \times X \to R$ such that for all $u, v \in X$,

$$d_v(u,v) = \sup_{t \in [0,A]} |u(t) - v(t)|^s. \qquad (24)$$

Clearly, d_v is a $b_v((v+1)^{s-1})$-metric space.
For some $r \in [0, A]$, we have

$|F(u(t), v(t)) - F(x(t), y(t))|^p$
$= |\int_0^A G(t,r) f(t,u(r),v(r)) dr + g(t) - \int_0^A G(t,r) f(t,x(r),y(r)) dr + g(t)|^p$
$\leq \int_0^A |G(t,r)|^p |f(t,u(r),v(r)) - f(t,x(r),y(r))|^p dr$
$\leq (\int_0^A |G(t,r)|^p dr) \lambda^p [\max\{|g(u(r)) - g(x(r))|^p, |g(v(r)) - g(y(r))|^p,$
$\mu|g(u(r)) - F(u(r),v(r))|^p, \mu|g(v(r)) - F(v(r),u(r))|^p,$
$\nu|g(x(r)) - F(x(r),y(r))|^p, \nu|g(y(r)) - F(y(r),x(r))|^p\}.$
$\leq (\int_0^A |G(t,r)|^p dr) \lambda^p [\max\{d_v(g(u), g(x)), d_v(g(v), g(y)), \mu d_v(g(u), F(u,v)), \mu d_v(g(v), F(v,u)),$
$\nu d_v(g(x), F(x,y)), \nu d_v(g(y), F(y,x))\}.$

Thus, using condition (iv), we have

$d_v(F(u,v), F(x,y)) = \sup_{t \in [0,A]} |F(u(t), v(t)) - F(x(t), y(t))|^p$
$\leq \lambda [\max\{d_v(g(u), g(x)), d_v(g(v), g(y)), \mu d_v(g(u), F(u,v)), \mu d_v(g(v), F(v,u)),$
$\nu d_v(g(x), F(x,y)), \nu d_v(g(y), F(y,x))\}.$

Thus, all the conditions of Corollary 1 are satisfied and so F has a unique coupled fixed point $(u', v') \in C([0, A] \times C([0, A])$, which is the unique solution of (22) and the sequence $(< gu_n(t) >, < gv_n(t) >)$ defined by (23) converges to the unique solution of (22). □

Example 1. Let $X = C[0,1]$ be the space of all continuous real valued functions defined on $[0,1]$ and define $d_3 \colon X \times X \to R$ such that, for all $u, v \in X$,

$$d_3(u,v) = sup_{t \in [0,1]} \mid u(t) - v(t) \mid^2. \tag{25}$$

Clearly, d_3 is a $b_2(3)$-metric. Now, consider the functions $f \colon [0,1] \times R \times R \to R$ given by $f(t,u,v) = t^2 + \frac{9}{20}u + \frac{8}{20}v$, $G \colon [0,1] \times [0,1] \to R$ given by $G(t,r) = \frac{\sqrt{45}(t+r)}{10}$, $K \in C([0,1])$ given by $K(t) = t$. Then, Equation (22) becomes

$$u(t) = t + \int_0^1 \frac{\sqrt{45}(t+r)}{10}(t^2 + \frac{9}{20}u(r) + \frac{8}{20}v(r))dr$$
$$v(t) = t + \int_0^1 \frac{\sqrt{45}(t+r)}{10}(t^2 + \frac{9}{20}v(r) + \frac{8}{20}u(r))dr. \tag{26}$$

Then,

$$\begin{aligned}
\mid f(t,u,v) - f(t,x,y) \mid^2 &= \mid \frac{9}{20}(u-x) + \frac{8}{20}(v-y) \mid^2 \\
&\leq \mid Max\{\frac{9}{10}(u-x), \frac{8}{10}(v-y)\} \mid^2 \\
&\leq \frac{81}{100} Max\{\mid u-x \mid^2, \mid v-y \mid^2\}.
\end{aligned}$$

In addition,

$$sup_{t \in [0,1]} \int_0^1 \mid G(t,r) \mid^2 dr = \int_0^1 \frac{45}{100}(t+r)^2 dr = 1.05.$$

We see that all the conditions of Theorem 4 are satisfied, with $\lambda = \frac{9}{10}, \mu = 0, \nu = 0, p = 2$ and $g = I_X$(Identity mapping). Hence, Theorem 4 ensures a unique solution of (26). Now, for $u_0(t) = 1$ and $v_0(t) = 0$, we construct the sequence $(<u_n(t)>, <v_n(t)>\}$ given by

$$u_n(t) = t + \int_0^1 \frac{\sqrt{45}(t+r)}{10}(t^2 + \frac{9}{20}u_{n-1}(r) + \frac{8}{20}v_{n-1}(r))dr$$
$$v_n(t) = t + \int_0^1 \frac{\sqrt{45}(t+r)}{10}(t^2 + \frac{9}{20}v_{n-1}(r) + \frac{8}{20}u_{n-1}(r))dr. \tag{27}$$

Using MATLAB, we see that above sequence converges to $\{0.6708t^3 + 0.3354t^2 + 2.2339t + 0.7677, 0.6708t^3 + 0.3354t^2 + 2.2339t + 0.7677\}$, and this is the unique solution of the system of nonlinear integral Equation (26). The convergence table is given in Table 1 below.

Table 1. Convergence of sequences $<u_n(t)>$ and $<v_n(t)>$.

n	$u_n(t) = t + \int_0^1 \frac{\sqrt{45}(t+r)}{10}(t^2 + \frac{9}{20}u_{n-1}(r) + \frac{8}{20}v_{n-1}(r))dr$	$v_n(t) = t + \int_0^1 \frac{\sqrt{45}(t+r)}{10}(t^2 + \frac{9}{20}v_{n-1}(r) + \frac{8}{20}u_{n-1}(r))dr$
1	$u_1(t) = t + 0.0167(2t+1)(20t^2+9))$	$v_1(t) = t + .0671(2t+1)(5t^2+2))$
2	$u_2(t) = 0.6708t^3 + 0.3354t^2 + 1.3t + 0.5007$	$v_2(t) = 0.6708t^3 + 0.3354t^2 + 1.29t + 0.5115$
3	$u_3(t) = 0.6708t^3 + 0.3354t^2 + 1.8210t + 0.5174$	$v_3(t) = 0.6708t^3 + 0.3354t^2 + 1.8208t + 0.5171$
4	$u_4(t) = 0.6708t^3 + 0.3354t^2 + 1.9734t + 0.6179$	$v_4(t) = 0.6708t^3 + 0.3354t^2 + 1.9734t + 0.6178$
5	$u_5(t) = 0.6708t^3 + 0.3354t^2 + 2.0743t + 0.6755$	$v_5(t) = 0.6708t^3 + 0.3354t^2 + 2.0743t + 0.6755$
6	$u_6(t) = 0.6708t^3 + 0.3354t^2 + 2.1359t + 0.7111$	$v_6(t) = 0.6708t^3 + 0.3354t^2 + 2.1359t + 0.7111$
7	$u_7(t) = 0.6708t^3 + 0.3354t^2 + 2.1737t + 0.73298$	$v_7(t) = 0.6708t^3 + 0.3354t^2 + 2.1737t + 0.73298$
8	$u_8(t) = 0.6708t^3 + 0.3354t^2 + 2.19699t + 0.7464$	$v_8(t) = 0.6708t^3 + 0.3354t^2 + 2.19699t + 0.7464$
9	$u_9(t) = 0.6708t^3 + 0.3354t^2 + 2.2113t + 0.7547$	$v_9(t) = 0.6708t^3 + 0.3354t^2 + 2.2113t + 0.7547$
10	$u_{10}(t) = 0.6708t^3 + 0.3354t^2 + 2.2200t + 0.7597$	$v_{10}(t) = 0.6708t^3 + 0.3354t^2 + 2.2200t + 0.7597$
11	$u_{11}(t) = 0.6708t^3 + 0.3354t^2 + 2.2254t + 0.7628$	$v_{11}(t) = 0.6708t^3 + 0.3354t^2 + 2.2254t + 0.7628$
12	$u_{12}(t) = 0.6708t^3 + 0.3354t^2 + 2.2287t + 0.7647$	$v_{12}(t) = 0.6708t^3 + 0.3354t^2 + 2.2287t + 0.7647$
13	$u_{13}(t) = 0.6708t^3 + 0.3354t^2 + 2.2308t + 0.7658$	$v_{13}(t) = 0.6708t^3 + 0.3354t^2 + 2.2308t + 0.7658$
14	$u_{14}(t) = 0.6708t^3 + 0.3354t^2 + 2.23199t + 0.7666$	$v_{14}(t) = 0.6708t^3 + 0.3354t^2 + 2.23199t + 0.7666$
15	$u_{15}(t) = 0.6708t^3 + 0.3354t^2 + 2.2328t + 0.7671$	$v_{15}(t) = 0.6708t^3 + 0.3354t^2 + 2.2328t + 0.7671$
16	$u_{16}(t) = 0.6708t^3 + 0.3354t^2 + 2.2333t + 0.7674$	$v_{16}(t) = 0.6708t^3 + 0.3354t^2 + 2.2333t + 0.7674$
17	$u_{17}(t) = 0.6708t^3 + 0.3354t^2 + 2.2336t + 0.7675$	$v_{17}(t) = 0.6708t^3 + 0.3354t^2 + 2.2336t + 0.7675$
18	$u_{18}(t) = 0.6708t^3 + 0.3354t^2 + 2.2338t + 0.7676$	$v_{18}(t) = 0.6708t^3 + 0.3354t^2 + 2.2338t + 0.7676$
19	$u_{19}(t) = 0.6708t^3 + 0.3354t^2 + 2.2339t + 0.7677$	$v_{19}(t) = 0.6708t^3 + 0.3354t^2 + 2.2339t + 0.7677$
20	$u_{20}(t) = 0.6708t^3 + 0.3354t^2 + 2.2339t + 0.7677$	$v_{20}(t) = 0.6708t^3 + 0.3354t^2 + 2.2339t + 0.7677$

Remark 4. *Condition (iv) of Theorem 4 above is weaker than the corresponding conditions used in similar theorems of [11,13,14].*

Remark 5. *In example 1 above, we see that* $\sup_{t \in [0,1]} \int_0^1 |G(t,r)|^2 dr = \int_0^1 \frac{45}{100}(t+r)^2 dr = 1.05 > 1$ *and thus condition (v) of Theorem 3.1 of [11], condition (30) of Theorem 3.1 of [13] and condition (iii) of Theorem 3.1 of [14] are not satisfied.*

Author Contributions: Investigation, R.G., Z.D.M., and S.R.; Methodology, R.G.; Software, Z.D.M.; Supervision, R.G., Z.D.M., and S.R. All authors have read and agreed to the published version of the manuscript.

Funding: This research received no external funding.

Acknowledgments: 1. The authors are thankful to the Deanship of Scientific Research at Prince Sattam bin Abdulaziz University, Al-Kharj, Kingdom of Saudi Arabia, for supporting this research. 2. The authors are thankful to the learned reviewers for their valuable comments which helped in improving this paper.

Conflicts of Interest: The authors declare no conflict of interest.

References

1. Bakhtin, I.A. The contraction mapping principle in quasimetric spaces. *Funct. Anal. Ulianowsk Gos. Ped. Inst.* **1989**, *30*, 26–37.
2. Branciari, A. A fixed point theorem of Banach-Caccioppoli type on a class of generalized metric spaces. *Publicationes Mathematicae Debrecen* **2000**, *57*, 31–37.

3. George, R.; Radenović, S.; Reshma, K.P.; Shukla, S. Rectangular b-metric spaces and contraction principle. *J. Nonlinear Sci. Appl.* **2015**, *8*, 1005–1013. [CrossRef]
4. Hincal, E.; Auwalu, A. A note on Banach contraction mapping principle in cone hexagonal metric space. *Br. J. Math. Comput. Sci.* **2016**, *16*, 1–12.
5. Auwalul, A.; Hincal, E. Kannan type fixed point theorem in cone pentagonal metric spaces. *Intern. J. Pure Appl. Math.* **2016**, *108*, 29–38.
6. Mitrović, Z.D.; Radenović, S. The Banach and Reich contractions in $b_v(s)$-metric spaces. *J. Fixed Point Theory Appl.* **2017**, *19*, 3087–3095. [CrossRef]
7. Mitrović, Z.D. A fixed point theorem for mappings with a contractive iterate in rectangular b-metric spaces. *Matematicki Vesnik* **2018**, *70*, 204–210.
8. George, R.; Mitrović, Z.D. On Reich contraction principle in rectangular cone b-metric space over Banach algebra. *J. Adv. Math. Stud.* **2018**, *11*, 10–16.
9. George, R.; Nabwey, H.A.; Rajagopalan, R.; Radenović, S.; Reshma, K.P. Rectangular cone b-metric spaces over Banach algebra and contraction principle. *Fixed Point Theory Appl.* **2017**, *2017*, 14. [CrossRef]
10. Gu, F. On some common coupled fixed point results in rectangular b-metric spaces. *J. Nonlinear Sci. Appl.* **2017**, *10*, 4085–4098. [CrossRef]
11. Kumar, D.R.; Pitchaimani, M. New coupled fixed point theorems in cone metric spaces with applications to integral equations and Markov process. *Trans. A. Razmadze Math. Inst.* **2018**. [CrossRef]
12. Hussain, N.; Salimi, P.; Al-Mezel, S. Coupled fixed point results on quasi-Banach spaces with application to a system of integral equations. *Fixed Point Theory Appl.* **2013**, *2013*, 261. [CrossRef]
13. Nashine, H.K.; Sintunavarat, W.; Kumam, P. Cyclic generalized contractions and fixed point results with applications to an integral equation. *Fixed Point Theory Appl.* **2012**, *2012*, 217. [CrossRef]
14. Hussain, N.; Roshan, J.R.; Parvaneh, V.; Abbas, M. Common fixed point results for weak contractive mappings in ordered b-dislocated metric spaces with applications. *J. Ineq. Appl.* **2013**, *2013*, 486. [CrossRef]
15. Garai, H.; Dey, L.K.; Mondal, P.; Radenović, S. Some remarks and fixed point results with an application in $b_v(s)$-metric spaces. *Nonlinear Anal. Model. Control* **2020**, *25*, 1015–1034. [CrossRef]

Publisher's Note: MDPI stays neutral with regard to jurisdictional claims in published maps and institutional affiliations.

© 2020 by the authors. Licensee MDPI, Basel, Switzerland. This article is an open access article distributed under the terms and conditions of the Creative Commons Attribution (CC BY) license (http://creativecommons.org/licenses/by/4.0/).

Article

New Results on Start-Points for Multi-Valued Maps

Yaé Ulrich Gaba [1,2], Erdal Karapınar [3,4,5,*], Adrian Petruşel [6,7] and Stojan Radenović [8]

1. Institut de Mathématiques et de Sciences Physiques (IMSP), Porto-Novo 01 BP 613, Benin; yaeulrich.gaba@gmail.com
2. African Center for Advanced Studies (ACAS), P.O. Box 4477, Yaoundé, Cameroon
3. Division of Applied Mathematics, Thu Dau Mot University, Thu Dau Mot City 820000, Binh Duong Province, Vietnam
4. Department of Mathematics, Çankaya University, Etimesgut, Ankara 06790, Turkey
5. Department of Medical Research, China Medical University Hospital, China Medical University, Taichung 40402, Taiwan
6. Faculty of Mathematics and Computer Science, Babeş-Bolyai University Cluj-Napoca, Str.Mihail Kogălniceanu 1, 400000 Cluj-Napoca, Romania; petrusel@math.ubbcluj.ro
7. Academy of Romanian Scientists, Str. Splaiul Independenţei 54, Sector 5, RO-050094 Bucharest, Romania
8. Faculty of Mechanical Engineering, University of Belgrade, Kraljice Marije 16, 11120 Beograd, Serbia; radens@beotel.rs
* Correspondence: erdalkarapinar@tdmu.edu.vn or erdalkarapinar@yahoo.com

Received: 3 November 2020; Accepted: 28 November 2020; Published: 3 December 2020

Abstract: In this manuscript we investigate the existence of start-points for the generalized weakly contractive multi-valued mappings in the setting of left K-complete quasi-pseudo metric space. We provide an example to support the given result.

Keywords: quasi-pseudometric; start-point; end-point; fixed point; weakly contractive

MSC: Primary 47H05; Secondary 47H09; 47H10; 54H25

1. Introduction and Preliminaries

The quasi-pseudo metric space, which is obtained by relaxing the symmetry condition, is one of the refinements of the notion of metric space. In the point view of fixed point theory, the lack of the symmetry axiom leads to consider the orientation in this new structure. Roughly speaking, fixed points for mappings are usually limits of the Picard sequence, which is constructed by the recursive iteration of the operator by starting with an arbitrarily chosen point. On the other hand, in this new structure, the distance function is not symmetric. Consequently, for an arbitrary initial value ξ_0, the value of the distance from its n-th iteration, $T^n x_0$, to its limit, say x^* (if exists), and the value of the distance from its limit, x^* (if exists), to its n-th iteration, $T^n x_0$, need not be equal. Under this motivation, the notions of start-point, end-point, ε-start-point, and ε-end-point were defined in [1]. In other words, fixed point has been investigated in the oriented structure, quasi-pseudo metric space, under the names of start-point and end-point. It is clear that, under the condition symmetry, the start-points and end-points coincide with the fixed points [2–5].

An initial result in the theory of start-point was given in [1] in order to extend the idea of fixed points for multi-valued mappings defined on quasi-pseudo metric spaces. A series of three papers, see [1,6,7], has given a more or less detailed introduction to the subject. The theory of start-point came to extend the idea of fixed points for multi-valued mappings that are defined on quasi-pseudo metric spaces. More detailed introduction to the subject can be read in [1,6–13].

In this paper, we investigate the existence of start-points and end-points for a class of mappings, which are known as generalized weakly contractive multi-valued maps, in the context of left K-complete quasi-pseudo metric space.

Intuitively, as we mentioned above, the appropriate framework for the theory of start-point is the quasi-metric setting. For the sake of completeness, we recollect, in the present manuscript, the necessary notations and fundamental concepts from the literature. We first recall the basic notions regarding quasi-metric spaces as well as some additional definitions that are related to multi-valued maps on these spaces[14–16]. For a general approach in metric fixed point theory for multi-valued operators, see [17–19].

Definition 1 (See [1]). *Let $q : X \times X \to [0, \infty)$ be a function where X is a non-empty set. The function is called a **quasi-pseudometric** (respectively, T_0-**quasi-metric**) on X if (q_1) and (q_2) (respectively, $(q_1)^*$ and (q_2)) hold, where*

(q_1) $q(\xi, \xi) = 0$ for all $\xi \in X$,
$(q_1)^*$ $q(\xi, \eta) = 0 = q(\eta, \xi)$ implies $\xi = y$, and
(q_2) $q(\xi, \zeta) \leq q(\xi, \eta) + q(\eta, \zeta)$ for all $\xi, \eta, \zeta \in X$.

Note that the condition $(q_1)^*$ is known as the T_0-condition. Furthermore, for a quasi-pseudo metric q on X, the function $q^{-1} : X \times X \to [0, \infty)$, which is defined by $q^{-1}(\xi, \eta) = q(\eta, \xi)$ for all $\xi, \eta \in X$, forms a quasi-pseudo metric on the same set X and is named as the conjugate of q. For a T_0-quasi-metric d on X, a distance function $d_q : X \times X \to [0, \infty)$, defined by $d_q(\xi, \eta) = \max\{q(\xi, \eta), q(\eta, \xi)\}$ for all $(\xi, \eta) \in X \times X$, becomes a metric on X.

Remark 1. *In some sources, the quasi-pseudo metric is called hemi-metric (see [20]). Moreover, T_0-quasi-metric is known also as a quasi-metric in the literature.*

In what follows, we consider three well-known examples in order to illustrate the validity of Definition 1.

Example 1 (Truncated difference). *Set $\mathbb{R}_0^+ := [0, \infty)$ and $\delta : \mathbb{R}_0^+ \times \mathbb{R}_0^+ \to \mathbb{R}_0^+$ be given, for any $\xi, \eta \in X$, by*

$$\delta(\xi, \eta) = \max\{0, \xi - \eta\}.$$

Under these conditions, δ forms a T_0-quasi-metric. Further, the pair $(\mathbb{R}_0^+, \delta)$ becomes a T_0-quasi-metric space.

Example 2 (cf. [21]). *Let A, B be two non-empty set, such that $A \cap B \neq \emptyset$. Set $X = A \cup B$ and $q : X \times X \to [0, \infty)$ be given, for any $a, b \in X$, by*

$$q(a, b) = \begin{cases} 0 & \text{if } a = b, \\ \frac{3}{2} & \text{if } a \in A, b \in B, \\ 2 & \text{if } b \in A, a \in B, \\ 1 & \text{otherwise.} \end{cases}$$

Under these conditions, q forms a T_0-quasi-metric. Further, the pair (X, q) becomes a T_0-quasi-metric space.

Example 3 (cf. [22]). *Set $\mathbb{I} := [0, 1]$, and define $\delta : \mathbb{I} \times \mathbb{I} \to \mathbb{R}_0^+$ be defined as*

$$\delta(\xi, \eta) = \begin{cases} \xi - \eta, & : \xi \geq \eta, \\ 1, & : \xi < \eta. \end{cases}$$

Under these conditions, δ forms a quasi-pseudo metric that is obviously not T_0.

For a quasi-pseudo metric space (X, q), we define an open ε-ball at a point ξ as follows: For $\xi \in X$ and $\varepsilon > 0$,

$$B_q(\xi, \varepsilon) = \{\eta \in X : q(\xi, \eta) < \varepsilon\}.$$

Let (X,q) be a quasi-pseudo metric space. We say that the sequence $\{\xi_n\}$ is q-convergent to ξ (or left-convergent to ξ), if
$$q(\xi_n,\xi) \longrightarrow 0,$$
and we denote this fact by $\xi_n \xrightarrow{q} \xi$. More precisely, $\{\xi_n\}$ converges to ξ with respect to $\tau(q)$.

In a similar manner, a sequence $\{\xi_n\}$ is q^{-1}-convergent to ξ (or right-convergent to ξ), if
$$q(\xi,\xi_n) \longrightarrow 0, \tag{1}$$

fact denoted by $\xi_n \xrightarrow{q^{-1}} \xi$. Actually, $\{\xi_n\}$ converges to ξ with respect to $\tau(q^{-1})$

A sequence $\{\xi_n\}$, in the setting of a quasi-pseudo metric space (X,q), is said to be d_q-convergent to ξ in the case the sequence converges to ξ from left and right, which is,

$$\xi_n \xrightarrow{q} \xi \text{ and } \xi_n \xrightarrow{q^{-1}} \xi.$$

Moreover, it is denoted as $\xi_n \xrightarrow{d_q} \xi$ (or, $\xi_n \longrightarrow \xi$, if there is no confusion).

Remark 2. *From the definition of d_q-convergence, we have*

$$d_q\text{-convergence implies } q\text{-convergence.}$$

The reverse implication does not hold in general, as demonstrated in the following example.

Example 4 (cf. [22]). *Set $\mathbb{I} := [0,1]$, and define $q : \mathbb{I} \times \mathbb{I} \to \mathbb{R}_0^+$ be defined as*

$$q(\xi,\eta) = \begin{cases} 0 & : \xi \leq y \\ 1 & : \xi > y \end{cases}$$

Subsequently, it is evident that (X,q) forms a quasi-pseudo metric space.

Consider
$$\xi_n = \begin{cases} \frac{1}{2} + 2^{-n} & : n \text{ is odd} \\ \frac{1}{3} + 3^{-n} & : n \text{ is even} \end{cases}$$

It is easy to see that the sequence $\{\xi_n\}$ is right-convergent (to $1/3$) and left-convergent (to 1), but not d_q-convergent.

Definition 2 (See e.g., [1]). *A sequence $\{\xi_n\}$ in a quasi-pseudo metric space (X,q) is called left K-Cauchy if for every $\varepsilon > 0$, there exists $n_0 \in \mathbb{N}$, such that*

$$\text{for all } n,k : n_0 \leq k \leq n \quad q(\xi_k,\xi_n) < \varepsilon;$$

Similarly, we define right K-Cauchy sequences and observe that a sequence is left K-Cauchy with respect to q if and only if it is right K-Cauchy with respect to q^{-1}.

Example 5 (See [8]). *Set $\mathbb{I} := (0,1)$, and define $\delta : \mathbb{O} \times \mathbb{O} \to \mathbb{R}_0^+$ be defined as*

$$q(\xi,\eta) = \begin{cases} \xi - \eta & : \xi \geq \eta \\ 1 & : \xi < \eta \end{cases}$$

Let us define the sequence $\{\xi_n\}$ given by $\xi_n = (n+1)^{-1}$. Subsequently,

$$q(\xi_r,\xi_s) < r^{-1}$$

for all $s > r$; hence, $\{\xi_n\}$ is left K-Cauchy. However, $\{\xi_n\}$ is not right K-Cauchy, since whenever $\xi \in X, q(\xi_m, \xi) = 1$ after a certain stage. On the other hand, if one considers the sequence $\{\eta_n\}$ where $\eta_n = 1 - (n+1)^{-1}$, one could easily see that it is right K-Cauchy.

Definition 3 (See [1,13]). *We say that (X,q) is left-K-complete if any left K-Cauchy sequence is q-convergent. Furthermore, we say that quasi-pseudo metric space (X,q) is Smyth complete if any left K-Cauchy sequence is d_q-convergent.*

It is easy to see that every Smyth-complete quasi-metric space is left K-complete [13], and the converse implication does not hold.

Definition 4 ([1]). *We say that a T_0-quasi-metric space (X,q) is said to be bicomplete if the corresponding metric d_q on X is complete.*

Example 6. Let us again consider Example 1. In that case, for any $\xi, \eta \in X = [0, \infty)$, we have that $d_q(\xi, \eta) = \max\{\xi - \eta, \eta - \xi\} = |\xi - \eta|$. We know that $(\mathbb{R}, |\,.\,|)$ is a complete metric space; hence, $([0, \infty), |\,.\,|)$ is an example of bicomplete T_0-quasi-metric space.

However, if we take the quasi-pseudo metric that is defined in Example 3, it is clear that (X, δ) is not bicomplete, since (X, δ) is not even T_0.

Definition 5 ([1]). *Let A be a subset of a quasi-pseudo metric space (X,q). We say that A is bounded if there exists a $\Delta > 0$, such that $q(\xi, \eta) < \Delta$ whenever $\xi, \eta \in A$.*

Example 7.

1. Let $X = \{a, b, c\}$. The map $q : X \times X \to [0, \infty)$ defined by $q(a,b) = q(a,c) = 0$, $q(b,a) = q(b,c) = 1$, $q(c,a) = q(c,b) = 2$ and $q(\xi, \xi) = 0$ for all $\xi \in X$ is a bounded T_0-quasi-metric on X. Indeed, for any $\xi, \eta \in X$, $q(\xi, \eta) \leq 2$.
2. The quasi-pseudo metric presented in Example 4 is bounded, as for any $\xi, \eta \in X$, $q(\xi, \eta) \leq 1$.

Let (X,q) be a quasi-pseudo metric space. We set $\mathscr{P}_0(X) := \mathscr{P}_0(X) \setminus \{\varnothing\}$, where $\mathscr{P}_0(X)$ denotes the power set of X.

$$\mathscr{P}_{cb}(X) := \{A \in \mathscr{P}_0(X) : A \text{ closed and bounded}\},$$
$$\mathscr{P}_k(X) := \{A \in \mathscr{P}_0(X) : A \text{ compact }\},$$
$$\mathscr{P}_c(X) := \{A \in \mathscr{P}_0(X) : A \text{ closed }\}.$$

For $\xi \in X$ and $A \in \mathscr{P}_0(X)$, we set:

$$q(\xi, A) := \inf\{q(\xi, a), a \in A\}, \quad q(A, \xi) := \inf\{q(a, \xi), a \in A\}.$$

We also define the map $H : \mathscr{P}_0(X) \times \mathscr{P}_0(X) \to [0, \infty]$ by

$$H(A,B) = \max\left\{\sup_{a \in A} q(a,B), \sup_{b \in B} q(A,b)\right\} \text{ whenever } A, B \in \mathscr{P}_0(X).$$

Subsequently, the distance function H is called the Hausdorff extended quasi-pseudo metric on $\mathscr{P}_0(X)$. Notice that, here, the word "extended" is use to emphasize that H can attain the value ∞ as it appears in the definition.

Finally, we recall some concepts that are related to the classical fixed point notions in the setting of a quasi-pseudo metric space.

Definition 6 (cf.[1]). *Let (X,q) be a quasi-pseudo metric space and $F: X \to \mathscr{P}_0(X)$ be a multi-valued map. Suppose that H is a Hausdorff quasi-pseudo metric on $\mathscr{P}_0(X)$. We say that $\xi \in X$ is*

(i) *a fixed point of F if $\xi \in F\xi$,*
(ii) *a strict fixed point if $F\xi = \{\xi\}$,*
(iii) *a start-point of F if $H(\{\xi\}, F\xi) = 0$, and*
(iv) *an end-point of F if $H(F\xi, \{\xi\}) = 0$.*

In this context, we can also write $H(\eta, F\eta) := H(\{\eta\}, F\eta)$, $\eta \in X$. Notice that $H(\{\eta\}, F\eta) = \sup_{\psi \in F\eta} q(\eta, \psi)$, while $H(F\eta, \{\eta\}) = \sup_{\psi \in F\eta} q(\psi, \eta)$.

2. Main Results

In this section, we give a new start-point theorem for a generalized weakly contractive multi-valued map.

As we dive into the topic, it could be very interesting to point out this known fact, which is always good to remember. That is, if ξ is both a start-point and an end-point of a multi-valued F, then ξ is a fixed point of F. In fact, $F\xi$ is a singleton. Observe that a fixed point of a multi-valued F need not be a start-point or an end-point. We provide the following three examples in order to illustrate that fact.

Example 8. *Consider the T_0-quasi-pseudo metric space (X,q), where $X = \{a,b,c\}$ and q defined by $q(a,b) = q(a,c) = 0$, $q(b,a) = q(b,c) = 2$, $q(c,a) = q(c,b) = 4$ and $q(\xi, \xi) = 0$ for $\xi = a, b, c$. The multi-valued map $F: X \to \mathscr{P}_0(X)$ is considered by $Fa = \{a,b\}$ and $F\xi = X \setminus \{\xi\}$ for $\xi = b, c$. Obviously, a is a fixed point for F. Moreover, since*

$$H(\{a\}, Fa) = \max\{q(a,a), q(a,b)\} = 0,$$

we derive that a is a start-point, but, since

$$H(Fa, \{a\}) = \max\{q(a,a), q(b,a)\} = 2 \neq 0,$$

we derive that a is not an end-point. Furthermore, there is no other start-point or end-point for F.

Example 9. *Consider the T_0-quasi-pseudo metric space (X,q), as defined in the previous example (Example 8). The multi-valued map $F: X \to \mathscr{P}_0(X)$ is considered by $F\xi = \{a,b\}$ for $\xi = a, b, c$. Obviously, a, b are fixed points for F. Again, a is a start-point, but not an end-point. Observe this time around that b is an end-point, but not a start-point.*

Example 10. *Consider the T_0-quasi-pseudo metric space (X,q), as defined in the previous example (Example 8). The multi-valued map $F: X \to \mathscr{P}_0(X)$ is considered by $Fa = \{b\}$, $Fb = \{c\}$, $Fc = \{a\}$. The map F does not have any fixed point. However, we can easily that a is the only start-point and c the only end-point for F.*

Remark 3. *So far in the examples, we have been obtaining fixed points. Let us observe what happens when we are in the presence of a strict fixed point.*

Example 11. *Consider the T_0-quasi-pseudo metric space (X,q), where $X = \{a,b,c\}$ and q defined by $q(a,b) = q(a,c) = q(b,c) = 0$, $q(b,a) = 2$, $q(c,a) = q(c,b) = 4$ and $q(\xi, \xi) = 0$ for $\xi = a, b, c$. We define, on X, the multi-valued map $F: X \to \mathscr{P}_0(X)$ by $Fa = \{a\}$ and $Fb = Fc = \{b,c\}$ for $\xi = b, c$.*

$$H(\{a\}, Fa) = q(a,a) = 0,$$

and

$$H(Fa, \{a\}) = q(a,a) = 0,$$

i.e., a is is both a start-point and an end-point for F.

The point b is both a fixed point (which is not strict) and end-point for F, while c is neither a (strict) fixed point nor a start-point nor an end-point for F.

In fact, the above example illustrates the following fact:

Lemma 1. *Let X be non-empty set and H the Hausdorff quasi-pseudo metric that is derived by a quasi-pseudo metric q. Let $F : X \to \mathscr{P}_0(X)$ be a multi-valued map. If $\xi \in X$ is a strict fixed point, then ξ is both a start-point and an end-point.*

Proof. The result is immediate, since, for $F\xi = \{\xi\}$, we have

$$H(\{\xi\}, F\xi) = q(\xi, \xi) = 0 = q(\xi, \xi) = H(F\xi, \{\xi\}) = 0.$$

□

We begin with the following intermediate result.

Lemma 2. *Let (X, q) be T_0-quasi-metric space and $A \subset X$. If A is a compact subset of (X, d_q), then it is a closed subset of (X, q). That is, $\mathscr{P}_k(X) \subset \mathscr{P}_c(X)$.*

Proof. Let $\{\xi_n\}$ be a sequence in A, such that $q(\xi, \xi_n) \to 0$ for some $\xi \in X$. Because A is a compact subset of (X, d_q), there exists a subsequence $\{\xi_{n_k}\}$ of $\{\xi_n\}$ and a point $\zeta \in A$, such that $d_q(\zeta, \xi_{n_k}) \to 0$. Thus, we have $q(\xi_{n_k}, \zeta) \to 0$. While using the triangle inequality, we have

$$q(\xi, \zeta) \leq q(\xi, \xi_{n_k}) + q(\xi_{n_k}, \zeta).$$

Letting $k \to \infty$ in above inequality, we obtain $\xi = \zeta$ and $\xi \in A$. Thus, A is a closed subset of (X, q). □

The concept of weakly contractive maps that appeared in [23] (Definition 1) is one of the generalizations of contractions on metric spaces. In [23], the authors defined such maps for single valued maps on Hilbert spaces and proved the existence of fixed points. Later, it was shown that most of the results of [23] still hold in any Banach space, see e.g., Rhoades[24–29]. As it is expected, this notion was extended to multi-valued maps and it was characterized in the setting of quasi-metric spaces.

In the literature, one of the useful auxiliary function is the comparison function that is initiated by [30], and, later, discussed and investigated densely by Rus [31] and many others. A function $\varphi : [0, \infty) \to [0, \infty)$ is called a comparison function [30,31] if it is increasing and $\varphi^n(t) \to 0$ as $n \to \infty$ for every $t \in [0, \infty)$, where φ^n is the n-th iterate of φ. A simple example of such mappings is $\varphi(t) = \frac{kt}{n}$, where $k \in [0, 1)$ and $n \in \{2, 3, \cdots\}$.

Let Γ be the family of functions $\gamma : [0, \infty) \to [0, \infty)$ satisfying the following conditions:

(Γ_1) γ is nondecreasing;

(Γ_2) $\sum_{n=1}^{+\infty} \gamma^n(t) < \infty$ for all $t > 0$.

Subsequently, a function $\phi \in \Gamma$ is called (c)-comparison function, see also [31,32].

Lemma 3 ([31]). *If $\gamma : [0, \infty) \to [0, \infty)$ is a comparison function, then*

1. *each iterate γ^k of γ, $k \geq 1$ is also a comparison function;*
2. *γ is continuous at 0; and,*
3. *$\gamma(t) < t$ for all $t > 0$.*

The listed properties above are also valid for (c)-comparison functions, since the class of (c)-comparison functions is a subclass of comparison functions.

For our own purpose, we introduce the $(c)^*$-comparison function, as follows:

Definition 7. *A function $\gamma : [0, \infty) \to [0, \infty)$ is called a $(c)^*$-comparison function if*

(γ_1) γ is nondecreasing with $\gamma(0) = 0$ and $0 < \gamma(t) < t$ for each $t > 0$; and,

(γ_2^*) for any sequence $\{t_n\}$ of $(0, \infty)$, $\sum_{n=1}^{\infty} \gamma(t_n) < \infty$ implies $\sum_{n=1}^{\infty} t_n < \infty$.

Definition 8. *Let (X, q) be T_0-quasi-metric space.*

1. A multi-valued map $F : X \to \mathscr{P}_0(X)$ is called weakly contractive if there exists a $(c)^*$-comparison function γ, such that, for each $\xi \in X$ there exists $\eta \in F\xi$ satisfying

$$H(\eta, F\eta) \leq q(\xi, \eta) - \gamma(q(\xi, \eta)). \quad (2)$$

2. A single-valued map $f : X \to X$ is called weakly contractive if there exists a $(c)^*$-comparison function γ, such that

$$q(f\xi, f\eta) \leq q(\xi, \eta) - \gamma(q(\xi, \eta)), \text{ for every } \xi, \eta \in X. \quad (3)$$

The following is the main result of the paper.

Theorem 1. *Let (X, q) be a left K-complete quasi-pseudo metric space, $F : X \to \mathscr{P}_{cb}(X)$ be a weakly contractive multi-valued mapping. Subsequently, F has a start-point in X.*

Proof. Let $\xi_0 \in X$ be arbitrary. By (2), there exists $\xi_1 \in F\xi_0$, such that, for every $\xi_2 \in F\xi_1$, we have

$$q(\xi_1, \xi_2) \leq H(\xi_1, F\xi_1) \leq q(\xi_0, \xi_1) - \gamma(q(\xi_0, \xi_1)).$$

Again, by (2), there exists an element $\xi_2 \in F\xi_1$, such that, for every $\xi_3 \in F\xi_2$, we have

$$q(\xi_2, \xi_3) \leq H(\xi_2, F\xi_2) \leq q(\xi_1, \xi_2) - \gamma(q(\xi_1, \xi_2)) \leq q(\xi_1, \xi_2) \leq H(\xi_1, F\xi_1).$$

Continuing this process, we can find a sequence $\{\xi_n\} \subset X$, such that, for $n \in \{0, 1, 2, \cdots\}$, we have

$$\xi_{n+1} \in F\xi_n$$

and

$$q(\xi_{n+1}, \xi_{n+2}) \leq H(\xi_{n+1}, F\xi_{n+1}) \leq q(\xi_n, \xi_{n+1}) - \gamma(q(\xi_n, \xi_{n+1})) \leq q(\xi_n, \xi_{n+1}) \leq H(\xi_n, F\xi_n).$$

Thus, the sequence $\{q(\xi_n, \xi_{n+1})\}$ is non-increasing and so we can conclude that $\lim_{n \to \infty} q(\xi_n, \xi_{n+1}) = l$ for some $l \geq 0$. We show that $l = 0$. Suppose that $l > 0$. Subsequently, we have

$$q(\xi_n, \xi_{n+1}) \leq q(\xi_{n-1}, \xi_n) - \gamma(q(\xi_{n-1}, \xi_n)) \leq q(\xi_{n-1}, \xi_n) - \gamma(l),$$

and so

$$q(\xi_{n+N}, \xi_{n+N+1}) \leq q(\xi_{n-1}, \xi_n) - N\gamma(l),$$

which is a contradiction for N large enough. Thus, we have

$$\lim_{n \to \infty} q(\xi_n, \xi_{n+1}) = 0.$$

For $m \in \mathbb{N}$ with $m \geq 3$, we have

$$q(\xi_{m-1},\xi_m) \leq q(\xi_{m-2},\xi_{m-1}) - \gamma(q(\xi_{m-2},\xi_{m-1})) \cdots$$
$$\leq q(\xi_1,\xi_2) - \gamma(q(\xi_1,\xi_2)) - \cdots - \gamma(q(\xi_{m-2},\xi_{m-1})).$$

Hence, we get

$$\sum_{k=1}^{m-2} \gamma(q(\xi_k,\xi_{k+1})) \leq q(\xi_1,\xi_2) - q(\xi_{m-1},\xi_m).$$

Letting $m \to \infty$ in above inequality, we obtain

$$\sum_{k=1}^{\infty} \gamma(q(\xi_k,\xi_{k+1})) \leq q(\xi_1,\xi_2) < \infty,$$

which implies, using (γ_2^*), that

$$\sum_{k=1}^{\infty} q(\xi_k,\xi_{k+1}) < \infty.$$

We conclude that $\{\xi_n\}$ is a left K-Cauchy sequence. On account of the left K-completeness, there exists $\xi^* \in X$, such that $\xi_n \xrightarrow{q} \xi^*$.

Given the function $h\xi := H(\xi, F\xi)$, observe that the sequence $\{h\xi_n\} = \{H(\xi_n, F\xi_n)\}$ is decreasing and it converges to 0. Recall that h is $\tau(q)$-lower semicontinuous (as supremum of $\tau(q)$-lower semicontinuous functions), which yields

$$0 \leq h\xi^* \leq \liminf_{n \to \infty} h\xi_n = 0.$$

Hence, $h\xi^* = 0$, i.e. $H(\{\xi^*\}, F\xi^*) = 0$. This completes the proof. □

Remark 4. *It is clear that, if we replace the condition (2) by the dual condition*

$$H(F\eta, \eta) \leq q(\eta, \xi) - \gamma(q(\eta, \xi)), \tag{4}$$

then the conclusion of Theorem 1 would be that the multi-valued function F possesses an end-point. Moreover for the multi-valued function F to admit a fixed point, it is enough that

$$H^{d_q}(F\eta, \eta) \leq \min\{q(\xi,\eta) - \gamma(q(\xi,\eta)), q(\eta,\xi) - \gamma(q(\eta,\xi))\}, \tag{5}$$

where

$$H^{d_q}(A,B) = \max\left\{\sup_{a \in A} d_q(a,B), \sup_{b \in B} d_q(A,b)\right\} \text{ whenever } A, B, \in \mathscr{P}_0(X).$$

If let $\gamma(t) = (1-k)t$ for $k \in [0,1)$ in Theorem 1, then we obtain the following version of Nadler's theorem in the setting of left K-complete quasi-pseudo metric space.

Theorem 2. *Let (X,q) be a left K-complete quasi-pseudo metric space and $F : X \to \mathscr{P}_{cb}(X)$ be a multi-valued mapping. If there exists $k \in [0,1)$, such that, for each $\xi \in X$, there exists $\eta \in F\xi$ satisfying*

$$H(\eta, F\eta) \leq kq(\xi, \eta),$$

then F possesses a start-point in X.

We conclude this part of the paper with the following illustrative example:

Example 12. Let
$$X = \left\{ \frac{1}{2^n} : n = 0, 1, 2, \cdots \right\} \cup \{0\}$$

and let
$$q(\xi, \eta) = \begin{cases} \eta - \xi, & \text{if } \eta \geq \xi, \\ 2(\xi - \eta), & \text{if } \xi > \eta. \end{cases}$$

Subsequently, (X, q) is a left K-complete T_0-quasi-metric space. Set $\gamma(t) = \frac{t}{2}$ for all $t \geq 0$. Let $F : X \to \mathcal{P}_{cb}(X)$ be a multi-valued map defined as

$$F\xi = \begin{cases} \left\{ \frac{1}{2^{n+1}}, 0 \right\} & \text{if } \xi = \frac{1}{2^n} : n = 0, 1, 2, \cdots, \\ \{0\}, & \text{if } \xi = 0. \end{cases}$$

We now show that F satisfies condition (2).

Case 1. $\xi = 0$, there exists $\eta = 0 \in F\xi = F0 = \{0\}$ such that
$$0 = H(\eta, F\eta) = H(0, F0) \leq q(0,0) - \gamma(q(0,0)) = 0.$$

Case 2. $\xi = \frac{1}{2^n}$, there exists $\eta = 0 \in F\xi = \left\{ \frac{1}{2^n}, 0 \right\}$, such that
$$0 = H(\eta, F\eta) = H(0, F0) \leq q\left(\frac{1}{2^n}, 0\right) - \gamma\left(q\left(\frac{1}{2^n}, 0\right)\right).$$

The map F satisfies the assumptions of Theorem 1, so it has a start-point, which, in this case, is 0.

In the case of a single-valued mapping, Theorem 1 produces the following existence result.

Theorem 3. *Let (X, q) be a left K-complete quasi-pseudo metric space and $f : X \to X$ be a weakly contractive single-valued mapping. Subsequently, f possesses at least one start-point in X, i.e., there exists $\xi^* \in X$, such that $q(\xi^*, f\xi^*) = 0$.*

We conclude the paper with a start-point result for a multi-valued mapping satisfying a stronger weakly contractive type condition. In this case, we can obtain a stability result for the start-point problem.

Definition 9. *Let (X, q) be T_0-quasi-metric space. A multi-valued mapping $F : X \to \mathcal{P}_0(X)$ is called s-weakly contractive if there exists a $(c)^*$-comparison function γ, such that, for each $\xi \in X$, there exists $\eta \in F\xi$ satisfying*

$$H(\eta, Fv) \leq q(\xi, v) - \gamma(q(\xi, v)), \text{ for every } v \in X. \quad (6)$$

Notice that any s-weakly contractive multi-valued mapping is weakly contractive, but not reversely.

The following existence and stability result holds for s-weakly contractive multi-valued mappings. For the sake of simplicity, we will present the result when $\gamma(t) = (1-k)t, t \in [0, \infty)$, with some $k \in [0, 1)$.

Theorem 4. *Let (X, q) be a left K-complete quasi-pseudo metric space and $F : X \to \mathcal{P}_{cb}(X)$ be a multi-valued mapping. Suppose that there exists $k \in [0, 1)$, such that, for each $\xi \in X$, there exists $\eta \in F\xi$ satisfying*

$$H(\eta, Fv) \leq kq(\xi, v), \text{ for every } v \in X.$$

Then:

(a) *F possesses a start-point in X; and,*

(b) *the start-point problem for F is Ulam–Hyers stable with respect to the end-point problem for F, in the sense that there exists $C > 0$, such that, for any $\varepsilon > 0$ and any $\rho^* \in X$ with $H(F\rho^*, \rho^*) \leq \varepsilon$, there exists a start-point $\xi^* \in X$ of F, such that $q(\xi^*, \rho^*) \leq C\varepsilon$.*

Proof.

(a) follows by Theorem 1. Denote, by $\xi^* \in X$, a start-point of F.

(b) For any $u \in F\xi^*$, we can write

$$q(\xi^*, \rho^*) \leq H(\xi^*, F\xi^*) + H(u, F\rho^*) + H(F\rho^*, \rho^*) = H(u, F\rho^*) + H(F\rho^*, \rho^*).$$

For $\xi^* \in X$, there exists $u^* \in F\xi^*$, such that $H(u^*, F\rho^*) \leq kq(\xi^*, \rho^*)$.

Thus,

$$q(\xi^*, \rho^*) \leq \frac{1}{1-k}\varepsilon.$$

□

Author Contributions: Conceptualization, Y.U.G., E.K., A.P. and S.R.; investigation, Y.U.G., E.K., A.P. and S.R.; writing—review and editing, Y.U.G., E.K., A.P. and S.R. All authors have read and agreed to the published version of the manuscript.

Funding: This research received no external funding.

Acknowledgments: Y.U.G. wishes to thank the Sefako Makgatho Health Sciences University (SU), in South Africa, which accepted him as a visitor in November 2020 and provided funding for his stay.

Conflicts of Interest: The authors declare no conflict of interest.

References

1. Gaba, Y.U. Start-points and (α, γ)-contractions in quasi-pseudo metric spaces. *J. Math.* **2014**, *2014*, 709253. [CrossRef]
2. Agarwal, R.P.; Karapinar, E.; O'Regan, D.; Roldán-López-de-Hierro, A.F. *Fixed Point Theory in Metric Type Spaces*; Springer International Publishing Switzerland: Cham, Switzerland, 2015.
3. Ćirić, L. *Some Recent Results in Metrical Fixed Point Theory*; University of Belgrade: Beograd, Serbia, 2003.
4. Kirk, W.; Shahzad, N. *Fixed Point Theory in Distance Spaces*; Springer International Publishing Switzerland: Berlin/Heidelberg, Germany, 2014.
5. Todorčević, V. *Harmonic Quasiconformal Mappings and Hyperbolic Type Metrics*; Springer: Cham, Switzerland, 2019.
6. Gaba, Y.U. Advances in start-point theory for quasi-pseudo metric spaces. *Bull. Allahabad Math. Soc.* **2015**, *30*, 119–146.
7. Gaba, Y.U. New Results in the start-point theory for quasi-pseudo metric spaces. *J. Oper.* **2014**, *2014*, 741818.
8. Alegre, C.; Marin, J. Modified w-distances on quasi-metric spaces and a fixed point theorem on complete quasi-metric spaces. *Topol. Appl.* **2016**, *203*, 32–41. [CrossRef]
9. Alegre, C.; Marin, J. A Caristi fixed point theorem for complete quasi-metric spaces by using mw-distances. *Fixed Point Theory* **2018**, *19*, 25–31. [CrossRef]
10. Künzi, H.-P. An introduction to quasi-uniform spaces. *Contemp. Math.* **2009**, *486*, 239–304.
11. Marín, J.; Romaguera, S.; Tirado, P. Generalized contractive set-valued maps on complete preordered quasi-metric spaces. *J. Funct. Spaces Appl.* **2013**, *2013*, 269246. [CrossRef]
12. Marín, J.; Romaguera, S.; Tirado, P. Q-functions on quasimetric spaces and fixed points for multivalued maps. *Fixed Point Theory Appl.* **2011**, *2011*, 603861. [CrossRef]
13. Romaguera, S.; Tirado, P. A characterization of Smyth complete quasi-metric spaces via Caristi's fixed point theorem. *Fixed Point Theory Appl.* **2015**, *2015*, 183. [CrossRef]

14. Mlaiki, N.; Kukić, K.; Gardaševixcx-Filipovixcx, M.; Aydi, H. On almost b-metric spaces related fixed points results. *Axioms* **2019**, *8*, 70. [CrossRef]
15. Mustafa, Z.; Huang, H.; Radenović, S. Some remarks on the paper "Some fixed point generalizations are not real generalizations". *J. Adv. Math. Stud.* **2016**, *9*, 110–116.
16. Romaguera, S.; Tirado, P. A characterization of quasi-metric completeness in terms of α-ψ-contractive mappings having fixed points. *Mathematics* **2020**, *8*, 16. [CrossRef]
17. Petruşel, A.; Petruşel, G. Multivalued Picard operators. *J. Nonlinear Conxex Anal.* **2012**, *13*, 157–171.
18. Petruşel, A.; Rus, I.A.; Şerban, M.A. Basic problems of the metric fixed point theory and the relevance of a metric fixed point theorem for multivalued operators. *J. Nonlinear Convex Anal.* **2014**, *15*, 493–513.
19. Petruşel, A.; Petruşel, G. Some variants of the contraction principle for multi-valued operators, generalizations and applications. *J. Nonlinear Convex Anal.* **2019**, *20*, 2187–2203.
20. Goubault-Larrecq, J. *Non-Hausdorff Topology and Domain Theory: Selected Topics in Point-Set Topology*; Cambridge University Press: Cambridge, UK, 2013; Volume 22.
21. Felhi, A.; Sahmim, S.; Aydi, H. Ulam-Hyers stability and well-posedness of fixed point problems for α-λ-contractions on quasi b-metric spaces. *Fixed Point Theory Appl.* **2016**, *2016*, 1. [CrossRef]
22. Reilly, I.L.; Subrahmanyam, P.V.; Vamanamurthy, M.K. Cauchy sequences in quasi-pseudo-metric spaces. *Monatshefte Math.* **1982**, *93*, 127–140. [CrossRef]
23. Alber, Y.I.; Guerre-Delabriere, S. Principles of weakly contractive maps in Hilbert spaces. In *New Results in Operator Theory. Operator Theory: Advances and Applications*; Gohberg, I., Lyubich, Y., Eds.; Birkhäuser: Basel, Switzerland, 1997; Volume 98, pp. 7–22._2. [CrossRef]
24. Aydi, H.; Karapinar, E.; Shatanawi, W. Coupled fixed point results for weakly contractive condition in ordered partial metric spaces. *Comput. Math. Appl.* **2011**, *62*, 4449–4460. [CrossRef]
25. Chi, K.P.; Karapinar, E.; Thanh, T.D. On the fixed point theorems in generalized weakly contractive mappings on partial metric spaces. *Bull. Iranian Math. Soc.* **2013**, *39*, 369–381.
26. Karapinar, E. Fixed point theory for cyclic weak ϕ-contraction. *Appl. Math. Lett.* **2011**, *24*, 822–825. [CrossRef]
27. Karapinar, E. Weak ϕ-contraction on partial contraction. *J. Comput. Anal. Appl.* **2012**, *14*, 206–210.
28. Karapinar, E. Best proximity points of Kannan type cylic weak phi-contractions in ordered metric spaces. *Analele Stiintifice Ale Univ. Ovidius Constan* **2012**, *20*, 51–64.
29. Rhoades, B.E. Some theorems on weakly contractive maps. *Nonlinear Anal.* **2001**, *47*, 2683–2693. [CrossRef]
30. Browder, F.E. On the convergence of successive approximations for nonlinear functional equations. *Nederl. Akad. Wetensch. Ser. A71 Indag. Math.* **1968**, *30*, 27–35. [CrossRef]
31. Rus, I.A. *Generalized Contractions and Applications*; Cluj University Press: Cluj-Napoca, Romania, 2001.
32. Berinde, V. Generalized contractions in quasi-metric spaces. In *Seminar on Fixed Point Theory*; Research Seminars No. 3; Babeş-Bolyai University: Cluj-Napoca, Romania, 1993; pp. 3–9.

Publisher's Note: MDPI stays neutral with regard to jurisdictional claims in published maps and institutional affiliations.

© 2020 by the authors. Licensee MDPI, Basel, Switzerland. This article is an open access article distributed under the terms and conditions of the Creative Commons Attribution (CC BY) license (http://creativecommons.org/licenses/by/4.0/).

Article

On a Viscosity Iterative Method for Solving Variational Inequality Problems in Hadamard Spaces

Kazeem Olalekan Aremu, Chinedu Izuchukwu, Hammed Anuolwupo Abass and Oluwatosin Temitope Mewomo *

School of Mathematics, Statistics and Computer Science, University of KwaZulu-Natal, Durban 4001, South Africa; 218081063@stu.ukzn.ac.za or aremukazeemolalekan@gmail.com (K.O.A.); izuchukwu_c@yahoo.com or izuchukwuc@ukzn.ac.za (C.I.); 216075727@stu.ukzn.ac.za (H.A.A.)
* Correspondence: mewomoo@ukzn.ac.za

Received: 24 October 2020; Accepted: 18 November 2020; Published: 16 December 2020

Abstract: In this paper, we propose and study an iterative algorithm that comprises of a finite family of inverse strongly monotone mappings and a finite family of Lipschitz demicontractive mappings in an Hadamard space. We establish that the proposed algorithm converges strongly to a common solution of a finite family of variational inequality problems, which is also a common fixed point of the demicontractive mappings. Furthermore, we provide a numerical experiment to demonstrate the applicability of our results. Our results generalize some recent results in literature.

Keywords: variational inequalities; inverse strongly monotone mappings; demicontractive mappings; fixed point problems; Hadamard spaces

MSC: 47H06; 47H09; 47J05; 47J25

1. Introduction

The classical variational inequality problem (VIP) is defined in a real Hilbert space setting as: find $x \in D$ such that

$$\langle Tx, y - x \rangle \geq 0 \; \forall \, y \in D, \qquad (1)$$

where T is a nonlinear operator defined on D and D is a nonempty subset of the Hilbert space. The theory of VIP combines concepts of nonlinear operators and convex analysis in such a way that it generalizes both and is used to model nonlinear problems of physical phenomena in economics, sciences and engineering (see [1] for details). The VIP (1) was first introduced in finite dimensional spaces by Stampacchia [2], and since then researchers have devoted a lot of attention to VIP in finite and infinite dimensional spaces (see [3–8] and other references therein). Another form of the VIP widely studied in real Hilbert space settings (see [9,10] and the references therein) is defined as: find $x \in D$ such that

$$\langle x - Tx, y - x \rangle \geq 0 \; \forall \, y \in D. \qquad (2)$$

Several algorithms have been developed for solving VIP and related optimization problems in Hilbert and Banach spaces (see [3,7,11–16] and other references therein). It is well known that many of the problems in practical applications of optimization are constrained optimization problems, where the constraints are nonlinear, non-convex and non-smooth. Hence, it is pertinent to extend the study of these optimization problems to the nonlinear space settings, due to its ability to see non-convex and non-smooth constrained optimization problems as convex, smooth and unconstrained problems.

For this reason, Németh [17] introduced and generalized the existence and uniqueness results of the classical VIP from Euclidean spaces to complete Riemannian manifolds. This development led to increasing interest from researchers in the study of VIPs and their generalizations in nonlinear spaces (see [18–27] and other references therein). Despite the increasing attention of researchers in this direction, little attention has been given to other more general nonlinear spaces apart from the Riemannian manifolds. In 2015, Khatibzadeh and Ranjbar [28] extended the study of VIP (2) to the framework of complete CAT(0) spaces. They formulated the VIP as follows:

$$\text{Find } x \in D \text{ such that } \langle \overrightarrow{Txx}, \overrightarrow{xy} \rangle \geq 0 \ \forall \ y \in D, \tag{3}$$

where D is a nonempty, closed and convex subset of an Hadamard space X and T is a nonexpansive mapping. They established the existence of solutions for the VIP (3) and also employed an inexact proximal point algorithm to approximate a fixed point of the nonexpansive mapping which is also a solution of (3). They obtained convergence result for the algorithm under suitable conditions on the control sequences. Very recently, Alizadeh-Dehghan-Moradlou [29] introduced the notion of inverse strongly monotone mappings in metric spaces as follows: Let D be a nonempty subset of a metric space X and $T : D \to X$ be a mapping. T is called α-inverse strongly monotone if there exists $\alpha > 0$ such that

$$d^2(x,y) - \langle \overrightarrow{TxTy}, \overrightarrow{xy} \rangle \leq \alpha \Phi_T(x,y), \ \forall \ x, y \in D, \tag{4}$$

where $\Phi_T(x,y) = d^2(x,y) + d^2(Tx, Ty) - 2\langle \overrightarrow{TxTy}, \overrightarrow{xy} \rangle$.

Additionally, in [29], the authors studied the VIP (3) in an Hadamard space, where T is an inverse strongly monotone mapping. They established the existence of solutions for the VIP (3) associated with an inverse strongly monotone mapping. Furthermore, they introduced the following iterative algorithm to solve the VIP (3): for arbitrary $x_1 \in D$, the sequence $\{x_n\}$ is generated by

$$\begin{cases} y_n = P_D[\beta_n x_n \oplus (1-\beta_n)Tx_n], \\ x_{n+1} = P_D[\alpha_n x_n \oplus (1-\alpha_n)Sx_n], n \geq 1, \end{cases} \tag{5}$$

where $\{\beta_n\}$ and $\{\alpha_n\} \in (0,1)$, P_D is a metric projection, T is inverse strongly monotone and S is nonexpansive mapping. They obtained that Algorithm (5) Δ-converges to a solution of the VIP (3), which is also a fixed point of the nonexpansive mapping S.

Very recently, Osisiogu et al. [30] proposed and studied the following Halpern-type algorithm in Hadamard spaces for approximating a common solution of a finite family of the VIP (3):

$$\begin{cases} y_n = \bigoplus_{i=1}^{N} \beta_i P_D T_{\lambda_i} x_n, \\ x_{n+1} = \alpha_n u \oplus \beta_n x_n \oplus \gamma_n S y_n \ \forall \ n \geq 1, \end{cases} \tag{6}$$

where $T_{\lambda_i} = (1-\lambda_i)x \oplus \lambda_i Tx$, $0 < \lambda_i < 2\alpha_i$, for each $i = 1, 2, \cdots, N$, $\{\beta_n\}, \{\alpha_n\}$ and $\{\gamma_n\} \in (0,1)$, T is an inverse strongly monotone mapping and S is a nonexpansive mapping. They obtained a strong convergence result of Algorithm (6) under some suitable conditions.

Motivated by the results of Khatibzadeh and Ranjbar [28], Alizadeh-Dehghan-Moradlou [29] and Osisiogu et al. [30], we propose and study a viscosity iterative algorithm (from the fact that viscosity-type algorithms converge faster than Halpern-type algorithms and also Halpern-type algorithms are particular cases of viscosity-type algorithms, see [31,32]) that comprises of a finite family of inverse strongly monotone mappings (3) and a finite family of Lipchitz demicontractive mappings in an Hadamard space. Additionally, we establish that the proposed algorithm converges strongly to a common solution of a finite family of VIPs, which is also a common fixed point of a finite family of Lipchitz demicontractive mappings in the framework of Hadamard spaces. Furthermore, we provide a numerical experiment to demonstrate the applicability of our results. Our result generalizes the works of Alizadeh-Dehghan-Moradlou [29] and Osisiogu et al. [30] and other similar works in literature.

2. Preliminaries

Let (X, d) be a metric space, $x, y \in X$ and $I = [0, d(x, y)]$ be an interval. A curve c (or simply a geodesic path) joining x to y is an isometry $c : I \to X$ such that $c(0) = x$, $c(d(x,y)) = y$ and $d(c(t), c(t')) = |t - t'|$ for all $t, t' \in I$. The image of a geodesic path is called the geodesic segment, which is denoted by $[x, y]$ whenever it is unique. We say that a metric space X is a geodesic space if for every pair of points $x, y \in X$, there is a minimal geodesic from x to y. A geodesic triangle $\Delta(x_1, x_2, x_3)$ in a geodesic metric space (X, d) consists of three vertices (points in X) with geodesic segments between each pair of vertices. For any geodesic triangle, there is a comparison (Alexandrov) triangle $\bar{\Delta} \subset \mathbb{R}^2$ such that $d(x_i, x_j) = d_{\mathbb{R}^2}(\bar{x}_i, \bar{x}_j)$ for $i, j \in \{1, 2, 3\}$. A geodesic space X is a CAT(0) space if the distance between arbitrary pair of points on a geodesic triangle Δ does not exceed the distance between its pair of corresponding points on its comparison triangle $\bar{\Delta}$. If Δ is a geodesic triangle and $\bar{\Delta}$ is its comparison triangle in X, then Δ is said to satisfy the CAT(0) inequality for all points x, y of Δ and $\bar{x}, \bar{y}$ of $\bar{\Delta}$, if

$$d(x, y) = d_{\mathbb{R}^2}(\bar{x}, \bar{y}). \tag{7}$$

Let x, y, z be points in X and y_0 be the midpoint of the segment $[y, z]$; then the CAT(0) inequality implies

$$d^2(x, y_0) \leq \frac{1}{2} d^2(x, y) + \frac{1}{2} d^2(x, z) - \frac{1}{4} d(y, z). \tag{8}$$

Inequality (8) is known as CN inequality of Bruhat and Tits [33]. A geodesic space X is said to be a CAT(0) space if all geodesic triangles satisfy the CN inequality. Equivalently, X is called a CAT(0) space if and only if it satisfies the CN inequality. Examples of CAT(0) spaces includes Hadamard manifold, $\mathbb{R}$-trees [34], pre-Hilbert spaces [35], hyperbolic metric [36] and Hilbert balls [37].

Let D be a nonempty subset of a metric space (X, d). A point $x \in X$ is called a fixed point of a nonlinear mapping $T : D \to X$, if $x = Tx$. We denote by $F(T)$ the set of fixed points of T. The mapping T is said to be:

1. L-Lipschitz, if there exists $L > 0$ such that

$$d(Tx, Ty) \leq L d(x, y), \ \forall \ x, y \in X;$$

if $L = 1$, then T is called nonexpansive;

2. Firmly nonexpansive (see [38]), if

$$d^2(Tx, Ty) \leq \langle \overrightarrow{TxTy}, \overrightarrow{xy} \rangle \ \forall \ x, y \in X;$$

3. Quasi-nonexpansive, if $F(T) \neq \emptyset$ and

$$d(Tx, p) \leq d(x, p) \ \forall \ x \in X \text{ and } p \in F(T);$$

4. k-demicontractive, if $F(T) \neq \emptyset$ and there exists $k \in [0, 1)$ such that

$$d^2(Tx, p) \leq d^2(x, p) + k d^2(x, Tx) \ \forall \ x, y \in X, \text{ and } p \in F(T).$$

Obviously, the class of quasi-nonexpansive are k-demicontractive mappings. However, the converse is not true (see [39] Example 1.1).

Definition 1. *[40] Let a pair $(a,b) \in X \times X$, denoted by $\overrightarrow{ab}$, be called a vector in $X \times X$. The quasilinearization map $\langle .,. \rangle : (X \times X) \times (X \times X) \to \mathbb{R}$ is defined by*

$$\langle \overrightarrow{ab}, \overrightarrow{cd} \rangle = \frac{1}{2}(d^2(a,d) + d^2(b,c) - d^2(a,c) - d^2(b,d)), \ \forall \, a,b,c,d \in X. \tag{9}$$

It is easy to see that $\langle \overrightarrow{ba}, \overrightarrow{cd} \rangle = -\langle \overrightarrow{ab}, \overrightarrow{cd} \rangle$, $\langle \overrightarrow{ab}, \overrightarrow{cd} \rangle = \langle \overrightarrow{ae}, \overrightarrow{cd} \rangle + \langle \overrightarrow{eb}, \overrightarrow{cd} \rangle$ and $\langle \overrightarrow{ab}, \overrightarrow{cd} \rangle = \langle \overrightarrow{cd}, \overrightarrow{ab} \rangle$ for all $a,b,c,d,e \in X$. Furthermore, a geodesic space X is said to satisfy the Cauchy–Schwarz inequality if

$$\langle \overrightarrow{ab}, \overrightarrow{cd} \rangle \leq d(a,b)d(c,d), \ \forall \, a,b,c,d \in X.$$

It is known from [41] that a geodesically connected metric space is a CAT(0) space if and only if it satisfies the Cauchy–Schwarz inequality.

We state some known and useful results which will be needed in the proof of our main results. In the sequel, we denote strong and Δ-convergence by "$\to$" and "$\rightharpoonup$" respectively.

Let $\{x_n\}$ be a bounded sequence in X and $r(., \{x_n\}) : X \to [0, \infty)$ be a continuous mapping defined by $r(x, \{x_n\}) = \limsup_{n \to \infty} d(x, x_n)$. The asymptotic radius of $\{x_n\}$ is given by $r(\{x_n\}) := \inf\{r(x, \{x_n\}) : x \in X\}$ while the asymptotic center of $\{x_n\}$ is the set $A(\{x_n\}) = \{x \in X : r(x, \{x_n\}) = r(\{x_n\})\}$. It is known that in an Hadamard space X, $A(\{x_n\})$ consists of exactly one point. A sequence $\{x_n\}$ in X is said to be Δ-convergent to a point $x \in X$, if $A(\{x_{n_k}\}) = \{x\}$ for every subsequence $\{x_{n_k}\}$ of $\{x_n\}$. In this case, we write $\Delta\text{-}\lim_{n \to \infty} x_n = x$ (see [42,43]).

Definition 2. *Let D be a nonempty, closed and convex subset of an Hadamard space X. The metric projection is a mapping $P_D : X \to D$ which assigns to each $x \in X$, the unique point $P_D x \in D$ such that*

$$d(x, P_D x) = \inf\{d(x,y) : y \in D\}.$$

Lemma 1. *[44] Let D be a nonempty, closed convex subset of an Hadamard space X, $x \in X$ and $u \in D$. Then $u = P_D x$ if and only if $\langle \overrightarrow{xu}, \overrightarrow{uy} \rangle \geq 0$, for all $y \in D$.*

Lemma 2. *[29] Let D be a nonempty convex subset of an Hadamard space X and $T : D \to X$ be an α-inverse strongly monotone mapping. Assume $\lambda \in [0,1]$ and define $T_\lambda : D \to X$ by $T_\lambda x = (1 - \lambda)x \oplus \lambda T x$. If $0 < \lambda < 2\alpha$, then T_λ is nonexpansive and $F(T_\lambda) = F(T)$.*

Lemma 3. *[29] Let D be a nonempty convex subset of an Hadamard space X and $T : D \to X$ be a mapping. Then*

$$VI(D,T) = VI(D, T_\lambda),$$

where $\lambda \in (0,1]$ and $T_\lambda : D \to X$ is a mapping defined by $T_\lambda x = (1-\lambda)x \oplus \lambda Tx$, for all $x \in D$.

Remark 1. *Observe from Lemma 2 that*

$$F(P_D T) = VI(D,T) = VI(D, T_\lambda) = F(P_D T_\lambda).$$

Lemma 4. *[29] Let D be a nonempty bounded closed convex subset of an Hadamard space X and $T : D \to X$ be an α-inverse-strongly monotone mapping. Then $VI(D,T)$ is nonempty, closed and convex.*

Lemma 5. *[41,45] Let X be an Hadamard space. Then for all $x,y,z \in X$ and all $t,s \in [0,1]$, we have*

(i) $d(tx \oplus (1-t)y, z) \leq td(x,z) + (1-t)d(y,z),$
(ii) $d^2(tx \oplus (1-t)y, z) \leq td^2(x,z) + (1-t)d^2(y,z) - t(1-t)d^2(x,y),$
(iii) $d^2(z, tx \oplus (1-t)y) \leq t^2 d^2(z,x) + (1-t)^2 d^2(z,y) + 2t(1-t)\langle \overrightarrow{zx}, \overrightarrow{zy} \rangle.$

Lemma 6. *[46] Let X be a CAT(0) space and $z \in X$. Let $x_1, \cdots, x_N \in X$ and $\gamma_1, \cdots, \gamma_N$ be real numbers in $[0,1]$ such that $\sum_{i=1}^{N} \gamma_i = 1$. Then the following inequality holds:*

$$d^2\left(\sum_{i=1}^{N} \oplus \gamma_i x_i, z\right) \leq \sum_{i=1}^{N} \gamma_i d^2(x_i, z) - \sum_{i,j=1, i \neq j}^{N} \gamma_i \gamma_j d^2(x_i, x_j).$$

Lemma 7. *[47] Every bounded sequence in an Hadamard space has a Δ-convergent subsequence.*

Lemma 8. *[48] Let X be an Hadamard space, $\{x_n\}$ be a sequence in X and $x \in X$. Then $\{x_n\}$ Δ-converges to x if and only if*

$$\limsup_{n \to \infty} \langle \overrightarrow{x_n x}, \overrightarrow{yx} \rangle \leq 0, \ \forall y \in X.$$

Definition 3. *Let D be a nonempty, closed and convex subset of an Hadamard space X. A mapping $T : D \to D$ is said to be Δ-demiclosed, if for any bounded sequence $\{x_n\}$ in X, such that $\Delta - \lim_{n \to \infty} x_n = x$ and $\lim_{n \to \infty} d(x_n, T x_n) = 0$, then $x = Tx$.*

Lemma 9. *[49] Let X be an Hadamard space and $T : X \to X$ be a nonexpansive mapping. Then T is Δ-demiclosed.*

Lemma 10. *[50,51] Let $\{a_n\}$ be a sequence of non-negative real numbers satisfying*

$$a_{n+1} \leq (1 - \alpha_n) a_n + \delta_n, \ n \geq 0,$$

where $\{\alpha_n\}$ and $\{\delta_n\}$ satisfy the following conditions:

(i) $\{\alpha_n\} \subset [0,1], \sum_{n=0}^{\infty} \alpha_n = \infty$,

(ii) $\limsup_{n \to \infty} \frac{\delta_n}{\alpha_n} \leq 0$ or $\sum_{n=0}^{\infty} |\delta_n| \leq \infty$.

Then $\lim_{n \to \infty} a_n = 0$.

Lemma 11. *[52] Let $\{a_n\}$ be a sequence of non-negative real numbers such that there exists a subsequence $\{n_j\}$ of $\{n\}$ with $a_{n_j} < a_{n_j+1}$ for all $j \in \mathbb{N}$. Then there exists a nondecreasing sequence $\{m_k\} \subset \mathbb{N}$ such that $m_k \to \infty$ and the following properties are satisfied by all (sufficiently large) numbers $k \in \mathbb{N}$:*

$$a_{m_k} \leq a_{m_k+1} \text{ and } a_k \leq a_{m_k+1}.$$

In fact, $m_k = \max\{i \leq k : a_i < a_{i+1}\}$.

3. Main Results

In this section, we present our strong convergence results.

Theorem 1. *Let X be an Hadamard space and D be a nonempty, closed and convex subset of X. Let $S_i : D \to D$ be a finite family of L_i-Lipschitz k_i-demicontractive mappings and Δ-demiclosed such that $L > 0, L = \max\{L_i, i = 1, 2, \cdots, N\}, k \in [0,1), k = \max\{k_i, i = 1, 2, \cdots, N\}, k_i \in [0,1), i = 1, 2, \cdots, N$. Let $T_i : D \to X$ be a finite family of α_i-inverse strongly monotone mappings and f be a contraction on D with coefficient $\theta \in (0,1)$. Suppose that $\Gamma := \cap_{i=1}^{N} VI(D, T_i) \cap \cap_{i=1}^{N} F(S_i) \neq \emptyset$. For arbitrary $x_1 \in D$, let the sequence $\{x_n\}$ be generated by*

$$\begin{cases} w_n = \gamma_n f(x_n) \oplus (1-\gamma_n)x_n, \\ y_n = \beta_{n,0}w_n \oplus \sum_{i=1}^{N} \oplus \beta_{n,i} P_D T_{\lambda_i} w_n, \\ x_{n+1} = \alpha_{n,0}y_n \oplus \sum_{i=1}^{N} \oplus \alpha_{n,i} S_i y_n, \ \forall\, n \geq 1, \end{cases} \qquad (10)$$

where $T_{\lambda_i} = (1-\lambda_i)x \oplus \lambda_i T_i x$, $0 < \lambda_i < 2\alpha_i$, for each $i = 1, 2, \cdots, N$, $\{\gamma_n\}, \{\beta_{n,i}\}$ and $\{\alpha_{n,i}\} \in (0,1)$ such that the following conditions are satisfied:

(A1) $\lim_{n \to \infty} \gamma_n = 0$;

(A2) $\sum_{n=1}^{\infty} \gamma_n = \infty$,

(A3) $0 < a \leq \beta_{n,i}, \alpha_{n,i} \leq b < 1$; $\sum_{i=0}^{N} \alpha_{n,i} = 1$ and $\sum_{i=0}^{N} \beta_{n,i} = 1$;

(A4) $0 < c \leq \alpha_{n,0} - k \leq d < 1$.

Then, the sequence $\{x_n\}$ converges strongly to an element $\bar{z} = P_\Gamma f(\bar{z})$, where P_Γ is the metric projection of X onto Γ.

Proof. Let $p \in \Gamma$; then by (10), condition (A3), Lemma 6 and the fact that $P_D T_\lambda$ is nonexpansive, we have

$$d^2(y_n, p) = d^2(\beta_{n,0}w_n \sum_{i=1}^{N} \oplus \beta_{n,i} P_D T_{\lambda_i} w_n, p)$$

$$\leq \beta_{n,0} d^2(w_n, p) + \sum_{i=1}^{N} \beta_{n,i} d^2(P_D T_{\lambda_i} w_n, p) - \sum_{i=1}^{N} \beta_{n,0}\beta_{n,i} d^2(P_D T_{\lambda_i} w_n, w_n)$$

$$\leq d^2(w_n, p) - \sum_{i=1}^{N} \beta_{n,0}\beta_{n,i} d^2(P_D T_{\lambda_i} w_n, w_n). \qquad (11)$$

Additionally, since S_i is demicontractive, we have from (10) and Lemma 6 that

$$d^2(x_{n+1}, p) \leq \alpha_{n,0} d^2(y_n, p) + \sum_{i=1}^{N} \alpha_{n,i} d^2(S_i y_n, p) - \sum_{i=1}^{N} \alpha_{n,0}\alpha_{n,i} d^2(y_n, S_i y_n)$$

$$\leq \alpha_{n,0} d^2(y_n, p) + \sum_{i=1}^{N} \alpha_{n,i}[d^2(y_n, p) + k d^2(S_i y_n, y_n)] - \sum_{i=1}^{N} \alpha_{n,0}\alpha_{n,i} d^2(y_n, S_i y_n)$$

$$= d^2(y_n, p) - \sum_{i=1}^{N} \alpha_{n,i}(\alpha_{n,0} - k) d^2(y_n, S_i y_n). \qquad (12)$$

From (11), (12) and condition (A4), we have

$$d(x_{n+1}, p) \leq \gamma_n d(f(x_n), p) + (1-\gamma_n) d(x_n, p)$$

$$\leq \gamma_n \theta d(x_n, p) + \gamma_n d(f(p), p) + (1-\gamma_n) d(x_n, p)$$

$$\leq (1 - \gamma_n(1-\theta)) d(x_n, p) + \gamma_n d(f(p), p).$$

$$\leq \max\{d(x_n, p), \frac{1}{1-\theta} d(f(p), p)\}$$

$$\vdots$$

$$\leq \max\{d(x_1, p), \frac{1}{1-\theta} d(f(p), p)\}.$$

Thus, $\{x_n\}$ is bounded. Consequently, $\{w_n\}, \{y_n\}, \{Sy_n\}$ and $\{P_D T_{\lambda_i} w_n\}$ are also bounded. Now we divide the rest of the proof into two cases:

Case 1: Assume that $\{d^2(x_n, p)\}$ is a monotonically non-increasing sequence. Then, $\{d^2(x_n, p)\}$ is convergent and
$$d^2(x_n, p) - d^2(x_{n+1}, p) \to 0, \text{ as } n \to \infty.$$

Hence, from (11) and (12), we have

$$\begin{aligned} d^2(x_{n+1}, p) &\leq d^2(w_n, p) - \sum_{i=1}^{N} \beta_{n,0}\beta_{n,i}d^2(P_D T_{\lambda_i} w_n, w_n) - \sum_{i=1}^{N} \alpha_{n,i}(\alpha_{n,0} - k)d^2(y_n, S_i y_n) \\ &\leq \gamma_n d^2(f(x_n), p) + (1 - \gamma_n)d^2(x_n, p) - \sum_{i=1}^{N} \beta_{n,0}\beta_{n,i}d^2(P_D T_{\lambda_i} w_n, w_n) - \sum_{i=1}^{N} \alpha_{n,i}(\alpha_{n,0} - k)d^2(y_n, S_i y_n), \end{aligned} \quad (13)$$

which implies

$$\sum_{i=1}^{N} \beta_{n,0}\beta_{n,i}d^2(P_D T_{\lambda_i} w_n, w_n) \leq \gamma_n(d^2(f(x_n), p) - d^2(x_n, p)) + d^2(x_n, p) - d^2(x_{n+1}, p).$$

By conditions (A1) and (A3), we obtain that

$$\lim_{n \to \infty} d(P_D T_{\lambda_i} w_n, w_n) = 0, \ i = 0, 1, 2, \cdots, N. \quad (14)$$

Similarly, from (13) and condition (A1), we have

$$\lim_{n \to \infty} d(S_i y_n, y_n) = 0, \ i = 0, 1, 2, \cdots, N. \quad (15)$$

Additionally,

$$d(w_n, x_n) \leq \gamma_n d(f(x_n), x_n) \to 0 \text{ as } n \to \infty. \quad (16)$$

Again from (10) and Lemma 6, we have

$$\begin{aligned} d^2(y_n, x_n) &\leq \beta_{n,0}d^2(w_n, x_n) + \sum_{i=1}^{N} \beta_{n,i}d^2(P_D T_{\lambda_i} w_n, x_n) \\ &\leq \beta_{n,0}d^2(w_n, x_n) + \sum_{i=1}^{N} \beta_{n,i}\big[d(P_D T_{\lambda_i} w_n, w_n) + d(w_n, x_n)\big]^2. \end{aligned}$$

Hence, from (14) and (16), we obtain

$$d(y_n, x_n) \to 0 \text{ as } n \to \infty. \quad (17)$$

Additionally, from (15) and (17), we have

$$d(S_i y_n, x_n) \leq d(S_i y_n, y_n) + d(y_n, x_n) \to 0, \text{ as } n \to \infty, \ i = 0, 1, 2, \cdots, N. \quad (18)$$

Since S_i is Lipschitz, then from (17) and (18), we have that

$$\begin{aligned} d(S_i x_n, x_n) &\leq d(S_i x_n, S_i y_n) + d(S_i y_n, x_n) \\ &\leq L d(x_n, y_n) + d(S_i y_n, x_n) \to 0, \text{ as } n \to \infty, \ i = 0, 1, 2, \cdots, N. \end{aligned} \quad (19)$$

Hence, from (14), (16) and Lemma 2, we obtain that

$$\begin{aligned} d(P_D T_{\lambda_i} x_n, x_n) &\leq d(P_D T_{\lambda_i} x_n, P_D T_{\lambda_i} w_n) + d(P_D T_{\lambda_i} w_n, w_n) + d(w_n, x_n) \\ &\leq d(x_n, w_n) + d(P_D T_{\lambda_i} w_n, w_n) + d(w_n, x_n) \to 0, \text{ as } n \to \infty, \ i = 0, 1, 2, \cdots, N. \end{aligned} \quad (20)$$

Since $\{x_n\}$ is bounded, by Lemma 7 there exists a subsequence $\{x_{n_k}\}$ of $\{x_n\}$ such that Δ-$\lim_{k\to\infty} x_{n_k} = z$ for some $z \in D$. Then, it follows from (16) that there exists a subsequence $\{w_{n_k}\}$ of $\{w_n\}$, such that Δ-$\lim_{k\to\infty} w_{n_k} = z$. Additionally, from (17), we have that Δ-$\lim_{k\to\infty} y_{n_k} = z$. Since S_i is Δ-demiclosed for each $i = 1, 2, \cdots, N$, it follows from (19) that $z \in \cap_{i=0}^{N} F(S_i)$. Additionally, $P_D T_{\lambda_i}$ is nonexpansive (by Lemma 2) for each $i = 1, 2, \cdots, N$, thus we obtain from (20) and Remark 1 that $z \in \cap_{i=0}^{N} F(P_D T_{\lambda_i}) = \cap_{i=0}^{N} F(P_D T_i)$. Hence, $z \in \Gamma$.

Next we show that $\{x_n\}$ converges strongly to $\bar{z} = P_\Gamma f(\bar{z})$. Since $\{x_n\}$ is bounded, we may choose without loss of generality, a subsequence $\{x_{n_k}\}$ of $\{x_n\}$ such that $\{x_{n_k}\}$ Δ-converges to z and

$$\limsup_{n\to\infty} \langle \overrightarrow{f(\bar{z})\bar{z}}, \overrightarrow{x_n\bar{z}} \rangle = \lim_{k\to\infty} \langle \overrightarrow{f(\bar{z})\bar{z}}, \overrightarrow{x_{n_k}\bar{z}} \rangle. \tag{21}$$

Thus, by (21) and Lemma 1, we obtain that

$$\limsup_{k\to\infty} \langle \overrightarrow{f(\bar{z})\bar{z}}, \overrightarrow{x_n\bar{z}} \rangle = \langle \overrightarrow{f(\bar{z})\bar{z}}, \overrightarrow{z\bar{z}} \rangle \leq 0. \tag{22}$$

From (12), Lemma 5 (iii) and quasilinearization properties in Definition 1, we have that

$$\begin{aligned}
d^2(x_{n+1}, \bar{z}) &\leq \gamma_n^2 d^2(f(x_n), \bar{z}) + (1-\gamma_n)^2 d^2(x_n, \bar{z}) + 2\gamma_n(1-\gamma_n) \langle \overrightarrow{f(x_n)\bar{z}}, \overrightarrow{x_n\bar{z}} \rangle \\
&\leq \gamma_n^2 d^2(f(x_n), \bar{z}) + (1-\gamma_n)^2 d^2(x_n, \bar{z}) + 2\gamma_n(1-\gamma_n)\left[\langle \overrightarrow{f(x_n)f(\bar{z})}, \overrightarrow{x_n\bar{z}} \rangle + \langle \overrightarrow{f(\bar{z})\bar{z}}, \overrightarrow{x_n\bar{z}} \rangle \right] \\
&\leq \gamma_n^2 d^2(f(x_n), \bar{z}) + (1-\gamma_n)^2 d^2(x_n, \bar{z}) + 2\gamma_n(1-\gamma_n)\left[\theta d^2(x_n, \bar{z}) + \langle \overrightarrow{f(\bar{z})\bar{z}}, \overrightarrow{x_n\bar{z}} \rangle\right] \\
&\leq (1 - 2\gamma_n + 2\gamma_n \theta) d^2(x_n, \bar{z}) + 2\gamma_n(1-\gamma_n)\langle \overrightarrow{f(\bar{z})\bar{z}}, \overrightarrow{x_n\bar{z}} \rangle + \gamma_n^2(d^2(f(x_n), \bar{z}) + d^2(x_n, \bar{z})) \\
&= 1 - 2\gamma_n(1-\theta) d^2(x_n, \bar{z}) + 2\gamma_n(1-\theta)\left[\frac{1-\gamma_n}{1-\theta}\langle \overrightarrow{f(\bar{z})\bar{z}}, \overrightarrow{x_n\bar{z}} \rangle + \frac{\gamma_n}{2(1-\theta)}(d^2(f(x_n), \bar{z}) + d^2(x_n, \bar{z}))\right].
\end{aligned}$$

That is,

$$d^2(x_{n+1}, \bar{z}) \leq 1 - 2\gamma_n(1-\theta) d^2(x_n, \bar{z}) + 2\gamma_n(1-\theta) M_n, \tag{23}$$

where

$$M_n = \left[\frac{1-\gamma_n}{1-\theta}\langle \overrightarrow{f(\bar{z})\bar{z}}, \overrightarrow{x_n\bar{z}} \rangle + \frac{\gamma_n}{2(1-\theta)}(d^2(f(x_n), \bar{z}) + d^2(x_n, \bar{z}))\right].$$

Thus from (22), (23) and condition (A1), we conclude by Lemma 10 that $\{x_n\}$ converges strongly to $\bar{z} = P_\Gamma f(\bar{z})$.

Case 2: Suppose there exists a subsequence $\{n_k\}$ of $\{n\}$ such that $d^2(x_{n_k}, p) \leq d^2(x_{k+1}, p)$ for all $k \in \mathbb{N}$. Then by Lemma 11, there exists a nondecreasing sequence $\{m_k\} \subset \mathbb{N}$ such that $m_k \to \infty$:

$$d(x_{m_k}, p) < d(x_{m_{k+1}}, p), \text{ and } d(x_k, p) < d(x_{k+1}, p) \; \forall \, k \in \mathbb{N}. \tag{24}$$

Therefore

$$\begin{aligned}
0 &\leq \lim_{k\to\infty} \left(d(x_{m_{k+1}}, p) - d(x_{m_k}, p) \right) \\
&\leq \limsup_{n\to\infty} \left(d(x_{n+1}, p) - d(x_n, p) \right) \\
&\leq \limsup_{n\to\infty} \left(\gamma_n d(f(x_n), p) + (1-\gamma_n) d(x_n, p) - d(x_n, p) \right) \\
&= \limsup_{n\to\infty} \left(\gamma_n \bigl(d(f(x_n), p) - d(x_n, p) \bigr) \right) = 0.
\end{aligned}$$

This implies that

$$\lim_{k\to\infty} \left(d(x_{m_{k+1}}, p) - d(x_{m_k}, p)\right) = 0. \tag{25}$$

Following the arguments as in Case 1, we get

$$\lim_{k\to\infty} \langle \overrightarrow{f(\bar{z})\bar{z}}, \overrightarrow{x_{m_k}\bar{z}} \rangle \leq 0. \tag{26}$$

Hence, from (23), we obtain that

$$d^2(x_{m_{k+1}}, \bar{z}) \leq 1 - 2\gamma_{m_k}(1-\theta)d^2(x_{m_k}, \bar{z}) + 2\gamma_{m_k}(1-\theta)M_{m_k}.$$

Additionally, from (24), we have that

$$d^2(x_{m_k}, \bar{z}) \leq M_{m_k},$$

which implies that

$$\lim_{k\to\infty} d^2(x_{m_k}, \bar{z}) = 0.$$

Thus, from cases 1 and 2, we conclude that $\{x_n\}$ converges to $\bar{z} = P_\Gamma f(\bar{z})$ which is an element of Γ. □

We present some consequences of our main results.

Now, by setting S_i to be a family of quasi-nonexpansive mappings in Theorem 1, we obtain the following result:

Corollary 2. *Let X be an Hadamard space and D be a nonempty, closed and convex subset of X. Let $S_i : D \to D$ be a finite family of quasi-nonexpansive mappings, $T_i : D \to X$ be a finite family of α_i-inverse strongly monotone mappings and f be a contraction on D with coefficient $\theta \in (0,1)$. Suppose that $\Gamma := \cap_{i=1}^N VI(D, T_i) \cap \cap_{i=1}^N F(S_i) \neq \emptyset$. For arbitrary $x_1, \in D$, let the sequence $\{x_n\}$ be generated by*

$$\begin{cases} w_n = \gamma_n f(x_n) \oplus (1-\gamma_n)x_n, \\ y_n = \beta_{n,0} w_n \oplus \sum_{i=1}^N \oplus \beta_{n,i} P_D T_{\lambda_i} w_n, \\ x_{n+1} = \alpha_{n,0} y_n \oplus \sum_{i=1}^N \oplus \alpha_{n,i} S_i y_n, \quad \forall n \geq 1, \end{cases} \tag{27}$$

where $T_{\lambda_i} = (1-\lambda_i)x \oplus \lambda_i T_i x$, $0 < \lambda_i < 2\alpha_i$, for each $i = 1, 2, \cdots, N$, $\{\gamma_n\}, \{\beta_{n_i}\}$ and $\{\alpha_{n_i}\} \in (0,1)$ such that conditions (A1)-(A3) of Theorem 1 are satisfied. Then, the sequence $\{x_n\}$ converges strongly to an element $\bar{z} = P_\Gamma f(\bar{z})$, where P_Γ is the metric projection of X onto Γ.

Proof. The proof follows from the proof of Theorem 1. □

By setting $N = 1$ in Corollary 2, we obtain the following result:

Corollary 3. *Let X be an Hadamard space and D be a nonempty, closed and convex subset of X. Let $S : D \to D$ be a quasi-nonexpansive mapping, $T : D \to X$ be an α-inverse strongly monotone mapping and f be a* contraction on D with coefficient $\theta \in (0,1)$. Suppose that $\Gamma := VI(D,T) \cap F(S) \neq \emptyset$. For arbitrary $x_1 \in D$, let the sequence $\{x_n\}$ be generated by

$$\begin{cases} w_n = \gamma_n f(x_n) \oplus (1-\gamma_n) x_n, \\ y_n = \beta_{n,0} w_n \oplus \beta_{n,1} P_D T_\lambda w_n, \\ x_{n+1} = \alpha_{n,0} y_n \oplus \alpha_{n,1} S y_n, \quad \forall n \geq 1, \end{cases} \quad (28)$$

where $T_\lambda = (1-\lambda)x \oplus \lambda T x$, $0 < \lambda < 2\alpha$, $\{\gamma_n\}$ and $\{\alpha_n\} \in (0,1)$ such that $0 < a \leq \beta_{n,i}, \alpha_{n,i} \leq b < 1$, for $i = 0,1$, $\sum_{i=0}^{1} \alpha_{n,i} = 1$, $\sum_{i=0}^{1} \beta_{n,i} = 1$ and conditions (A1)-(A2) of Theorem 1 are satisfied. Then, the sequence $\{x_n\}$ converges strongly to an element $\bar{z} = P_\Gamma f(\bar{z})$, where P_Γ is the metric projection of X onto Γ.

Suppose $u = f(x)$ for arbitrary but fixed $u \in X$ and for all $x \in X$ in Theorem 1, we obtain the following result:

Corollary 4. *Let X be an Hadamard space and D be a nonempty, closed and convex subset of X. Let $S_i : D \to D$ be a finite family of L_i-Lipschitz demicontractive mappings and Δ-demiclosed such that $L > 0$, $L = \max\{L_i, i = 1,2, \cdots, N\}$, $k \in [0,1)$, $k = \max\{k_i, i = 1,2, \cdots, N\}$, $k_i \in [0,1)$, $i = 1,2, \cdots, N$. Let $T_i : D \to X$ be a finite family of α_i-inverse strongly monotone mappings and suppose that $\Gamma := \cap_{i=1}^{N} VI(D, T_i) \cap \cap_{i=1}^{N} F(S_i) \neq \emptyset$. For arbitrary $x_1, u \in D$, let the sequence $\{x_n\}$ be generated by*

$$\begin{cases} w_n = \gamma_n u \oplus (1-\gamma_n) x_n, \\ y_n = \beta_{n,0} w_n \oplus \sum_{i=1}^{N} \oplus \beta_{n,i} P_D T_{\lambda_i} w_n, \\ x_{n+1} = \alpha_{n,0} y_n \oplus \sum_{i=1}^{N} \oplus \alpha_{n,i} S_i y_n, \quad \forall n \geq 1, \end{cases} \quad (29)$$

where $T_{\lambda_i} = (1-\lambda_i)x \oplus \lambda_i T_i x$, $0 < \lambda_i < 2\alpha_i$, for each $i = 1,2, \cdots, N$, $\{\gamma_n\}, \{\beta_{n_i}\}$ and $\{\alpha_{n_i}\} \in (0,1)$ such that conditions (A1)-(A4) of Theorem 1 are satisfied. Then, the sequence $\{x_n\}$ converges strongly to an element $\bar{z} \in \Gamma$ which is the nearest point to u.

By setting $S_i \equiv I$ for all $i = 1,2, \cdots, N$ in Theorem 1, we obtain the following result:

Corollary 5. *Let X be an Hadamard space and D be a nonempty, closed and convex subset of X. Let $T_i : D \to X$ be a finite family of α_i-inverse strongly monotone mappings and f be a contraction on D with coefficient $\theta \in (0,1)$. Suppose that $\Gamma := \cap_{i=1}^{N} VI(D, T_i) \neq \emptyset$. For arbitrary $x_1 \in D$, let the sequence $\{x_n\}$ be generated by*

$$\begin{cases} w_n = \gamma_n f(x_n) \oplus (1-\gamma_n) x_n, \\ y_n = \beta_{n,0} w_n \oplus \sum_{i=1}^{N} \oplus \beta_{n,i} P_D T_{\lambda_i} w_n \\ x_{n+1} = \alpha_{n,0} y_n \oplus \sum_{i=i}^{N} \oplus \alpha_{n,i} y_n \quad \forall n \geq 1, \end{cases} \quad (30)$$

where $T_{\lambda_i} = (1-\lambda_i)x \oplus \lambda_i T_i x$, $0 < \lambda_i < 2\alpha_i$, for each $i = 1,2, \cdots, N$, $\{\gamma_n\}, \{\beta_{n_i}\}$ and $\{\alpha_{n_i}\} \in (0,1)$ such that conditions (A1)-(A3) of Theorem 1 are satisfied. Then, the sequence $\{x_n\}$ converges strongly to an element $\bar{z} = P_\Gamma f(\bar{z})$, where P_Γ is the metric projection of X onto Γ.

4. Numerical Example

In this section, we give a numerical experiment to show the applicability of our result.

Example 1. *[29] Let $X = \mathbb{R}^2$ be an R-tree with radial metric d_r, where $d_r(x,y) = d(x,y)$ if x and y are situated on a Euclidean straight line passing through the origin and $d_r(x,y) = d(x,\mathbf{0}) + d(y,\mathbf{0})$ otherwise. We put $p = (0,1), q = (1,0)$ and $D = A \cup B \cup C$, where*

$$A = \{(0,t) : t \in [2/3,1]\}, \quad B = \{(t,0) : t \in [2/3,1]\}, \quad C = \{(t,s) : t+s = 1, t \in (0,1)\}.$$

Define $T : D \to X$ by

$$Tx = \begin{cases} q \text{ if } x \in A, \\ p \text{ if } x \in B, \\ x \text{ if } x \in C, \end{cases} \tag{31}$$

then T is $\frac{1}{4}$-inverse strongly monotone in (X, d_r).

Now, define $S : D \to D$ by $Sx = \frac{5}{8}x$. We make the following choices of parameters: $\lambda = \frac{1}{4}$, $\gamma_n = \frac{1}{n+1}$, $\alpha_{n,0} = \frac{2n}{5n+7}$, $\alpha_{n,1} = \frac{3n+7}{5n+7}$, $\beta_{n,0} = \frac{n}{3n+2}$, $\beta_{n,1} = \frac{2n+2}{3n+2}$ $\forall n \geq 1$ and $f(x) = \frac{1}{2}x$ $\forall x \in X$; then the conditions (A1)–(A3) of Theorem 2 are satisfied. Therefore, for $x_1 \in X$, Algorithm (28) becomes

$$\begin{cases} w_n = \frac{1}{n+1} f(x_n) \oplus (1 - \frac{1}{n+1}) x_n, \\ y_n = \beta_{n,0} w_n \oplus \beta_{n,1} P_D T_\lambda w_n, \\ x_{n+1} = \alpha_{n,0} y_n \oplus \alpha_{n,1} S y_n \quad \forall n \geq 1. \end{cases} \tag{32}$$

We now consider the following 3 cases for our numerical experiments given in Figure 1 above.

Case 1: $x_1 = (-0.5, 0.5)^T$.
Case 2: $x_1 = (0.5, -0.5)^T$.
Case 2: $x_1 = (1, 2)^T$.

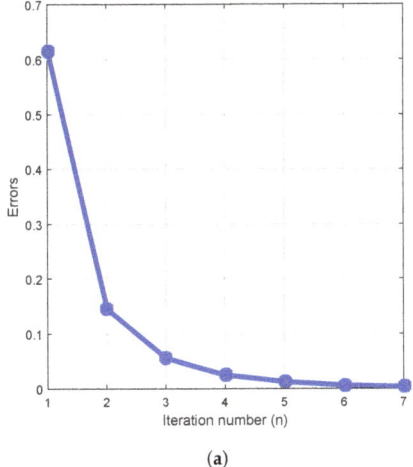
(a)

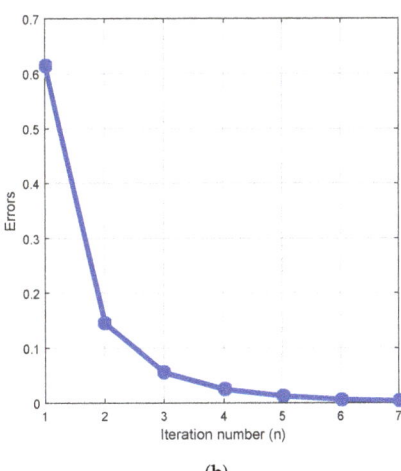
(b)

Figure 1. *Cont.*

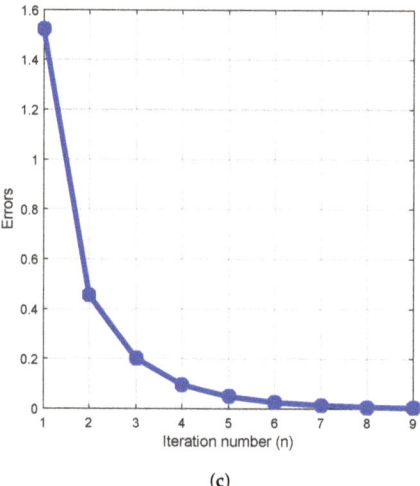

(c)

Figure 1. Errors vs. iteration numbers (n) for Example 1: case 1 (**a**); case 2 (**b**); case 3 (**c**).

Author Contributions: Conceptualization of the article was given by K.O.A., C.I. and O.T.M., methodology by H.A.A., K.O.A., and C.I., software by C.I., validation by O.T.M. and C.I., formal analysis, investigation, data curation, and writing–original draft preparation by K.O.A., H.A.A. and C.I. resources by K.O.A., C.I., H.A.A. and O.T.M. writing–review and editing by K.O.A., C.I. and O.T.M., visualization by C.I. and O.T.M., project administration by O.T.M., Funding acquisition by O.T. All authors have read and agreed to the published version of the manuscript.

Funding: C.I. is funded by National Research Foundation (NRF) South Africa (S&F-DSI/NRF Free Standing Postdoctoral Fellowship; grant number: 120784); H.A.A. is funded by Department of Science and Technology and National Research Foundation, Republic of South Africa Center of Excellence in Mathematical and Statistical Sciences (DST-NRF COE-MaSS) and O.T.M. is funded by National Research Foundation (NRF) of South Africa Incentive Funding for Rated Researchers (grant number 119903).

Acknowledgments: The authors sincerely thank the anonymous reviewers for their careful reading, constructive comments and fruitful suggestions that substantially improved the manuscript.

Conflicts of Interest: The authors declare no conflict of interest.

References

1. Oden, J.T.; Kikuchi, N. Theory of variational inequalities with applications to problems of flow through porous media. *Int. J. Eng. Sci.* **1980**, *18*, 1173–1284. [CrossRef]
2. Stampacchia, G. Variational inequalities. In *Theory and Applications of Monotone Operators*; Ghizzetti, A., Ed.; Edizioni Oderisi: Gubbio, Italy, 1969; pp. 102–192.
3. Alakoya, T.O.; Jolaoso, L.O.; Mewomo, O.T. Modified inertia subgradient extragradient method with self adaptive stepsize for solving monotone variational inequality and fixed point problems. *Optimization* **2020**. [CrossRef]
4. Alakoya, T.O.; Jolaoso, L.O.; Mewomo, O.T. Two modifications of the inertial Tseng extragradient method with self-adaptive step size for solving monotone variational inequality problems. *Demonstr. Math.* **2020**, *53*, 208–224. [CrossRef]
5. Allen, G. Variational inequalities, complementarity problems, and duality theorems. *J. Math. Anal. Appl.* **1977**, *58*, 1–10. [CrossRef]
6. Izuchukwu, C.; Mebawondu, A.A.; Mewomo, O.T. A New Method for Solving Split Variational Inequality Problems without Co-coerciveness. *J. Fixed Point Theory Appl.* **2020**, *22*, 1–23. [CrossRef]
7. Jolaoso, L.O.; Taiwo, A.; Alakoya, T.O.; Mewomo, O.T. Strong convergence theorem for solving pseudo-monotone variational inequality problem using projection method in a reflexive Banach space. *J. Optim. Theory Appl.* **2020**, *185*, 744–766. [CrossRef]

8. Khan, S.H.; Alakoya, T.O.; Mewomo, O.T. Relaxed projection methods with self-adaptive step size for solving variational inequality and fixed point problems for an infinite family of multivalued relatively nonexpansive mappings in Banach spaces. *Math. Comput. Appl.* **2020**, *25*, 54. [CrossRef]
9. Marino, G.; Xu, H.K. Explicit hierarchical fixed point approach to variational inequalities. *J. Optim. Theory. Appl.* **2011**, *149*, 61–78. [CrossRef]
10. Moudafi, A. Viscosity approximation methods for fixed-point problems. *J. Math. Anal. Appl.* **2000**, *241*, 46–55. [CrossRef]
11. Alakoya, T.O.; Jolaoso, L.O.; Mewomo, O.T. A self adaptive inertial algorithm for solving split variational inclusion and fixed point problems with applications. *J. Ind. Manag. Optim.* **2020**. [CrossRef]
12. Gibali, A.; Jolaoso, L.O.; Mewomo, O.T.; Taiwo, A. Fast and simple Bregman projection methods for solving variational inequalities and related problems in Banach spaces. *Results Math.* **2020**, *75*, 1–36. [CrossRef]
13. Izuchukwu, C.; Ogwo, G.N.; Mewomo, O.T. An Inertial Method for solving Generalized Split Feasibility Problems over the solution set of Monotone Variational Inclusions. *Optimization* **2020**. [CrossRef]
14. Jolaoso, L.O.; Alakoya, T.O.; Taiwo, A.; Mewomo, O.T. Inertial extragradient method via viscosity approximation approach for solving Equilibrium problem in Hilbert space. *Optimization* **2020**. [CrossRef]
15. Oyewole, O.K.; Abass, H.A.; Mewomo, O.T. A Strong convergence algorithm for a fixed point constrainted split null point problem. *Rend. Circ. Mat. Palermo II* **2020**. [CrossRef]
16. Taiwo, A.; Owolabi, A.O.-E.; Jolaoso, L.O.; Mewomo, O.T.; Gibali, A. A new approximation scheme for solving various split inverse problems. *Afr. Mat.* **2020**. [CrossRef]
17. Németh, S.Z. Variational inequalities on Hadamard manifolds. *Nonlinear Anal.* **2003**, *52*, 1491–1498. [CrossRef]
18. Aremu, K.O.; Abass, H.A.; Izuchukwu, C.; Mewomo, O.T. A viscosity-type algorithm for an infinitely countable family of (f,g)-generalized k-strictly pseudononspreading mappings in CAT(0) spaces. *Analysis* **2020**, *40*, 19–37. [CrossRef]
19. Aremu, K.O.; Izuchukwu, C.; Ogwo, G.N.; Mewomo, O.T. Multi-step Iterative algorithm for minimization and fixed point problems in p-uniformly convex metric spaces. *J. Ind. Manag. Optim.* **2020**. [CrossRef]
20. Aremu, K.O.; Izuchukwu, C.; Ugwunnadi, G.C.; Mewomo, O.T. On the proximal point algorithm and demimetric mappings in CAT(0) spaces. *Demonstr. Math.* **2018**, *51*, 277–294. [CrossRef]
21. Bento, G.C.; Ferreira, O.P.; Oliveira, P.R. Proximal point method for a special class of non convex functions on Hadamard manifolds. *Optimization* **2015**, *64*, 289–319. [CrossRef]
22. Chen, S.L.; Huang, N.J. Vector variational inequalities and vector optimization problems on Hadamard manifolds. *Optim. Lett.* **2016**, *10*, 753–767. [CrossRef]
23. Dehghan, H.; Izuchukwu, C.; Mewomo, O.T.; Taba, D.A.; Ugwunnadi, G.C. Iterative algorithm for a family of monotone inclusion problems in CAT(0) spaces. *Quaest. Math.* **2020**, *43*, 975–998. [CrossRef]
24. Ogwo, G.N.; Izuchukwu, C.; Aremu, K.O.; Mewomo, O.T. A viscosity iterative algorithm for a family of monotone inclusion problems in an Hadamard space. *Bull. Belg. Math. Soc. Simon Stevin* **2020**, *27*, 127–152. [CrossRef]
25. Ogwo, G.N.; Izuchukwu, C.; Aremu, K.O.; Mewomo, O.T. On θ-generalized demimetric mappings and monotone operators in Hadamard spaces. *Demonstr. Math.* **2020**, *53*, 95–111. [CrossRef]
26. Taiwo, A.; Jolaoso, L.O.; Mewomo, O.T. Viscosity approximation method for solving the multiple-set split equality common fixed-point problems for quasi-pseudocontractive mappings in Hilbert Spaces. *J. Ind. Manag. Optim.* **2017**. [CrossRef]
27. Ugwunnadi, G.C.; Izuchukwu, C.; Mewomo, O.T. Strong convergence theorem for monotone inclusion problem in CAT(0) spaces. *Afr. Mat.* **2019**, *31*, 151–169. [CrossRef]
28. Khatibzadeh, H.; Ranjbar, S. A variational inequality in complete CAT(0) spaces. *J. Fixed Point Theory Appl.* **2015**, *17*, 557–574. [CrossRef]
29. Alizadeh, S.; Dehghan, H.; Moradlou, F. Δ-convergence theorem for inverse strongly monotone mapping in CAT(0) spaces. *Fixed Point Theory* **2018**, *19*, 45–56. [CrossRef]
30. Osisiogu, U.A.; Adum, F.L.; Efor, T.E. Strong convergence results for variational inequality problem in CAT(0) spaces. *Adv. Nonlinear Var. Inequal.* **2020**, *23*, 84–101.
31. Izuchukwu, C.; Okeke, C.C.; Isiogugu, F.O. Viscosity iterative technique for split variational inclusion problem and fixed point problem between Hilbert space and Banach space. *J. Fixed Point Theory Appl.* **2018**, *20*, 1–25. [CrossRef]

32. Song, Y.; Liu, X. Convergence comparison of several iteration algorithms for the common fixed point problems. *Fixed Point Theory Appl.* **2009**, *2009*, 824374. [CrossRef]
33. Bruhat, F.; Tits, J. Groupes réductits sur un cor local. In *Donneés Radicielles Valueés*; Institut des Hautes Études Scientifiques: Bures-sur-Yvette, France, 1972; Volume 41.
34. Kirk, W.A. Fixed point theorems in CAT(0) spaces and $\mathbb{R}$-trees. *Fixed Point Theory Appl.* **2004**, *2004*, 309–316. [CrossRef]
35. Bridson, M.R.; Haefliger, A. Metric Spaces of Non-Positive Curvature. In *Fundamental of Mathematical Sciences*; Springer: Berlin, Germany, 1999; Volume 319.
36. Reich, S.; Shafrir, I. Nonexpansive iterations in hyperbolic spaces. *Nonlinear Anal.* **1990**, *15*, 537–558. [CrossRef]
37. Goebel, K.; Reich, S. *Uniform Convexity, Hyperbolic Geometry and Nonexpansive Mappings*; Marcel Dekker: New York, NY, USA, 1984.
38. Khatibzadeh, H.; Ranjbar, S. Monotone operators and the proximal point algorithm in complete CAT(0) metric spaces. *J. Aust. Math. Soc.* **2017**, *103*, 70–90. [CrossRef]
39. Aremu, K.O.; Jolaoso, L.O.; Izuchukwu, C.; Mewomo, O.T. Approximation of common solution of finite family of monotone inclusion and fixed point problems for demicontractive mappings in CAT(0) spaces. *Ric. Mat.* **2020**, *69*, 13–34. [CrossRef]
40. Berg, I.D.; Nikolaev, I.G. Quasilinearization and curvature of Alexandrov spaces. *Geom. Dedicata* **2008**, *133*, 195–218. [CrossRef]
41. Dhompongsa, S.; Panyanak, B. On Δ-convergence theorems in CAT(0) spaces. *Comp. Math. Appl.* **2008**, *56*, 2572–2579. [CrossRef]
42. Dhompongsa, S.; Kirk, W.A.; Sims, B. Fixed points of uniformly Lipschitzian mappings. *Nonlinear Anal.* **2006**, *65*, 762–772. [CrossRef]
43. Kirk, W.A.; Panyanak, B. A concept of convergence in geodesic spaces. *Nonlinear Anal.* **2008**, *68*, 3689–3696. [CrossRef]
44. Dehghan, H.; Rooin, J. A characterization of metric projection in CAT(0) spaces. *arXiv* **2013**, arXiv:1311.4174VI.
45. Dehghan, H.; Rooin, J. Metric projection and convergence theorems for nonexpansive mapping in Hadamard spaces. *arXiv* **2014**, arXiv:1410.1137VI.
46. Chidume, C.E.; Bello, A.U.; Ndambomve, P. Strong and Δ-convergence theorems for common fixed points of a finite family of multivalued demicontractive mappings in CAT(0) soaces. *Abstr. Appl. Anal.* **2014**, *2014*, 805168. [CrossRef]
47. Leustean, L. Nonexpansive iterations in uniformly convex W-hyperbolic spaces. *arXiv* **2013**, arXiv:0810.4117.
48. Kakavandi, B.A.; Amini, M. Duality and subdifferential for convex functions on complete CAT(0) metric spaces. *Nonlinear Anal.* **2010**, *73*, 3450–3455. [CrossRef]
49. Dhompongsa, S.; Kirk, W.A.; Panyanak, B. Nonexpansive set-valued mappings in metric and Banach spaces. *J. Nonlinear Convex Anal.* **2007**, *8*, 35–45.
50. Taiwo, A.; Alakoya, T.O.; Mewomo, O.T. Halpern-type iterative process for solving split common fixed point and monotone variational inclusion problem between Banach spaces. *Numer. Algorithms* **2020**. [CrossRef]
51. Xu, H.K. Iterative algorithms for nonlinear operators. *J. Lond. Math. Soc.* **2002**, *66*, 240–256. [CrossRef]
52. Taiwo, A.; Jolaoso, L.O.; Mewomo, O.T.; Gibali, A. On generalized mixed equilibrium problem with α-β-μ bifunction and μ-τ monotone mapping. *J. Nonlinear Convex Anal.* **2020**, *21*, 1381–1401.

Publisher's Note: MDPI stays neutral with regard to jurisdictional claims in published maps and institutional affiliations.

© 2020 by the authors. Licensee MDPI, Basel, Switzerland. This article is an open access article distributed under the terms and conditions of the Creative Commons Attribution (CC BY) license (http://creativecommons.org/licenses/by/4.0/).

Article
Iterative Sequences for a Finite Number of Resolvent Operators on Complete Geodesic Spaces

Kengo Kasahara * and Yasunori Kimura

Department of Information Science, Toho University, Miyama, Funabashi, Chiba 274-8510, Japan; yasunori@is.sci.toho-u.ac.jp
* Correspondence: 7518001k@st.toho-u.jp

Abstract: We consider Halpern's and Mann's types of iterative schemes to find a common minimizer of a finite number of proper lower semicontinuous convex functions defined on a complete geodesic space with curvature bounded above.

Keywords: geodesic space; convex minimization problem; resolvent; common fixed point; iterative scheme

1. Introduction

We consider finding a common fixed point of a finite number of resolvents operators for proper lower semicontinuous convex functions on a geodesic space. To find this point, we often use iterative schemes. We focus on Mann's [1] and Halpern's [2] iterative schemes. We know many authors have considered these schemes by using nonexpansive mappings. In a Banach space, Reich [3] proved weak convergence of Mann-type iteration, and Takahashi and Tamura [4] proved that by using two nonexpansive mappings. In a Hilbert space, Wittmann [5] proved strong convergence of the Halpern-type iteration.

We also know many researchers have proved iterative schemes on geodesic spaces. In a CAT(0) space, Dhompongsa and Panyanak [6] proved Δ-convergence of Mann's iterative scheme, and Saejung [7] also proved convergence of Halpern's iterative scheme. We know a large number of results by using Mann's and Halpern's iterative schemes in a CAT(1) space. Piątek [8] considered Halpern's iterative scheme by using a nonexpansive mapping in CAT(1) space. Kimura and Satô [9] proved that by using a strongly quasi-nonexpansive and Δ-demiclosed mapping in a complete CAT(1) space. Kimura, Saejung, and Yotkaew [10] also proved convergence of Halpern's iterative schemes under the same setting. Kimura and Kohsaka [11] proved convergence of Mann and Halpern types of iterative schemes with a sequence of resolvent operators for a single proper lower semicontinuous convex function. We are particularly interested in these results [9–11], and obtain Theorems 1 and 2 with a finite number of resolvent operators in a complete CAT(1) space.

In a Hilbert space, the resolvent operator J_f is defined as follows. Let f be a proper lower semicontinuous convex function from a Hilbert space H to $]-\infty, +\infty]$. Then, J_f is defined by

$$J_f x = \operatorname*{argmin}_{y \in H} \{f(y) + \frac{1}{2}\|y - x\|^2\}$$

for all $x \in H$. We know the resolvent J_f is a single-valued mapping from H to H and it is nonexpansive. For a proper lower semicontinuous convex function f from a complete CAT(0) space X into $]-\infty, +\infty]$, Jost [12] and Mayer [13] defined the resolvent R_f of f by

$$R_f x = \operatorname*{argmin}_{y \in X} \{f(y) + \frac{1}{2}d(y, x)^2\}$$

for all $x \in X$. We also know the resolvent R_f is a single-valued mapping from X to X and it is nonexpansive. In this paper, we use the resolvent in a complete CAT(1) space defined by Kimura and Kohsaka [11,14].

2. Preliminaries

Let (X, d) be a metric space. For $x, y \in X$, a geodesic between x and y is an isometric mapping $c \colon [0, d(x,y)] \to X$ with $c(0) = x$ and $c(d(x,y)) = y$. We say X is an r-geodesic space for $r > 0$ if a geodesic exists for every pair of points in X satisfying $d(x,y) < r$. Further, a metric space X is said to be r-uniquely geodesic if such a geodesic is unique for each pair of points satisfying $d(x,y) < r$. The image of a unique geodesic between x and y is denoted by $[x, y]$.

For an r-uniquely geodesic space X, the convex combination between $x, y \in X$ with $d(x,y) < r$ is naturally defined. That is, for $\alpha \in [0,1]$, we denote by $\alpha x \oplus (1-\alpha)y$ the point $c((1-\alpha)d(x,y))$, where c is a geodesic between x and y. It follows that

$$d(\alpha x \oplus (1-\alpha)y, x) = (1-\alpha)d(x,y) \text{ and } d(\alpha x \oplus (1-\alpha)y, y) = \alpha d(x,y).$$

A subset C of X is said to be r-convex if $\alpha x \oplus (1-\alpha)y \in C$ for every $x, y \in C$ with $d(x,y) < r$ and $\alpha \in [0,1]$.

If X is r-geodesic for any $r > 0$, then X is simply called a geodesic space. A uniquely geodesic space and a convex subset are also defined in the same way.

Let X be a uniquely geodesic space and $x, y, z \in X$. For a triangle $\triangle(x, y, z) = [y, z] \cup [z, x] \cup [x, y] \subset X$ satisfying $d(y,z) + d(z,x) + d(x,y) < 2\pi$, we define its comparison triangle $\triangle(\bar{x}, \bar{y}, \bar{z})$ in the two-dimensional unit sphere $\mathbb{S}^2$ by the triangle such that each corresponding edge has the same length as that of the original triangle. Using this notion, we call X a CAT(1) space if for every $x, y, z \in X$, $p, q \in \triangle(x, y, z)$, and their corresponding points $\bar{p}, \bar{q} \in \mathbb{S}^2$, the following relation is satisfied,

$$d(p,q) \leq d_{\mathbb{S}^2}(\bar{x}, \bar{y}),$$

where $d_{\mathbb{S}^2}$ is the spherical metric on $\mathbb{S}^2$.

The following results are fundamental and important for our work.

Lemma 1 (Kimura-Satô [15]). *Let X be a CAT(1) space. Then, for every $x, y, z \in X$ with $d(x,y) + d(y,z) + d(z,x) < 2\pi$ and $\alpha \in [0,1]$, the following inequality holds,*

$$\cos d(x, w) \sin d(y, z) \geq \cos d(x,y) \sin(\alpha d(y,z)) + \cos d(x,z) \sin((1-\alpha)d(y,z)),$$

where $w = \alpha y \oplus (1-\alpha)z$.

Lemma 2 (Kimura-Satô [9]). *Let X be a CAT(1) space. Then, for every $x, y, z \in X$ with $d(x,y) + d(y,z) + d(z,x) < 2\pi$ and $\alpha \in [0,1]$, the following inequality holds,*

$$\cos d(x, w) \geq \alpha \cos d(x,y) + (1-\alpha) \cos d(x,z),$$

where $w = \alpha y \oplus (1-\alpha)z$.

Lemma 3 (Kimura-Satô [9]). *Let X be a CAT(1) space such that $d(v, v') < \pi$ for every $v, v' \in X$. Let $\alpha \in [0,1]$ and $u, y, z \in X$. Then,*

$$1 - \cos d(\alpha u \oplus (1-\alpha)y, z)$$

$$\leq (1-\beta)(1 - \cos d(y,z)) + \beta\left(1 - \frac{\cos d(u,z)}{\sin d(u,y) \tan(\frac{\alpha}{2}d(u,y)) + \cos d(u,y)}\right),$$

where

$$\beta = \begin{cases} 1 - \dfrac{\sin((1-\alpha)d(u,y))}{\sin d(u,y)} & (u \neq y), \\ \alpha & (u = y). \end{cases}$$

Let $\{x_n\} \subset X$ be a bounded sequence. We say a point $z \in X$ is an asymptotic center of $\{x_n\}$ if it is a minimizer of the function $\limsup_{n\to\infty} d(x_n, \cdot)$, that is,

$$\limsup_{n\to\infty} d(x_n, z) \leq \limsup_{n\to\infty} d(x_n, y)$$

for every $y \in X$. If $z \in X$ is the unique asymptotic center of all subsequences of $\{x_n\}$, then we say $\{x_n\}$ is Δ-convergent to a Δ-limit z. We know that in a CAT(1) space, every sequence $\{x_n\}$ satisfying $\inf_{y\in X} \limsup_{n\to\infty} d(x_n, y) < \pi/2$ has a unique asymptotic center and a Δ-convergent subsequence.

Let X be a CAT(1) space and $T: X \to X$. The set of all fixed points of T is denoted by $F(T)$. Namely, $F(T) = \{z \in X : z = Tz\}$. T is said to be quasi-nonexpansive if $F(T) \neq \emptyset$ and $d(Tx, z) \leq d(x, z)$ for every $x \in X$ and $z \in F(T)$. A quasi-nonexpansive mapping T is said to be strongly quasi-nonexpansive if $\lim_{n\to\infty} d(x_n, Tx_n) = 0$ whenever $\{x_n\} \subset X$ satisfies $\sup_{n\in\mathbb{N}} d(x_n, p) < \pi/2$ and $\lim_{n\to\infty} (\cos d(x_n, p)/\cos d(Tx_n, p)) = 1$ for every $p \in F(T)$.

A mapping T is said to be Δ-demiclosed if $z \in F(T)$ whenever $\{x_n\}$ is Δ-convergent to z and $\lim_{n\to\infty} d(x_n, Tx_n) = 0$.

Following [16], we define the notions of a strongly quasi-nonexpansive sequence and a Δ-demiclosed sequence on CAT(1) spaces as follows. Let $\{T_n\}$ be a sequence of mappings from X to X. $\{T_n\}$ is said to be a strongly quasi-nonexpansive sequence if each T_n is quasi-nonexpansive and $\lim_{n\to\infty} d(x_n, T_n x_n) = 0$ whenever $\sup_{n\in\mathbb{N}} d(x_n, p) < \pi/2$ and $\lim_{n\to\infty} (\cos d(x_n, p)/\cos d(T_n x_n, p)) = 1$ for every $p \in \bigcap_{n=1}^{\infty} F(T_n)$. $\{T_n\}$ is said to be a Δ-demiclosed sequence if $z \in \bigcap_{n=1}^{\infty} F(T_n)$ whenever $\{x_n\}$ is Δ-convergent to z and $\lim_{n\to\infty} d(x_n, T_n x_n) = 0$.

Let X be a complete CAT(1) space and $C \subset X$ a nonempty closed π-convex subset such that $d(x, C) = \inf_{y\in C} d(x, y) < \pi/2$ for every $x \in X$. Then, for each $x \in X$, there exists a unique point $y_x \in C$ satisfying $d(x, y_x) = \inf_{y\in C} d(x, y)$. Using this point, we define a metric projection $P_C \colon X \to C$ by $P_C x = y_x$ for $x \in X$.

Let X be a complete CAT(1) space such that $d(v, v') < \pi/2$ for every $v, v' \in X$. Let $f: X \to]-\infty, +\infty]$ be a proper lower semicontinuous convex function. The resolvent R_f of f is defined by

$$R_f x = \operatorname*{argmin}_{y\in X}(f(y) + \tan d(y, x) \sin d(y, x))$$

for all $x \in X$; (see in [14]). We know that R_f is a single-valued mapping from X to X. We also know that the resolvent R_f is strongly quasi-nonexpansive and Δ-demiclosed such that $F(R_f) = \operatorname{argmin}_{x\in X} f$ (see [11,14]).

We recall some lemmas useful for our results.

Lemma 4 (Kimura-Satô [17]). *Let X be a complete CAT(1) space such that $d(u, v) < \pi/2$ for all $u, v \in X$. Let S, T be quasi-nonexpansive mappings from X to X with $F(S) \cap F(T) \neq \emptyset$. Then, for every $\alpha \in\,]0, 1[$, $F(S) \cap F(T) = F(\alpha S \oplus (1-\alpha)T)$ and the mapping $\alpha S \oplus (1-\alpha)T$ is quasi-nonexpansive.*

Lemma 5 (He-Fang-López-Li [18]). *Let X be a complete CAT(1) space and $p \in X$. If a sequence $\{x_n\}$ in X satisfies that $\limsup_{n\to\infty} d(x_n, p) < \pi/2$ and that $\{x_n\}$ is Δ-convergent to $x \in X$, then $d(x, p) \leq \liminf_{n\to\infty} d(x_n, p)$.*

Lemma 6 (Saejung-Yotkaew [19], Aoyama-Kimura-Kohsaka [20]). *Let $\{s_n\}$ and $\{t_n\}$ be sequences of real numbers such that $s_n \geq 0$ for every $n \in \mathbb{N}$. Let $\{\beta_n\}$ be a sequence in $]0,1[$ such that $\sum_{n=0}^{\infty} \beta_n = \infty$. Suppose that $s_{n+1} \leq (1-\beta_n)s_n + \beta_n t_n$ for every $n \in \mathbb{N}$. If $\limsup_{k \to \infty} t_{n_k} \leq 0$ for every nondecreasing sequence $\{n_k\}$ of $\mathbb{N}$ satisfying $\liminf_{k \to \infty}(s_{n_k+1} - s_{n_k}) \geq 0$, then $\lim_{n \to \infty} s_n = 0$.*

3. Lemmas for a Finite Number of Resolvent Operators

In this section, we prove some lemmas by using a finite number of resolvent operators for iterative schemes. Throughout this section, let X be a CAT(1) space such that $d(v, v') < \pi/2$ for every $v, v' \in X$.

Lemma 7. *For a given real number $a \in \left]0, \tfrac{1}{2}\right]$, let $\sigma \in [a, 1-a]$. For given points $y, y^0, y^1 \in X$, define $w \in X$ by*
$$w = \sigma y^0 \oplus (1-\sigma) y^1.$$

Then,
$$\cos d(w, y) \cos(ad(y^0, y^1)) \geq \min\{\cos d(y^0, y), \cos d(y^1, y)\}.$$

Proof. If $y^0 = y^1$, it is obvious. Otherwise, by Lemma 1, we have

$\cos d(w, y) \sin d(y^0, y^1)$
$\geq \cos d(y^0, y) \sin(\sigma d(y^0, y^1)) + \cos d(y^1, y) \sin((1-\sigma) d(y^0, y^1))$
$\geq \min\{\cos d(y^0, y), \cos d(y^1, y)\} (\sin(\sigma d(y^0, y^1)) + \sin((1-\sigma) d(y^0, y^1)))$
$= 2 \min\{\cos d(y^0, y), \cos d(y^1, y)\} \sin \dfrac{d(y^0, y^1)}{2} \cos \dfrac{(2\sigma - 1) d(y^0, y^1)}{2}.$

Dividing above by $2 \sin(d(y^0, y^1)/2)$, we have

$\cos d(w, y) \cos \dfrac{d(y^0, y^1)}{2}$
$\geq \min\{\cos d(y^0, y), \cos d(y^1, y)\} \cos \dfrac{(2\sigma - 1) d(y^0, y^1)}{2}$
$\geq \min\{\cos d(y^0, y), \cos d(y^1, y)\} \cos \dfrac{(1 - 2a) d(y^0, y^1)}{2}.$

Moreover, dividing above by $\cos((1-2a) d(y^0, y^1)/2)$, we have

$\min\{\cos d(y^0, y), \cos d(y^1, y)\}$
$\leq \cos d(w, y) \dfrac{\cos \dfrac{(1-2a) d(y^0, y^1)}{2} \cos(ad(y^0, y^1)) - \sin \dfrac{(1-2a) d(y^0, y^1)}{2} \sin(ad(y^0, y^1))}{\cos \dfrac{(1-2a) d(y^0, y^1)}{2}}$
$\leq \cos d(w, y) \cos(ad(y^0, y^1)).$

This completes the proof. $\square$

Lemma 8. *For a given real number $a \in \left]0, \tfrac{1}{2}\right]$, let $\sigma^l \in [a, 1-a]$ for every $l = 0, 1, \ldots, N-1$. For given points $y, y^k \in X$ for every $k = 0, 1, \ldots, N$, define $w^l \in X$ by*
$$w^N = y^N \text{ and } w^l = \sigma^l y^l \oplus (1 - \sigma^l) w^{l+1}$$

for every $l = 0, 1, \ldots, N-1$. Then,
$$\cos d(w^0, y) \cos(ad(y^0, w^1)) \geq \min_{k \in \{0,1,\ldots,N\}} \cos d(y^k, y).$$

Proof. By Lemma 7,

$$\cos d(w^0, y) \cos(ad(y^0, w^1)) \geq \min\{\cos d(y^0, y), \cos d(w^1, y)\}.$$

We also have

$$\cos d(w^l, y) \geq \cos d(w^l, y) \cos(ad(y^l, w^{l+1}))$$
$$\geq \min\{\cos d(y^l, y), \cos d(w^{l+1}, y)\}$$

for $l = 1, 2, \ldots, N-1$. Therefore, $\cos d(w^0, y) \cos(ad(y^0, w^1)) \geq \min_{k \in \{0,1,\ldots,N\}} \cos d(y^k, y)$. This completes the proof. □

Corollary 1. *Let T^k be a quasi-nonexpansive mapping from X to X for every $k = 0, 1, \ldots, N$. For a given real number $a \in \left]0, \frac{1}{2}\right]$, let $\sigma^l \in [a, 1-a]$ for every $l = 0, 1, \ldots, N-1$. Define $U^l : X \to X$ by*

$$U^N = T^N \text{ and } U^l = \sigma^l T^l \oplus (1 - \sigma^l) U^{l+1}$$

for every $l = 0, 1, \ldots, N-1$. Let $x \in X$ and $p \in \bigcap_{k=0}^{N} F(T^k)$. Then,

$$\cos d(U^0 x, p) \cos(ad(T^0 x, U^1 x)) \geq \cos d(x, p).$$

Next, we show several properties of a sequence of resolvents. Let f be a proper lower semicontinuous convex function from X into $]-\infty, +\infty]$ such that $\operatorname{argmin}_X f \neq \emptyset$ and let $\{\lambda_n\}$ be a real sequence such that $\inf \lambda_n > 0$. Then we know that $\{R_{\lambda_n f}\}$ is a strongly quasi-nonexpansive sequence and Δ-demiclosed sequence (see [11]). Therefore, we obtain the following results, using Lemma 4.

Lemma 9. *Let f^k be a proper lower semicontinuous convex function from X into $]-\infty, +\infty]$ for every $k = 0, 1, \ldots, N$ such that $\bigcap_{k=0}^{N} \operatorname{argmin}_X f^k \neq \emptyset$. For a given real number $a \in \left]0, \frac{1}{2}\right]$, let $\sigma^l \in [a, 1-a]$ for every $l = 0, 1, \ldots, N-1$ and $\lambda^k \in [a, +\infty[$ for every $k = 0, 1, \ldots, N$. Let $R_{\lambda^k f^k}$ be the resolvent of $\lambda^k f^k$ for every $k = 0, 1, \ldots, N$. Define $U^l : X \to X$ by*

$$U^N = R_{\lambda^N f^N} \text{ and } U^l = \sigma^l R_{\lambda^l f^l} \oplus (1 - \sigma^l) U^{l+1}$$

for every $l = 0, 1, \ldots, N-1$. Then

$$F(U^0) = \bigcap_{k=0}^{N} \operatorname{argmin}_X f^k.$$

Lemma 10. *Let $\{T_n\}$ be a strongly quasi-nonexpansive sequence. Let f be a proper lower semicontinuous convex function from X into $]-\infty, +\infty]$ such that $\bigcap_{n=1}^{\infty} F(T_n) \cap \operatorname{argmin}_X f \neq \emptyset$. For a given real number $a \in \left]0, \frac{1}{2}\right]$, let $\{\sigma_n\} \subset [a, 1-a]$ and $\{\lambda_n\} \subset [a, +\infty[$. Let $R_{\lambda_n f}$ be the resolvent of $\lambda_n f$ for every $n \in \mathbb{N}$. Then $\{\sigma_n R_{\lambda_n f} \oplus (1 - \sigma_n) T_n\}$ is a strongly quasi-nonexpansive sequence.*

Proof. Let $V_n = \sigma_n R_{\lambda_n f} \oplus (1 - \sigma_n) T_n$ for every $n \in \mathbb{N}$. From Lemma 4, V_n is a quasi-nonexpansive mapping for every $n \in \mathbb{N}$. From Corollary 1, for $\{x_n\} \subset X$ and $p \in \bigcap_{n=1}^{\infty} F(T_n) \cap \operatorname{argmin}_X f$ such that $\lim_{n \to \infty} \cos d(x_n, p) / \cos d(V_n x_n, p) = 1$ and $\sup_{n \in \mathbb{N}} d(x_n, p) < \pi/2$, we have

$$\cos d(V_n x_n, p) \cos(ad(R_{\lambda_n f} x_n, T_n x_n)) \geq \cos d(x_n, p)$$

and thus

$$\cos(ad(R_{\lambda_n f} x_n, T_n x_n)) \geq \frac{\cos d(x_n, p)}{\cos d(V_n x_n, p)} \to 1.$$

That is, $\lim_{n\to\infty} d(R_{\lambda_n f} x_n, T_n x_n) = 0$. Therefore, we have

$$\lim_{n\to\infty} d(T_n x_n, V_n x_n) = \lim_{n\to\infty} \sigma_n d(R_{\lambda_n f} x_n, T_n x_n) = 0.$$

As $1 = \lim_{n\to\infty} \cos d(x_n, p) / \cos d(V_n x_n, p) = \lim_{n\to\infty} \cos d(x_n, p) / \cos d(T_n x_n, p)$, we have

$$\lim_{n\to\infty} d(T_n x_n, x_n) = 0.$$

Thus, we obtain

$$d(V_n x_n, x_n) \leq d(V_n x_n, T_n x_n) + d(T_n x_n, x_n) \to 0.$$

This completes the proof. □

Corollary 2. *Let f^k be the same as in Lemma 9 for $k = 0, 1, \ldots, N$. For a given real number $a \in \left]0, \frac{1}{2}\right]$, let $\{\sigma_n^l\} \subset [a, 1-a]$ for every $l = 0, 1, \ldots, N-1$ and $\{\lambda_n^k\} \subset [a, +\infty[$ for every $k = 0, 1, \ldots, N$. Let $R_{\lambda_n^k f^k}$ be the resolvent of $\lambda_n^k f^k$ for every $k = 0, 1, \ldots, N$ and $n \in \mathbb{N}$. Define $U_n^l : X \to X$ by*

$$U_n^N = R_{\lambda_n^N f^N} \text{ and } U_n^l = \sigma_n^l R_{\lambda_n^l f^l} \oplus (1 - \sigma_n^l) U_n^{l+1}$$

for every $l = 0, 1, \ldots, N-1$ and $n \in \mathbb{N}$. Then, $\{U_n^0\}$ is a strongly quasi-nonexpansive sequence.

Lemma 11. *Let $\{T_n\}$ be a quasi-nonexpansive and Δ-demiclosed sequence. Let f be a proper lower semicontinuous convex function from X into $]-\infty, +\infty]$ such that $\bigcap_{n=1}^{\infty} F(T_n) \cap \mathrm{argmin}_X f \neq \emptyset$. For a given real number $a \in \left]0, \frac{1}{2}\right]$, let $\{\sigma_n\} \subset [a, 1-a]$ and $\{\lambda_n\} \subset [a, +\infty[$. Let $R_{\lambda_n f}$ be the resolvent of $\lambda_n f$ for every $n \in \mathbb{N}$. Then $\{\sigma_n R_{\lambda_n f} \oplus (1 - \sigma_n) T_n\}$ is a Δ-demiclosed sequence.*

Proof. Let $V_n = \sigma_n R_{\lambda_n f} \oplus (1 - \sigma_n) T_n$ for every $n \in \mathbb{N}$. Let $p \in \bigcap_{n=1}^{\infty} F(T_n) \cap \mathrm{argmin}_X f$, $\{x_n\} \subset X$, and $z \in X$ such that $\lim_{n\to\infty} d(V_n x_n, x_n) = 0$ and suppose that $\{x_n\}$ is Δ-convergent to z. Then,

$$\cos d(V_n x_n, p) \cos(a d(R_{\lambda_n f} x_n, T_n x_n)) \geq \cos d(x_n, p)$$

and thus

$$1 \geq \cos(a d(R_{\lambda_n f} x_n, T_n x_n)) \geq \frac{\cos d(x_n, p)}{\cos d(V_n x_n, p)}$$

$$\geq \frac{\cos(d(x_n, V_n x_n) + d(V_n x_n, p))}{\cos d(V_n x_n, p)} \to 1.$$

Therefore, $\lim_{n\to\infty} d(R_{\lambda_n f} x_n, T_n x_n) = 0$. Thus, we have

$$d(R_{\lambda_n f} x_n, V_n x_n) = (1 - \sigma_n) d(R_{\lambda_n f} x_n, T_n x_n)$$
$$\leq (1 - a) d(R_{\lambda_n f} x_n, T_n x_n) \to 0.$$

Since $R_{\lambda_n f}$ is a Δ-demiclosed sequence, we have $R_{\lambda_n f} z = z$. Similarly,

$$d(T_n x_n, V_n x_n) = \sigma_n d(R_{\lambda_n f} x_n, T_n x_n)$$
$$\leq (1 - a) d(R_{\lambda_n f} x_n, T_n x_n) \to 0.$$

Since $\{T_n\}$ is a Δ-demiclosed sequence, we have $T_n z = z$. Thus, $V_n z = z$. This completes the proof. □

Corollary 3. *Let f^k, $\{\sigma_n^l\}$, $\{\lambda_n^k\}$ and $\{U_n^l\}$ be the same as in Corollary 2 for $k = 0, 1, \ldots, N$ and $l = 0, 1, \ldots, N-1$. Then $\{U_n^0\}$ is a Δ-demiclosed sequence.*

4. Iterative Schemes for a Finite Resolvents Operators

We prove convergence of Mann and Halpern types of iterative sequences for finitely many convex functions by using the properties of a sequence of the resolvents in CAT(1) space.

Theorem 1. *Let X be a complete CAT(1) space such that $d(v,v') < \pi/2$ for every $v,v' \in X$. Let f^k be a proper lower semicontinuous convex function from X into $]-\infty, +\infty]$ for every $k = 0, 1, \ldots, N$ such that $F = \bigcap_{k=0}^{N} \arg\min_X f^k \neq \emptyset$. For a given real number $a \in \left]0, \frac{1}{2}\right]$, let $\{\sigma_n^l\} \subset [a, 1-a]$ for every $l = 0, 1, \ldots, N-1$ and $\{\lambda_n^k\} \subset [a, +\infty[$ for every $k = 0, 1, \ldots, N$. Let $R_{\lambda_n^k f^k}$ be the resolvent of $\lambda_n^k f^k$ for every $k = 0, 1, \ldots, N$ and $n \in \mathbb{N}$. Define $U_n^l : X \to X$ by*

$$U_n^N = R_{\lambda_n^N f^N} \text{ and } U_n^l = \sigma_n^l R_{\lambda_n^l f^l} \oplus (1 - \sigma_n^l) U_n^{l+1}$$

for every $l = 0, 1, \ldots, N-1$ and $n \in \mathbb{N}$. Let $\{\alpha_n\}$ be a real sequence in $[a, 1-a]$. For a given point $x_1 \in X$, let $\{x_n\}$ be the sequence in X generated by

$$x_{n+1} = \alpha_n x_n \oplus (1 - \alpha_n) U_n^0 x_n$$

for $n \in \mathbb{N}$. Then, $\{x_n\}$ Δ-converges to a point of F.

Proof. Let $z \in F$. As U_n^0 is a quasi-nonexpansive mapping, it follows from Lemma 2 that

$$\cos d(x_{n+1}, z) \geq \alpha_n \cos d(x_n, z) + (1 - \alpha_n) \cos d(U_n^0 x_n, z)$$
$$\geq \cos d(x_n, z).$$

Thus we have $d(x_{n+1}, z) \leq d(x_n, z)$ for $n \in \mathbb{N}$. There exists $D = \lim_{n \to \infty} d(x_n, z) \leq d(x_1, z) < \pi/2$. From Lemma 1, we get

$$\cos d(x_{n+1}, z) \sin d(x_n, U_n^0 x_n)$$
$$\geq \cos d(x_n, z) \sin \alpha_n d(x_n, U_n^0 x_n) + \cos d(U_n^0 x_n, z) \sin(1 - \alpha_n) d(x_n, U_n^0 x_n)$$
$$\geq 2 \cos d(x_n, z) \sin \frac{d(x_n, U_n^0 x_n)}{2} \cos \frac{(2\alpha_n - 1) d(x_n, U_n^0 x_n)}{2}.$$

If $d(x_n, U_n^0 x_n) \neq 0$, we obtain

$$\cos d(x_{n+1}, z) \cos \frac{d(x_n, U_n^0 x_n)}{2} \geq \cos d(x_n, z) \cos \frac{(2\alpha_n - 1) d(x_n, U_n^0 x_n)}{2}.$$

As $\{\alpha_n\} \subset [a, 1-a]$, we get

$$1 > \frac{\cos \frac{d(x_n, U_n^0 x_n)}{2}}{\cos \frac{(1-2a) d(x_n, U_n^0 x_n)}{2}} \geq \frac{\cos \frac{d(x_n, U_n^0 x_n)}{2}}{\cos \frac{(2\alpha_n - 1) d(x_n, U_n^0 x_n)}{2}} \geq \frac{\cos d(x_n, z)}{\cos d(x_{n+1}, z)}.$$

As $D = \lim_{n \to \infty} d(x_n, z) \leq d(x_1, z) < \pi/2$, we have

$$\lim_{n \to \infty} \frac{\cos \frac{d(x_n, U_n^0 x_n)}{2}}{\cos \frac{(1-2a) d(x_n, U_n^0 x_n)}{2}} = 1$$

and thus $\lim_{n \to \infty} d(x_n, U_n^0 x_n) = 0$. Let x_0 be an asymptotic center of $\{x_n\}$ and y an asymptotic center of any subsequence $\{x_{n_k}\} \subset \{x_n\}$. There exists $\{x_{n_{k_l}}\} \subset \{x_{n_k}\}$ such that $\{x_{n_{k_l}}\}$

Δ-converges to w. As $\{U_{n_{k_l}}^0\}$ is a Δ-demiclosed sequence and $\lim_{n\to\infty} d(U_{n_{k_l}}^0 x_{n_{k_l}}, x_{n_{k_l}}) = 0$, we obtain $w \in F$. Since there exists $\lim_{n\to\infty} d(x_{n_k}, w)$, we have

$$\limsup_{k\to\infty} d(x_{n_k}, w) = \lim_{k\to\infty} d(x_{n_k}, w) = \lim_{l\to\infty} d(x_{n_{k_l}}, w)$$
$$\leq \limsup_{l\to\infty} d(x_{n_{k_l}}, y) \leq \limsup_{k\to\infty} d(x_{n_k}, y).$$

Therefore, we obtain $y = w \in F$. Similarly, we get $x_0 = y$. Therefore, $\{x_n\}$ Δ-converges to $x_0 \in F$. □

Theorem 2. *Let X, f^k, $\{\sigma_n^l\}$, $\{\lambda_n^k\}$ and $\{U_n^l\}$ be the same as in Theorem 1 for $k = 0, 1, \ldots, N$ and $l = 0, 1, \ldots, N-1$. Let $\{\alpha_n\}$ be a real sequence in $]0, 1[$ such that $\lim_{n\to\infty} \alpha_n = 0$ and $\sum_{n=0}^\infty \alpha_n = \infty$. For given points $u, x_1 \in X$, let $\{x_n\}$ be the sequence in X generated by*

$$x_{n+1} = \alpha_n u \oplus (1 - \alpha_n) U_n^0 x_n$$

for $n \in \mathbb{N}$. Suppose that one of the following conditions holds:
(a) $\sup_{v,v' \in X} d(v, v') < \pi/2$;
(b) $d(u, P_F u) < \pi/4$ and $d(u, P_F u) + d(x_0, P_F u) < \pi/2$;
(c) $\sum_{n=0}^\infty \alpha_n^2 = \infty$.
Then, $\{x_n\}$ converges to $P_F u$.

To prove this theorem, we also employ the technique proposed in [9]. Note that $F = \bigcap_{k=0}^N \arg\min_X f^k$.

Proof. Let $p = P_F u$ and let

$$s_n = 1 - \cos d(x_n, p),$$
$$t_n = 1 - \frac{\cos d(u, p)}{\sin d(u, U_n^0 x_n) \tan(\frac{\alpha_n}{2} d(u, U_n^0 x_n)) + \cos d(u, U_n^0 x_n)},$$
$$\beta_n = \begin{cases} 1 - \dfrac{\sin((1-\alpha_n)d(u, U_n^0 x_n))}{\sin d(u, U_n^0 x_n)} & (u \neq U_n^0 x_n), \\ \alpha_n & (u = U_n^0 x_n) \end{cases}$$

for $n \in \mathbb{N}$. Since U_n^0 is a quasi-nonexpansive mapping, it follows from Lemma 3 that

$$s_{n+1} \leq (1 - \beta_n)(1 - \cos d(U_n^0 x_n, p)) + \beta_n t_n \leq (1 - \beta_n) s_n + \beta_n t_n$$

for $n \in \mathbb{N}$. By Lemma 2, we have

$$\cos d(x_{n+1}, p) = \cos d(\alpha_n u \oplus (1 - \alpha_n) U_n^0 x_n, p)$$
$$\geq \alpha_n \cos d(u, p) + (1 - \alpha_n) \cos d(U_n^0 x_n, p)$$
$$\geq \alpha_n \cos d(u, p) + (1 - \alpha_n) \cos d(x_n, p)$$
$$\geq \min\{\cos d(u, p), \cos d(x_n, p)\}$$

for $n \in \mathbb{N}$. So we have

$$\cos d(x_n, p) \geq \min\{\cos d(u, p), \cos d(x_0, p)\} = \cos \max\{d(u, p), d(x_0, p)\} > 0$$

for $n \in \mathbb{N}$. Hence $\sup_{n \in \mathbb{N}} d(x_n, p) \leq \max\{d(u, p), d(x_0, p)\} < \pi/2$. Next, we will show for each of the conditions (a–c) imply that $\sum_{n=0}^\infty \beta_n = \infty$. For the conditions (a) and (b), let

$M = \sup_{n \in \mathbb{N}} d(u, U_n^0 x_n)$. Thus, we will show $M < \pi/2$. In case (a), it is obvious. In case (b), as $\sup_{n \in \mathbb{N}} d(x_n, p) \leq \max\{d(u, p), d(x_0, p)\}$, we have

$$M \leq \sup_{n \in \mathbb{N}}(d(u, p) + d(U_n^0 x_n, p))$$
$$\leq \sup_{n \in \mathbb{N}}(d(u, p) + d(x_n, p))$$
$$\leq \max\{2d(u, p), d(u, p) + d(x_0, p)\} < \pi/2.$$

Thus, for cases (a) and (b), we have

$$\beta_n \geq 1 - \frac{\sin((1 - \alpha_n)M)}{\sin M}$$
$$= \frac{2}{\sin M} \sin\left(\frac{\alpha_n}{2} M\right) \cos\left(\left(1 - \frac{\alpha_n}{2}\right) M\right)$$
$$\geq \alpha_n \cos M$$

for $n \in \mathbb{N}$. As $\sum_{n=0}^{\infty} \alpha_n = \infty$, each of the conditions (a) and (b) implies that $\sum_{n=0}^{\infty} \beta_n = \infty$. In the case (c), we have

$$\beta_n \geq 1 - \sin \frac{(1 - \alpha_n)\pi}{2} = 1 - \cos \frac{\alpha_n}{2} \geq \frac{\alpha_n^2 \pi^2}{16}$$

for $n \in \mathbb{N}$. Hence the condition (c) also implies that $\sum_{n=0}^{\infty} \beta_n = \infty$. For $\{s_{n_i}\} \subset \{s_n\}$ with a nondecreasing real sequence $\{n_i\} \subset \mathbb{N}$ such that $\liminf_{i \to \infty}(s_{n_i+1} - s_{n_i}) \geq 0$, we have

$$0 \leq \liminf_{i \to \infty}(s_{n_i+1} - s_{n_i})$$
$$= \liminf_{i \to \infty}(\cos d(x_{n_i}, p) - \cos d(x_{n_i+1}, p))$$
$$\leq \liminf_{i \to \infty}(\cos d(x_{n_i}, p) - (\alpha_{n_i} \cos d(u, p) + (1 - \alpha_{n_i}) \cos d(U_{n_i}^0 x_{n_i}, p)))$$
$$= \liminf_{i \to \infty}(\cos d(x_{n_i}, p) - \cos d(U_{n_i}^0 x_{n_i}, p))$$
$$\leq \limsup_{i \to \infty}(\cos d(x_{n_i}, p) - \cos d(U_{n_i}^0 x_{n_i}, p)) \leq 0.$$

Hence $\lim_{i \to \infty}(\cos d(x_{n_i}, p) - \cos d(U_{n_i}^0 x_{n_i}, p)) = 0$. Since $\sup_{n \in \mathbb{N}} d(U_n^0 x_n, p) < \pi/2$, we have $\lim_{i \to \infty}(\cos d(x_{n_i}, p) / \cos d(U_{n_i}^0 x_{n_i}, p)) = 1$. As $\{U_{n_i}^0\}$ is a strongly quasi-nonexpansive sequence, it follows that $\lim_{i \to \infty} d(x_{n_i}, U_{n_i}^0 x_{n_i}) = 0$. Let $\{x_{n_j}\} \subset \{x_{n_i}\}$ be a Δ-convergent subsequence such that $\lim_{j \to \infty} d(u, x_{n_j}) = \liminf_{i \to \infty} d(u, x_{n_i})$. Since $\{U_n^0\}$ is a Δ-demiclosed sequence and $\lim_{j \to \infty} d(x_{n_j}, U_{n_j}^0 x_{n_j}) = 0$, the Δ-limit $z \in \{x_{n_j}\}$ belongs to F. By Lemma 5, we have

$$\liminf_{i \to \infty} d(u, U_{n_i}^0 x_{n_i}) = \liminf_{i \to \infty} d(u, x_{n_i}) = \lim_{j \to \infty} d(u, x_{n_j}) \geq d(u, z) \geq d(u, p).$$

Hence

$$\limsup_{i \to \infty} t_{n_i} = \limsup_{i \to \infty} \left(1 - \frac{\cos d(u, p)}{\sin d(u, U_{n_i}^0 x_{n_i}) \tan(\frac{\alpha_{n_i}}{2} d(u, U_{n_i}^0 x_{n_i}) + \cos d(u, U_{n_i}^0 x_{n_i})}\right)$$
$$= \limsup_{i \to \infty} \left(1 - \frac{\cos d(u, p)}{\cos d(u, U_{n_i}^0 x_{n_i})}\right) \leq 0.$$

From Lemma 6, we have $\lim_{n \to \infty} s_n = 0$. Therefore, $\{x_n\}$ converges to p. This completes the proof. □

5. Applications to the Image Recovery Problem

At the end of this work, we apply our results to the problem of finding a point of the intersection of a finite family of closed convex subsets. This problem is also known as the image recovery problem. See the works in [21,22] and references therein.

Let C be a nonempty closed convex subset of a complete CAT(1) space such that $d(v, v') < \pi/2$ for every $v, v' \in X$. Then, the indicator function $i_C \colon C \to X$ of C defined by

$$i_C(x) = \begin{cases} 0 & (x \in C), \\ \infty & (x \notin C) \end{cases}$$

is proper, lower semicontinuous, and convex. As is mentioned in [14], the resolvent R_{i_C} of this function coincides with the metric projection P_C. Using this fact, we obtain the following results for the image recovery problem. The first result can be proved by using Theorem 1.

Theorem 3. *Let X be a complete CAT(1) space such that $d(v, v') < \pi/2$ for every $v, v' \in X$. Let $\{C_0, C_1, \ldots, C_N\}$ be a finite family of nonempty closed convex subsets of X such that $C = \bigcap_{k=0}^{N} C_K \neq \emptyset$. For a given real number $a \in \left]0, \frac{1}{2}\right]$, let $\{\sigma_n^l\} \subset [a, 1-a]$ for $l = 0, 1, \ldots, N-1$ and $n \in \mathbb{N}$. Let P_{C_k} be the metric projection onto C_k for $k = 0, 1, \ldots, N$. Define $U_n^l \colon X \to X$ by*

$$U_n^N = P_{C_N} \text{ and } U_n^l = \sigma_n^l P_{C_l} \oplus (1 - \sigma_n^l) U_n^{l+1}$$

for every $l = 0, 1, \ldots, N-1$ and $n \in \mathbb{N}$. Let $\{\alpha_n\}$ be a real sequence in $[a, 1-a]$. For a given point $x_1 \in X$, let $\{x_n\}$ be the sequence in X generated by

$$x_{n+1} = \alpha_n x_n \oplus (1 - \alpha_n) U_n^0 x_n$$

for $n \in \mathbb{N}$. Then, $\{x_n\}$ Δ-converges to a point of C.

Note that this theorem is a generalization of the result by [21] in the setting of Hilbert spaces, to complete CAT(1) spaces.

On the other hand, by using Thoerem 2, we can also prove the following theorem which was obtained by the authors of [23].

Theorem 4 (Kasahara-Kimura [23]). *Let X be a complete CAT(1) space such that $d(v, v') < \pi/2$ for every $v, v' \in X$. Let $\{C_0, C_1, \ldots, C_N\}$ be a finite family of nonempty closed convex subsets of X such that $C = \bigcap_{k=0}^{N} C_K \neq \emptyset$. For a given real number $a \in \left]0, \frac{1}{2}\right]$, let $\{\sigma_n^l\} \subset [a, 1-a]$ for $l = 0, 1, \ldots, N-1$ and $n \in \mathbb{N}$. Let P_{C_k} be the metric projection onto C_k for $k = 0, 1, \ldots, N$. Define $U_n^l \colon X \to X$ by*

$$U_n^N = P_{C_N} \text{ and } U_n^l = \sigma_n^l P_{C_l} \oplus (1 - \sigma_n^l) U_n^{l+1}$$

for every $l = 0, 1, \ldots, N-1$ and $n \in \mathbb{N}$. Let $\{\alpha_n\}$ be a real sequence in $]0,1[$ such that $\lim_{n \to \infty} \alpha_n = 0$ and $\sum_{n=0}^{\infty} \alpha_n = \infty$. For given points $u, x_1 \in X$, let $\{x_n\}$ be the sequence in X generated by

$$x_{n+1} = \alpha_n u \oplus (1 - \alpha_n) U_n^0 x_n$$

for $n \in \mathbb{N}$. Suppose that one of the following conditions holds:

(a) $\sup_{v, v' \in X} d(v, v') < \pi/2$;
(b) $d(u, P_C u) < \pi/4$ and $d(u, P_C u) + d(x_0, P_C u) < \pi/2$;
(c) $\sum_{n=0}^{\infty} \alpha_n^2 = \infty$.

Then $\{x_n\}$ converges to $P_C u$.

6. Conclusions

We proposed a new type of iterative scheme for the problem of finding a common minimizer of finitely many convex functions defined on a complete CAT(1) space. We considered the resolvent operators for proper lower semicontinuous convex functions defined on a complete CAT(1) space and their convex combination. As the convex combination on a CAT(1) space is defined only between two points, we need to take it repeatedly for three or more points.

In the first result (Theorem 1), we adopted a Mann-type sequence defined by the following iterative formula: $x_1 \in X$ is given and

$$x_{n+1} = \alpha_n x_n \oplus (1 - \alpha_n) U_n^0 x_n$$

for $n \in \mathbb{N}$, where a mapping U_n^0 is defined by the convex combination of finitely many resolvents. Then, $\{x_n\}$ is Δ-convergent to a solution to our problem.

In the second result (Theorem 2), we used a Halpern-type sequence defined as follows: $u, x_1 \in X$ is given and

$$x_{n+1} = \alpha_n u \oplus (1 - \alpha_n) U_n^0 x_n$$

for $n \in \mathbb{N}$. Then, it converges to $P_F u$, the nearest point of the solution set F to u.

Further, we showed that these results can be applied to the image recovery problem.

Author Contributions: The authors have contributed to this work on an equal basis. All authors read and approved the final manuscript.

Funding: This research received no external fundings.

Acknowledgments: The authors are grateful to anonymous referees for their valuable comments and suggestions.

Conflicts of Interest: The authors declare no conflict of interest.

References

1. Mann, W.R. Mean value methods in iteration; *Proc. Am. Math. Soc.* **1953**, *4*, 506–510. [CrossRef]
2. Halpern, B. Fixed points of nonexpanding maps. *Bull. Am. Math. Soc.* **1967**, *73*, 957–961. [CrossRef]
3. Reich, S. Weak convergence theorems for nonexpansive mappings in Banach spaces. *J. Math. Anal. Appl.* **1979**, *67*, 274–276. [CrossRef]
4. Takahashi, W.; Tamura, T. Convergence theorems for a pair of nonexpansive mappings. *J. Convex Anal.* **1998**, *5*, 45–56.
5. Wittmann, R. Approximation of fixed points of nonexpansive mappings. *Arch. Math.* **1992**, *58*, 486–491. [CrossRef]
6. Dhompongsa, S.; Panyanak, B. On Δ-convergence theorems in CAT(0) spaces. *Comput. Math. Appl.* **2008**, *56*, 2572–2579. [CrossRef]
7. Saejung, S. Halpern's iteration in CAT(0) spaces. *Fixed Point Theory Appl.* **2010**, *2010*, 1471781. [CrossRef]
8. Piątek, B. Halpern iteration in CAT(κ) spaces. *Acta Math. Sin. Engl. Ser.* **2011**, *27*, 635–646. [CrossRef]
9. Kimura, Y.; Satô, K. Halpern iteration for strongly quasinonexpansive mappings on a geodesic space with curvature bounded above by one. *Fixed Point Theory Appl.* **2013**, *2013*, 7. [CrossRef]
10. Kimura, Y.; Saejung, S.; Yotkaew, P. The Mann algorithm in a complete geodesic space with curvature bounded above. *Fixed Point Theory Appl.* **2013**, *2013*, 336. [CrossRef]
11. Kimura, Y.; Kohsaka, F. Two modified proximal point algorithms in geodesic spaces with curvature bounded above. *Rend. Circ. Mat. Pallemo II Ser.* **2019**, *68*, 83–104. [CrossRef]
12. Jost, J. Convex functionals and generalized harmonic maps into spaces of nonpositive curvature. *Comment. Math. Helv.* **1995**, *70*, 659–673. [CrossRef]
13. Mayer, U.F. Gradient flows on nonpositively curved metric spaces and harmonic maps. *Comm. Anal. Geom.* **1998**, *6*, 199–253. [CrossRef]
14. Kimura, Y.; Kohsaka, F. Spherical nonspreadingness of resolvents of convex functions in geodesic spaces. *J. Fixed Point Theory Appl.* **2016**, *18*, 93–115. [CrossRef]
15. Kimura Y.; Satô, K. Convergence of subsets of a complete geodesic space with curvature bounded above. *Nonlinear Anal.* **2012**, *75*, 5079–5085. [CrossRef]
16. Aoyama, K.; Kimura, Y. Strong convergence theorems for strongly nonexpansive sequences. *Appl. Math. Comput.* **2011**, *217*, 7537–7545. [CrossRef]
17. Kimura, Y.; Satô, K. Image recovery problem on a geodesic space with curvature bound above by one. In Proceedings of the Third Asian Conference on Nonlinear Analysis and Optimization, Matsue, Japan, 2–6 September 2012; pp. 165–172.

18. He, J.S.; Fang, D.H.; López, G.; Li, C. Mann's algorithm for nonexpansive mappings in CAT(κ) spaces. *Nonlinear Anal.* **2012**, *75*, 445–452. [CrossRef]
19. Saejung, S.; Yotkaew, P. Approximation of zeros of inverse strongly monotone operators in Banach spaces. *Nonlinear Anal.* **2012**, *75*, 742–750. [CrossRef]
20. Aoyama, K.; Kimura, Y.; Kohsaka, F. Strong convergence theorems for strongly relatively nonexpansive sequences and applications. *J. Nonlinear Anal. Optim.* **2012**, *3*, 67–77.
21. Crombez, G. Image recovery by convex combinations of projections. *J. Math. Anal. Appl.* **1991**, *155*, 413–419. [CrossRef]
22. Kitahara, S.; Takahashi, W. Image recovery by convex combinations of sunny nonexpansive retractions. *Topol. Methods Nonlinear Anal.* **1993**, *2*, 333–342. [CrossRef]
23. Kasahara, K.; Kimura, Y. An iterative sequence for a finite number of metric projections on a complete geodesic space, Nolinear analysis and convex analysis. *RIMS Kôkyûroku* **2019**, *2114*, 120–126.

Article

A Strong Convergence Theorem for Split Null Point Problem and Generalized Mixed Equilibrium Problem in Real Hilbert Spaces

Olawale Kazeem Oyewole [1,2] and Oluwatosin Temitope Mewomo [1,*]

1. School of Mathematics, Statistics and Computer Science, University of KwaZulu-Natal, Durban 4001, South Africa; 217079141@stu.ukzn.ac.za
2. DST-NRF Center of Excellence in Mathematical and Statistical Sciences (CoE-MaSS), Johannesburg 2001, South Africa
* mewomoo@ukzn.ac.za

Abstract: In this paper, we study a schematic approximation of solutions of a split null point problem for a finite family of maximal monotone operators in real Hilbert spaces. We propose an iterative algorithm that does not depend on the operator norm which solves the split null point problem and also solves a generalized mixed equilibrium problem. We prove a strong convergence of the proposed algorithm to a common solution of the two problems. We display some numerical examples to illustrate our method. Our result improves some existing results in the literature.

Keywords: split feasibility problem; null point problem; generalized mixed equilibrium problem; monotone mapping; strong convergence; Hilbert space

MSC: 47H06; 47H09; 47H10; 46N10; 47J25

1. Introduction

Let C be a nonempty, closed and convex subset of a real Hilbert space H. The Equilibrium Problem (EP) in the sense of Blum and Oettli [1] is to find a point $x \in C$, such that

$$F(x,y) \geq 0, \ y \in C, \tag{1}$$

where $F : C \times C \to \mathbb{R}$ is a bifunction. The EP unify many important problems, such as variational inequalities, fixed point problems, optimization problems, saddle point (minmax) problems, Nash equilibria problems and complimentarity problems [2–7]. It also finds applications in other fields of studies like physics, economics, engineering and so on [1,2,8–10]. The Generalized Mixed Equilibrium Problem (GMEP) (see e.g., [11]) is to find $x \in C$, such that

$$F(x,y) + \langle g(x), y - x \rangle + \phi(y) - \phi(x) \geq 0, \ \forall \ y \in C, \tag{2}$$

where $g : C \to H$ is a nonlinear mapping and $\phi : C \to \mathbb{R} \cup \{+\infty\}$ is a proper lower semicontinuous convex function. The solution set of (2) will be denoted $GMEP(F, g, \phi)$.

The GMEP includes as special cases, minimization problem, variational inequality problem, fixed point problem, nash equilibrium etc. GMEP (2) and these special cases have been studied in Hilbert, Banach, Hadamard and p-uniformly convex metric spaces , see [11–21].

For a real Hilbert space H, the Variational Inclusion Problem (VIP) consists of finding a point $x^* \in H$ such that

$$0 \in Ax^*, \tag{3}$$

where $A : H \to 2^H$ is a multivalued operator. If A is a maximal monotone operator, then the VIP reduces to the Monotone Inclusion Problem (MIP). The MIP provides a general framework for the study of many important optimization problems, such as convex programming, variationa inequalities and so on.

For solving Problem (3), Martinet [22] introduced the Proximal Point Algorithm (PPA), which is given as follows: $x_0 \in H$ and

$$x_{n+1} = J^A_{r_n} x_n, \qquad (4)$$

where $\{r_n\} \subset (0, +\infty)$ and $J^A_{r_n} = (I + r_n A)^{-1}$ is the resolvent of the maximal monotone operator A corresponding to the control sequence $\{r_n\}$. Several iterative algorithms have been proposed by authors in the literature for solving Problem (3) and related optimization problems, see [23–37].

Censor and Elfving [38] introduced the notion of Split Feasibility Problem (SFP). The SFP consists of finding a point

$$x^* \in C \quad \text{such that} \quad Lx^* \in Q, \qquad (5)$$

where C and Q are nonempty closed convex subsets of $\mathbb{R}^n$ and $\mathbb{R}^m$ respectively and L is an $m \times n$ matrix. The SFP has been studied by researchers due to its applications in various field of science and technology, such as signal processing, intensity-modulated radiation therapy and medical image construction, for details, see [39,40]. In solving (5), Byrne [39] introduced the following iterative algorithm: let $x_0 \in \mathbb{R}^n$ be arbitrary,

$$x_{n+1} = P_C(x_n - \gamma L^*(I - P_Q)Lx_n), \qquad (6)$$

where $\gamma \in (0, 2/||L||^2)$, L^* is the transpose of the matrix L, P_C and P_Q are nearest point mappings onto C and Q respectively. Lopez et al. [41] suggested the use of a stepsize γ_n in place of γ in Algorithm (6), where the stepsize does not depend on operator L. The stepsize γ_n is given as:

$$\gamma_n := \frac{\theta_n ||(I - P_Q)Lx_n||^2}{2||L^*(I - P_Q)Lx_n||^2}, \qquad (7)$$

where $\theta_n \in (0, 4)$ and $L^*(I - P_Q)Lx_n \neq 0$. They proved a weak convergence theorem of the proposed algorithm. The authors in [41] noted that for L with higher dimensions, it may be hard to compute the operator norm and this may have effect on the iteration process. Instances of this effect can be observed in the CPU time. The algorithm with stepsizes improves the performance of the Byrne algorithm.

The Split Null Point Problem (SNPP) was introduced in 2012 by Byrne et al. [42]. These authors combined the concepts of VIP and SFP and defined SNPP as follows: Find $x^* \in H_1$ such that

$$0 \in A_1(x^*) \quad \text{and} \quad Lx^* \in H_2 \text{ such that} \quad 0 \in A_2(Lx^*), \qquad (8)$$

where $A_i : H_i \to 2^{H_i}$, $i = 1, 2$ are maximal monotone operators, H_1 and H_2 are real Hilbert spaces. For solving (8), Byrne et al. [42] proposed the following iterative algorithm: For $r > 0$ and an arbitrary $x_0 \in H_1$,

$$x_{n+1} = J^{A_1}_r(x_n - \gamma L^*(I - J^{A_2}_r)Lx_n), \qquad (9)$$

where $\gamma \in (0, 2/||L||^2)$. They prove a weak convergence of (9) to a solution of (8).

One of our aim in this work is to consider a generalization of Problem (3) in the following form: Find $x^* \in H$ such that

$$0 \in \bigcap_{i=1}^{N} A_i(x^*), \tag{10}$$

where A_i is a finite family of maximal monotone operators. There have been some iterative algorithms for approximating the solution of (10) in the literature, (see [37] and the references therein).

In this study, we consider the problem of finding the common solution of the GMEP (2) and the SNPP for a finite family of intersection of maximal monotone operator in the frame work of real Hilbert spaces. We consider the following generalization of the SNPP: Find $x^* \in C$ such that $x^* \in GMEP(F, g, \phi)$ and

$$x^* \in \bigcap_{i=1}^{N} A_i^{-1}(0) \quad \text{such that} \quad Lx^* \bigcap_{i=1}^{N} B_i^{-1}(0). \tag{11}$$

In our quest to obtain a common element in the solution set of problems (2) and (11), the following two research questions arise.

(1) Can we obtain an iterative algorithm which solves problem (11), without depending on the operator norm?
(2) Can we obtain a strong convergence theorem for the proposed algorithm to the solution of problem (11) ?

In this work, we give an affirmative answer to the questions above by introducing an iterative algorithm which solves (11). Further, we prove a strong convergence theorem of the proposed algorithm to the common solution of problem given by (11).

2. Preliminaries

In this section, we give some important definitions and Lemmas which are useful in establishing our main results.

From now, we denote by H a real Hilbert space, C a nonempty closed convex subset of H with inner product and norm denoted by $\langle \cdot, \cdot \rangle$ and $|| \cdot ||$ respectively. We denote by $x_n \rightharpoonup x$ and $x_n \to x$ respectively the weak and strong convergence of a sequence $\{x_n\} \subset H$ to a point $x \in H$.

The nearest point mapping $P_C : H \to C$ is defined by $P_C x := \{x \in C : ||x - y|| = d_C(x), \forall y \in H\}$, where $d_C : H \to \mathbb{R}$ is the distance function of C. The mapping P_C is known to satisfy the inequality

$$\langle x - P_C x, y - P_C x \rangle \leq 0, \ \forall \, x \in H \ \text{ and } \ y \in C, \tag{12}$$

see e.g., [9,10] for details.

A point $x \in C$ is said to be a fixed point of a mapping $T : H \to H$, if $x = Tx$. We denote by $F(T)$ the set of fixed point of T. A mapping $f : C \to C$ is said to be a contraction, if there exists a constant $c \in (0,1)$, such that

$$||f(x) - f(y)|| \leq c||x - y||, \ \forall \ x, y \in C. \tag{13}$$

If $c = 1$, then f is called nonexpansive.

A mapping $T : H \to H$ is said to be firmly nonexpansive if, for all $x, y \in H$, the following holds

$$||Tx - Ty||^2 \leq ||x - y||^2 - ||(I - T)x - (I - T)y||^2, \tag{14}$$

where I is an identity mapping on H.

Lemma 1 ([43]). *Let $T : H \to H$ be a mapping. Then the following are equivalent:*

(i) T is firmly nonexpansive,
(ii) $I - T$ is firmly nonexpansive,
(iii) $2T - I$ is nonexpansive,
(iv) $\langle x - y, Tx - Ty \rangle \geq ||Tx - Ty||^2$,
(v) $\langle (I - T)x - (I - T)y, Tx - Ty \rangle \geq 0$.

A multivalued mapping $A : H \to 2^H$ is called monotone if for all $x, y \in H$, $u \in Ax$ and $v \in Ay$, we have

$$\langle x - y, u - v \rangle \geq 0. \tag{15}$$

A monotone mapping A is said to be maximal if its graph $G(A) := \{(x, u) \in H \times H : u \in Ax\}$ is not properly contained in the graph of any other monotone operator.

Let $A : H \to H$ be a single-valued mapping, then for a positive real number β, A is said to be β-inverse strongly monotone (β-ism), if

$$\langle x - y, Ax - Ay \rangle \geq \beta ||Ax - Ay||^2, \quad \forall \ x, y \ \in H. \tag{16}$$

This class of monotone mapping have been widely studied in literature (see [44,45]) for more details. If A is a monotone operator, then we can define, for each $r > 0$, a nonexpansive single-valued mapping $J_r^A : R(I + rA) \to D(A)$ by $J_r^A := (I + rA)^{-1}$ which is generally known as the resolvent of A, (see [46,47]). It is also known that $A^{-1}(0) = F(J_r^A)$, where $A^{-1}(0) = \{x \in H : 0 \in Ax\}$ and $F(J_r^A) = \{x \in H : J_r^A x = x\}$.

Lemma 2 ([6,48]). *Let H be a real Hilbert space. Then the following hold:*

(i) $||x + y||^2 \leq ||x||^2 + 2\langle y, x + y \rangle, \forall x, y \in H$,
(ii) $||x + y||^2 = ||x||^2 + 2\langle x, y \rangle + ||y||^2, x, y \in H$,
(iii) $||\lambda x + (1 - \lambda)y||^2 = \lambda ||x||^2 + (1 - \lambda)||y||^2 - \lambda(1 - \lambda)||x - y||^2, \forall x, y \in H$ and $\lambda \in [0, 1]$.

The bifunction $F : C \times C \to \mathbb{R}$ will be assumed to admit the following restrictions:

(C1) $F(x, x) = 0$ for all $x \in C$;
(C2) F is monotone, i.e., $F(x, y) + F(y, x) \leq 0$ for all $x, y \in C$;
(C3) for each $x, y, z \in C$, $\lim_{t \downarrow 0} F(tz + (1 - t)x, y) \leq F(x, y)$;
(C4) for each $x \in C$, $y \mapsto F(x, y)$ is convex and lower semicontinuous.

Lemma 3 ([11]). *Let C be a nonempty closed convex subset of real Hilbert space H. Let F be a real valued bifunction on $C \times C$ admitting restrictions $C1 - C4$, $g : C \to H$ be a nonlinear mapping and let $\phi : C \to \mathbb{R} \cup \{+\infty\}$ be a proper lower senicontinuous convex function. For any given $r > 0$ and $x \in H$, define a mapping $K_r^F : H \to C$ as*

$$K_r^F x = \{z \in C : F(z, y) + \langle g(z), y - z \rangle + \phi(y) - \phi(z) + \frac{1}{r}\langle y - z, z - x \rangle \geq 0, \ \forall \ y \in \ C\}, \tag{17}$$

for all $x \in H$. Then the following conclusions hold:

(i) for each $x \in H$, $K_r^F x \neq \emptyset$,
(ii) K_r^F is single valued,
(iii) K_r^F is firmly nonexpansive, i.e., for any $x, y \in H$

$$||K_r^F x - K_r^F y||^2 \leq \langle K_r^F x - K_r^F y, x - y \rangle,$$

(iv) $F(K_r^F(I - rg)) = GMEP(F, g, \phi)$,
(v) $GMEP(F, g, \phi)$ is closed and convex.

Lemma 4 ([49,50]). *Let $\{a_n\}$ be a sequence of nonnegative real numbers satisfying the following relation:*

$$a_{n+1} \leq (1-b_n)a_n + b_n c_n + d_n, \quad n \in \mathbb{N}, \tag{18}$$

where $\{b_n\}, \{c_n\}$ and $\{d_n\}$ are sequences of real numbers satisfying

(i) $\{b_n\} \subset [0,1], \sum_{n=1}^{\infty} b_n = \infty$;

(ii) $\limsup_{n \to \infty} c_n \leq 0$;

(iii) $d_n \geq 0, \sum_{n=0}^{\infty} d_n < \infty$.

Then, $\lim_{n=\infty} a_n = 0$.

3. Main Result

Throughout, we let $\Phi_{\lambda_{N,n}}^{A_N} = J_{\lambda_{N,n}}^{A_N} \circ J_{\lambda_{N-1,n}}^{A_{N-1}} \circ \cdots \circ J_{\lambda_{1,n}}^{A_1}$, where $\Phi_{\lambda_{0,n}}^{A_0} = I$. Define the stepsize γ_n by

$$\gamma_n = \begin{cases} \dfrac{\theta_n \|(I - \Phi_{\lambda_{i,n}}^{B_i})Lu_n\|^2}{\|L^*(I - \Phi_{\lambda_{i,n}}^{B_i})Lu_n\|^2}, & \text{if } L^*(I - \Phi_{\lambda_{i,n}}^{B_i})Lu_n \neq 0, \\ \gamma, & \text{otherwise,} \end{cases} \tag{19}$$

where γ_n depends on $\theta_n \in [a,b] \subset (0,1)$ and γ is any nonnegative number.

Lemma 5. *Let H be a real Hilbert space and $A : H \to 2^H$ be a monotone mapping. Then for $0 < s \leq r$, we have*

$$\|x - J_s^A x\| \leq 2\|x - J_r^A x\|.$$

Proof: Notice that $\frac{1}{s}(x - J_s^A) \in AJ_s^A x$ and $\frac{1}{r}(x - J_r^A x) \in AJ_r^A x$. Using the monotonicity of A, we have

$$\langle \frac{1}{s}(x - J_s^A x) - \frac{1}{r}(x - J_r^A x), J_s^A x - J_r^A x \rangle \geq 0.$$

That is

$$\langle x - J_s^A x - \frac{s}{r}(x - J_r^A x), J_s^A x - J_r^A x \rangle \geq 0,$$

which implies that

$$\langle x - J_s^A x, J_s^A x - J_r^A x \rangle \geq \frac{s}{r}\langle x - J_r^A x, J_s^A x - J_r^A x \rangle.$$

Using Lemma 2 (ii), we obtain

$$\frac{1}{2}(\|x - J_r^A x\|^2 - \|x - J_s^A x\|^2 - \|J_s^A x - J_r^A x\|^2) \geq \frac{s}{2r}(\|x - J_r^A x\|^2 + \|J_s^A x - J_r^A x\|^2 - \|x - J_s^A x\|^2),$$

that is

$$-\left(\frac{1}{2} + \frac{s}{2r}\right)\|J_s^A x - J_r^A x\|^2 \geq -\left(\frac{1}{2} - \frac{s}{2r}\right)\|x - J_r^A x\|^2 - \left(\frac{s}{2r} - \frac{1}{2}\right)\|x - J_s^A x\|^2$$

and

$$\left(\frac{r+s}{2r}\right)\|J_s^A x - J_r^A x\|^2 \leq \left(\frac{r-s}{2r}\right)\|x - J_r^A x\|^2 - \left(\frac{r-s}{2r}\right)\|x - J_s^A x\|^2.$$

Since $0 < s \leq r$, we obtain

$$||J_s^A x - J_r^A x||^2 \leq \left(\frac{r-s}{r+s}\right)||x - J_r^A x||^2,$$

which implies

$$||J_s^A x - J_r^A x|| \leq ||x - J_r^A x||. \quad (20)$$

Now, since $||x - J_s^A x|| \leq ||x - J_r^A x|| + ||J_r^A x - J_s^A x||$, by (20), we obtain

$$\begin{aligned} ||x - J_s^A x|| &\leq ||x - J_r^A x|| + ||x - J_r^A x|| \\ &= 2||x - J_r^A x||. \end{aligned}$$

Lemma 6. *Let C and Q be nonempty, closed and convex subsets of real Hilbert spaces H_1 and H_2 respectively and $L : H_1 \to H_2$ be a bounded linear operator. Assume F is a real valued bifunction on $C \times C$ which admits condition C1-C4. Let $\phi : H_1 \to \mathbb{R} \cup \{+\infty\}$ be a proper, lower semicontinuous convex function, g be a β-inverse strongly monotone mapping and $f : H_1 \to R$ be a differentiable function, such that ∇f is a contraction with coefficient $c \in (0,1)$. For $i = 1, 2 \cdots, N$, let $A_i : H_1 \to 2^{H_1}$ and $B_i : H_2 \to 2^{H_2}$ be finite families of monotone mappings. Assume $\Omega = GMEP(F, g, \phi) \cap \Gamma \neq \emptyset$, where $\Gamma = \{x^* \in H_1 : 0 \in \cap_{i=1}^N A_i(x^*) \text{ and } Lx^* \in H_2 : 0 \in \cap_{i=1}^N B_i(Lx^*)\}$. For an arbitrary $x_0 \in H_1$, let $\{x_n\} \subset H_1$ be a sequence defined iteratively by*

$$\begin{cases} F(u_n, y) + \langle g(u_n), y - u_n \rangle + \phi(y) - \phi(u_n) + \frac{1}{r_n}\langle y - u_n, u_n - x_n \rangle \geq 0, \ y \in H_1, \\ z_n = u_n - \gamma_n L^*(I - \Phi_{\lambda_{i,n}}^{B_i}) L u_n, \\ x_{n+1} = \alpha_n \nabla f(z_n) + (1 - \alpha_n) \Phi_{\lambda_{i,n}}^{A_i} z_n, \end{cases} \quad (21)$$

where $\{r_n\}$ is a nonnegative sequence of real numbers, $\{\alpha_n\}$ and $\{\lambda_{i,n}\}$ are sequences in $(0,1)$, γ_n is a nonnegative sequence defined by (19), satisfying the following restrictions:

(i) $\sum\limits_{n=1}^\infty \alpha_n = \infty$, $\lim\limits_{n \to \infty} \alpha_n = 0$;
(ii) $0 < \lambda_i \leq \lambda_{i,n}$;
(iii) $0 < a \leq r_n \leq b < 2\beta$.

Then $\{x_n\}$, $\{z_n\}$ and $\{u_n\}$ are bounded.

Proof. Observe that u_n can be rewritten as $u_n = K_{r_n}^F(x_n - r_n g(x_n))$ for each n. Fix $p \in \Omega$. Since $p = K_{r_n}^F(p - r_n p)$, g is β-inverse strongly monotone and $r_n \in (0, 2\beta)$, for any $n \in \mathbb{R}$, we have from (21) and Lemma 2 (ii) that

$$\begin{aligned} ||u_n - p||^2 &= ||K_{r_n}^F(x_n - r_n g(x_n)) - K_{r_n}^F(p - r_n g(p))||^2 \\ &\leq ||x_n - r_n g(x_n) - (p - r_n g(p))||^2 \\ &= ||(x_n - p) - r_n(g(x_n) - g(p))||^2 \\ &= ||x_n - p||^2 - 2r_n \langle x_n - p, g(x_n) - g(p) \rangle + r_n^2 ||g(x_n) - g(p)||^2 \quad (22) \\ &\leq ||x_n - p||^2 - 2\beta r_n ||g(x_n) - g(p)||^2 + r_n^2 ||g(x_n) - g(p)||^2 \\ &= ||x_n - p||^2 - r_n(2\beta - r_n)||g(x_n) - g(p)||^2 \\ &\leq ||x_n - p||^2. \end{aligned}$$

Also by Lemma 2, we have

$$
\begin{aligned}
||z_n - p||^2 &= ||u_n - \gamma_n L^*(I - \Phi^{B_i}_{\lambda_{i,n}})Lu_n - p||^2 \\
&= ||u_n - p - \gamma_n L^*(I - \Phi^{B_i}_{\lambda_{i,n}})Lu_n||^2 \\
&= ||u_n - p||^2 - 2\gamma_n \langle u_n - p, L^*(I - \Phi^{B_i}_{\lambda_{i,n}})Lu_n \rangle + \gamma_n^2 ||L^*(I - \Phi^{B_i}_{\lambda_{i,n}})Lu_n||^2 \\
&= ||u_n - p||^2 - 2\gamma_n \langle Lu_n - Lp, (I - \Phi^{B_i}_{\lambda_{i,n}})Lu_n \rangle + \gamma_n^2 ||L^*(I - \Phi^{B_i}_{\lambda_{i,n}})Lu_n||^2 \quad (23) \\
&\leq ||u_n - p||^2 - 2\gamma_n ||(I - \Phi^{B_i}_{\lambda_{i,n}})Lu_n||^2 + \gamma_n^2 ||L^*(I - \Phi^{B_i}_{\lambda_{i,n}})Lu_n||^2 \\
&\leq ||u_n - p||^2 - \gamma_n ||(I - \Phi^{B_i}_{\lambda_{i,n}})Lu_n||^2 + \gamma_n^2 ||L^*(I - \Phi^{B_i}_{\lambda_{i,n}})Lu_n||^2 \\
&= ||u_n - p||^2 - \gamma_n [||(I - \Phi^{B_i}_{\lambda_{i,n}})Lu_n||^2 - \gamma_n ||L^*(I - \Phi^{B_i}_{\lambda_{i,n}})Lu_n||^2].
\end{aligned}
$$

Using the definition of γ_n, we obtain

$$||z_n - p||^2 \leq ||u_n - p||^2, \qquad (24)$$

hence, $||z_n - p|| \leq ||u_n - p|| \leq ||x_n - p||$.

Further, we obtain that

$$
\begin{aligned}
||x_{n+1} - p|| &= ||\alpha_n \nabla f(z_n) + (1 - \alpha_n)\Phi^{A_i}_{\lambda_{i,n}} z_n - p|| \\
&= ||\alpha_n (\nabla f(z_n) - p) + (1 - \alpha_n)(\Phi^{A_i}_{\lambda_{i,n}} z_n - p)|| \\
&\leq \alpha_n ||\nabla f(z_n) - p|| + (1 - \alpha_n)||\Phi^{A_i}_{\lambda_{i,n}} z_n - p|| \\
&\leq \alpha_n ||\nabla f(z_n) - \nabla f(p)|| + \alpha_n ||\nabla f(p) - p|| + (1 - \alpha_n)||z_n - p|| \quad (25) \\
&\leq \alpha_n c ||z_n - p|| + \alpha_n ||\nabla f(p) - p|| + ||(1 - \alpha_n)||z_n - p|| \\
&= (1 - \alpha_n(1 - c))||z_n - p|| + \alpha_n ||\nabla f(p) - p|| \\
&\leq (1 - \alpha_n(1 - c))||x_n - p|| + \frac{\alpha_n(1-c)}{1-c}||\nabla f(p) - p||.
\end{aligned}
$$

Let $K = \max\{||x_0 - p||, \frac{||\nabla f(p) - p||}{1-c}\}$. We show that $||x_n - p|| \leq K$ for all $n \geq 0$. Indeed, we see that $||x_0 - p|| \leq K$. Now suppose $||x_j - p|| \leq K$ for some $j \in \mathbb{N}$. Then, we have that

$$
\begin{aligned}
||x_{j+1} - p|| &\leq (1 - \alpha_j(1-c))||x_j - p|| + \frac{\alpha_j(1-c)||\nabla f(p) - p||}{1-c} \\
&\leq (1 - \alpha_j(1-c))K + \alpha_j(1-c)K \qquad (26) \\
&\leq K.
\end{aligned}
$$

By induction, we obtain that $||x_n - p|| \leq K$ for all n. Therefore $\{x_n\}$ is bounded, consequently $\{z_n\}$ and $\{u_n\}$ are bounded. □

Theorem 1. *Let C and Q be nonempty, closed and convex subsets of real Hilbert spaces H_1 and H_2, respectively and $L : H_1 \to H_2$ be a bounded linear operator. Assume F is a real valued bifunction on $C \times C$ which admits condition C1-C4. Let $\phi : H_1 \to \mathbb{R} \cup \{+\infty\}$ be a proper, lower semicontinuous function, g be a β-inverse strongly monotone mapping and $f : H_1 \to \mathbb{R}$ be a differentiable function, such that ∇f is a contraction with coefficient $c \in (0,1)$. For $i = 1, 2 \cdots, N$, let $A_i : H_1 \to 2^{H_1}$ and $B_i : H_2 \to 2^{H_2}$ be finite families of monotone mappings. Assume $\Omega = GMEP(F, g, \phi) \cap \Gamma \neq \emptyset$, where $\Gamma = \{p \in H_1 : 0 \in \bigcap_{i=1}^N A_i(p) \text{ and } Lp \in H_2 : 0 \in \bigcap_{i=1}^N B_i(Lp)\}$. For an arbitrary $x_0 \in H_1$, let $\{x_n\} \subset H_1$ be a sequence defined iteratively by (21) satisfying the conditions of Lemma 6. Then $\{x_n\}$ converges strongly to $p \in \Omega$, where $p = P_\Omega \nabla f(p)$.*

Proof. We observe from (21), that

$$\begin{aligned}
||x_{n+1}-p||^2 &= \langle \alpha_n \nabla f(z_n) + (1-\alpha_n)(\Phi^{A_i}_{\lambda_{i,n}} z_n - p), x_{n+1}-p \rangle \\
&= \alpha_n \langle \nabla f(z_n), x_{n+1}-p \rangle + (1-\alpha_n)\langle \Phi^{A_i}_{\lambda_{i,n}} z_n - p, x_{n+1}-p \rangle \\
&= \alpha_n \langle \nabla f(z_n) - \nabla f(p), x_{n+1}-p \rangle + \alpha_n \langle \nabla f(p) - p, x_{n+1}-p \rangle + (1-\alpha_n)\langle \Phi^{A_i}_{\lambda_{i,n}} z_n - p, x_{n+1}-p \rangle \\
&\leq \alpha_n ||\nabla f(z_n) - \nabla f(p)|| \cdot ||x_{n+1}-p|| + (1-\alpha_n)||\Phi^{A_i}_{\lambda_{i,n}} z_n - p|| \cdot ||x_{n+1}-p|| \\
&\quad + \alpha_n \langle \nabla f(p) - p, x_{n+1}-p \rangle \\
&\leq \frac{\alpha_n}{2}\left(||\nabla f(z_n) - \nabla f(p)||^2 + ||x_{n+1}-p||^2 \right) + \left(\frac{1-\alpha_n}{2}\right)\left(||\Phi^{A_i}_{\lambda_{i,n}} z_n - p||^2 + ||x_{n+1}-p||^2 \right) \\
&\quad + \alpha_n \langle \nabla f(p) - p, x_{n+1}-p \rangle \\
&\leq \frac{\alpha_n c^2}{2}||z_n-p||^2 + \frac{\alpha_n}{2}||x_{n+1}-p||^2 + \frac{(1-\alpha_n)}{2}||z_n-p||^2 + \frac{(1-\alpha_n)}{2}||x_{n+1}-p||^2 \\
&\quad + \alpha_n \langle \nabla f(p) - p, x_{n+1}-p \rangle \\
&\leq \frac{[1-\alpha_n(1-c^2)]}{2}||z_n-p||^2 + \frac{1}{2}||x_{n+1}-p||^2 + \alpha_n \langle \nabla f(p) - p, x_{n+1}-p \rangle \\
&\leq \frac{[1-\alpha_n(1-c^2)]}{2}||u_n-p||^2 + \frac{1}{2}||x_{n+1}-p||^2 + \alpha_n \langle \nabla f(p) - p, x_{n+1}-p \rangle,
\end{aligned} \quad (27)$$

that is

$$\begin{aligned}
||x_{n+1}-p||^2 &\leq [1-\alpha_n(1-c^2)]||u_n-p||^2 + \alpha_n(1-c^2)\left(\frac{2}{(1-c^2)}\langle \nabla f(p)-p, x_{n+1}-p\rangle\right) \\
&\leq [1-\alpha_n(1-c^2)]||x_n-p||^2 + \alpha_n(1-c^2)\left(\frac{2}{(1-c^2)}\langle \nabla f(p)-p, x_{n+1}-p\rangle\right). \quad (28)
\end{aligned}$$

From now the rest of the proof shall be divide into two cases.

Case 1: Suppose that there exists $n_0 \in \mathbb{N}$ such that $\{||x_n - p||\}$ is not monotonically increasing. Then by Lemma 6, we have that $\{||x_n - p||\}$ is convergent. From (21), we have by Lemma 2 that

$$\begin{aligned}
||x_{n+1}-p||^2 &= ||\alpha_n \nabla f(z_n) + (1-\alpha_n)\Phi^{A_i}_{\lambda_{i,n}} z_n - p||^2 \\
&= ||\alpha_n \nabla(f(z_n)-p) + (1-\alpha_n)(\Phi^{A_i}_{\lambda_{i,n}} z_n - p)||^2 \\
&= \alpha_n ||\nabla f(z_n)-p||^2 + (1-\alpha_n)||\Phi^{A_i}_{\lambda_{i,n}} z_n - p||^2 - \alpha_n(1-\alpha_n)||\nabla f(z_n) - \Phi^{A_i}_{\lambda_{i,n}} z_n||^2 \\
&\leq \alpha_n ||\nabla f(z_n)-p||^2 + (1-\alpha_n)||z_n - p||^2.
\end{aligned} \quad (29)$$

Thus,

$$||z_n - p||^2 \geq ||x_{n+1}-p||^2 - \alpha_n(||\nabla f(z_n) - p||^2 - ||z_n-p||^2). \quad (30)$$

From (23), we have that

$$\begin{aligned}
\gamma_n[||(I-\Phi^{B_i}_{\lambda_{i,n}})Lu_n||^2 - \gamma_n ||L^*(I-\Phi^{B_i}_{\lambda_{i,n}})Lu_n||^2] &\\
&\leq ||u_n-p||^2 - ||z_n-p||^2 \\
&\leq ||u_n-p||^2 - ||x_{n+1}-p||^2 + \alpha_n(||\nabla f(z_n)-p||^2 - ||z_n-p||^2) \\
&\leq ||x_n-p||^2 - ||x_{n+1}-p||^2 + \alpha_n(||\nabla f(z_n)-p||^2 - ||z_n-p||^2),
\end{aligned}$$

by using restriction (i) in Lemma 6, we have

$$\lim_{n\to\infty} \gamma_n[||(I-\Phi^{B_i}_{\lambda_{i,n}})Lu_n||^2 - \gamma_n ||L^*(I-\Phi^{B_i}_{\lambda_{i,n}})Lu_n||^2] = 0. \quad (31)$$

Using (19), we have that

$$[||(I - \Phi^{B_i}_{\lambda_{i,n}})Lu_n||^2 - \gamma_n ||L^*(I - \Phi^{B_i}_{\lambda_{i,n}})Lu_n||^2] = \theta_n(1-\theta_n) \frac{||(I - \Phi^{B_i}_{\lambda_{i,n}})Lu_n||^4}{||L^*(I - \Phi^{B_i}_{\lambda_{i,n}})Lu_n||^2}, \quad (32)$$

thus by (31), we obtain

$$\theta_n(1-\theta_n) \frac{||(I - \Phi^{B_i}_{\lambda_{i,n}})Lu_n||^4}{||L^*(I - \Phi^{B_i}_{\lambda_{i,n}})Lu_n||^2} \to 0, \text{ as } n \to \infty.$$

Therefore, since $\theta_n \in (0,1)$, we obtain

$$\lim_{n \to \infty} \frac{||(I - \Phi^{B_i}_{\lambda_{i,n}})Lu_n||^2}{||L^*(I - \Phi^{B_i}_{\lambda_{i,n}})Lu_n||} = 0. \quad (33)$$

Notice that $||L^*(I - \Phi^{B_i}_{\lambda_{i,n}})Lu_n|| \leq ||L^*|| \cdot ||(I - \Phi^{B_i}_{\lambda_{i,n}})Lu_n||$, which implies

$$||(I - \Phi^{B_i}_{\lambda_{i,n}})Lu_n|| \leq \frac{||L^*|| \cdot ||(I - \Phi_{\lambda_{i,n}})Lu_n||^2}{||L^*(I - \Phi^{B_i}_{\lambda_{i,n}})Lu_n||},$$

by (33), we obtain

$$\lim_{n \to \infty} ||(I - \Phi^{B_i}_{\lambda_{i,n}})Lu_n|| = 0, \quad (34)$$

consequently,

$$\lim_{n \to \infty} ||L^*(I - \Phi^{B_i}_{\lambda_{i,n}})Lu_n|| = 0. \quad (35)$$

From (21), we see that

$$\begin{aligned} ||z_n - u_n|| &= ||u_n - \gamma_n L^*(I - \Phi^{B_i}_{\lambda_{i,n}})Lu_n - u_n|| \\ &\leq \gamma_n ||L^*(I - \Phi^{B_i}_{\lambda_{i,n}})Lu_n||. \end{aligned}$$

By (35), we get that

$$\lim_{n \to \infty} ||z_n - u_n|| = 0. \quad (36)$$

Furthermore, we have from (21),

$$\begin{aligned} ||x_{n+1} - z_n|| &= ||\alpha_n \nabla f(z_n) + (1-\alpha_n)\Phi^{A_i}_{\lambda_{i,n}} z_n - z_n|| \\ &= ||\alpha_n(\nabla f(z_n) - z_n) + (1-\alpha_n)(\Phi^{A_i}_{\lambda_{i,n}} z_n - z_n)|| \quad (37) \\ &\leq \alpha_n ||\nabla f(z_n) - z_n|| + (1-\alpha_n)||\Phi^{A_i}_{\lambda_{i,n}} z_n - z_n|| \\ &\leq \alpha_n ||\nabla f(z_n) - \nabla f(p)|| + \alpha_n ||\nabla f(p) - z_n|| + (1-\alpha_n)||z_n - \Phi^{A_i}_{\lambda_{i,n}} z_n|| \\ &\leq \alpha_n c ||z_n - p|| + \alpha_n ||\nabla f(p) - z_n|| + (1-\alpha_n)||z_n - \Phi^{A_i}_{\lambda_{i,n}} z_n||. \end{aligned}$$

Observe from (21), that

$$||\Phi^{A_i}_{\lambda_{i,n}} z_n - p|| \geq ||x_{n+1} - p|| - \alpha_n ||\nabla f(z_n) - \Phi^{A_i}_{\lambda_{i,n}} z_n||,$$

using the nonexpansivity of $\Phi^{A_i}_{\lambda_{i,n}}$, we obtain that

$$\begin{aligned} 0 &\leq ||z_n - p|| - ||\Phi^{A_i}_{\lambda_{i,n}} z_n - \Phi^{A_i}_{\lambda_{i,n}} p|| \\ &= ||z_n - p|| - ||\Phi^{A_i}_{\lambda_{i,n}} z_n - p|| \\ &\leq ||x_n - p|| - ||x_{n+1} - p|| + \alpha_n ||\nabla f(z_n) - \Phi^{A_i}_{\lambda_{i,n}} z_n||. \end{aligned}$$

Using restriction (i) in Lemma 6, the boundedness of $\{z_n\}$ and the convergence of $\{||x_n - p||\}$, we have that $||z_n - p|| - ||\Phi^{A_i}_{\lambda_{i,n}} z_n - p|| \to 0$ as $n \to \infty$. Thus by the strong nonexpansivity of $\Phi^{A_i}_{\lambda_{i,n}}$, we get that

$$\lim_{n \to \infty} ||z_n - \Phi^{A_i}_{\lambda_{i,n}} z_n|| = 0.$$

Using this and restriction (i) of Lemma 6 in (38), we get

$$\lim_{n \to \infty} ||x_{n+1} - z_n|| = 0. \tag{38}$$

Observe from (28), that

$$-||u_n - p||^2 \leq -||x_{n+1} - p||^2 - \alpha_n(1 - c^2)||u_n - p||^2 + 2\alpha_n \langle \nabla f(p) - p, x_{n+1} - p \rangle, \tag{39}$$

since $||u_n - x_n||^2 \leq ||x_n - p||^2 - ||u_n - p||^2$, using (39), we have that

$$||u_n - x_n||^2 \leq ||x_n - p||^2 - ||x_{n+1} - p||^2 - \alpha_n(1 - c^2)||u_n - p||^2 + 2\alpha_n \langle \nabla f(p) - p, x_{n+1} - p \rangle,$$

thus, by restriction (i) in Lemma 6, we obtain

$$\lim_{n \to \infty} ||u_n - x_n|| = ||K^F_{r_n} x_n - x_n|| = 0. \tag{40}$$

Combining (36) and (40), we obtain

$$\lim_{n \to \infty} ||z_n - x_n|| = 0. \tag{41}$$

Moreover, since

$$||x_{n+1} - x_n|| \leq ||x_{n+1} - z_n|| + ||z_n - x_n||,$$

we have that

$$||x_{n+1} - x_n|| \to 0 \text{ as } n \to \infty. \tag{42}$$

Furthermore,

$$\begin{aligned} ||x_n - \Phi^{A_i}_{\lambda_{i,n}} x_n|| &\leq ||x_n - x_{n+1}|| + ||x_{n+1} - \Phi^{A_i}_{\lambda_{i,n}} z_n|| + ||\Phi^{A_i}_{\lambda_{i,n}} - \Phi^{A_i}_{\lambda_{i,n}} x_n|| \\ &\leq ||x_n - x_{n+1}|| + ||x_{n+1} - \Phi^{A_i}_{\lambda_{i,n}} z_n|| + ||z_n - x_n||, \end{aligned} \tag{43}$$

but

$$\begin{aligned} ||x_{n+1} - \Phi^{A_i}_{\lambda_{i,n}} z_n|| &= ||\alpha_n \nabla f(z_n) + (1 - \alpha_n) \Phi^{A_i}_{\lambda_{i,n}} z_n - \Phi^{A_i}_{\lambda_{i,n}} z_n|| \\ &= \alpha_n ||\nabla f(z_n) - \Phi^{A_i}_{\lambda_{i,n}} z_n|| \to 0, \text{ as } n \to \infty. \end{aligned}$$

Hence, by substituting this, (41) and (42) into (43), we obtain

$$\lim_{n \to \infty} ||(I - \Phi^{A_i}_{\lambda_{i,n}}) x_n|| = 0.$$

Since $0 < \lambda_i \leq \lambda_{i,n}$, we have by Lemma 5, that

$$\lim_{n \to \infty} ||(I - \Phi^{A_i}_{\lambda_i})x_n|| = 0. \tag{44}$$

Now, since $\{x_n\}$ is bounded in H_1, there exists a subsequence $\{x_{n_j}\}$ of $\{x_n\}$ such that $x_{n_j} \rightharpoonup x^* \in H_1$. First, we show that $x^* \in \cap_{i=1}^N A_i^{-1}(0)$. Consider for each $j \in \mathbb{N}$,

$$||(I - \Phi^{A_i}_{\lambda_i})x^*||^2 \leq \langle (I - \Phi^{A_i}_{\lambda_i})x^*, x^* - x_{n_j}\rangle + \langle (I - \Phi^{A_i}_{\lambda_i})x^*, x_{n_j} - \Phi^{A_i}_{\lambda_i}x_{n_j}\rangle$$
$$+ \langle (I - \Phi^{A_i}_{\lambda_i})x^*, \Phi^{A_i}_{\lambda_i}x_{n_j} - \Phi^{A_i}_{\lambda_i}x^*\rangle. \tag{45}$$

Since $\{x_{n_j}\} \subset \{x_n\}$, as a consequence of (44), we have

$$\lim_{j \to \infty} ||x_{n_j} - \Phi^{A_i}_{\lambda_i}x_{n_j}|| = 0. \tag{46}$$

Therefore, using $x_{n_j} \rightharpoonup x^*$ and (46) in (45), we have

$$\lim_{n \to \infty} ||(I - \Phi^{A_i}_{\lambda_i})x^*|| = 0. \tag{47}$$

Thus, $x^* = \Phi^{A_i}_{\lambda_i}x^*$ and hence $x^* \in \cap_{i=1}^N A_i^{-1}(0)$.

Secondly, we show that $Lx^* \in \cap_{i=1}^N B_i^{-1}(0)$. Consider again for each $j \in \mathbb{N}$,

$$||(I - \Phi^{B_i}_{\lambda_i})Lx^*||^2 \leq \langle (I - \Phi^{B_i}_{\lambda_i})Lx^*, Lx^* - Lz_{n_j}\rangle + \langle (I - \Phi^{B_i}_{\lambda_i})Lx^*, Lz_{n_j} - \Phi^{B_i}_{\lambda_i}Lz_{n_j}\rangle$$
$$+ \langle (I - \Phi^{B_i}_{\lambda_i})Lx^*, \Phi^{B_i}_{\lambda_i}Lz_{n_j} - \Phi^{B_i}_{\lambda_i}Lx^*\rangle, \tag{48}$$

observe that,

$$||(I - \Phi^{B_i}_{\lambda_{i,n}})Lu_n|| \leq ||(I - \Phi^{B_i}_{\lambda_{i,n}})Lz_n - (I - \Phi^{B_i}_{\lambda_{i,n}})Lu_n|| + ||(I - \Phi^{B_i}_{\lambda_{i,n}})Lu_n||$$
$$\leq ||Lz_n - Lu_n|| + ||\Phi^{B_i}_{\lambda_{i,n}}Lz_n - \Phi^{B_i}_{\lambda_{i,n}}Lu_n|| + ||(I - \Phi^{B_i}_{\lambda_{i,n}})Lu_n||$$
$$\leq 2||L||||z_n - u_n|| + ||(I - \Phi^{B_i}_{\lambda_{i,n}})Lu_n||,$$

which by (34) and (36), implies

$$\lim_{n \to \infty} ||(I - \Phi^{B_i}_{\lambda_{i,n}})Lz_n|| = 0.$$

Again, since $0 < \lambda_i \leq \lambda_{i,n}$, we have by Lemma 5, that

$$\lim_{n \to \infty} ||(I - \Phi^{B_i}_{\lambda_i})Lz_n|| = 0. \tag{49}$$

So for any subsequence $\{z_{n_j}\} \subset \{z_n\}$, we also have that

$$\lim_{n \to \infty} ||(I - \Phi^{B_i}_{\lambda_i})Lz_{n_j}|| = 0. \tag{50}$$

Thus, by the linearity and continuity of L, $Lx_{n_j} \rightharpoonup Lx^*$ as $j \to \infty$ and $||z_n - x_n|| \to 0$ as $n \to \infty$ implies $Lz_{n_j} \rightharpoonup Lx^*$ as $j \to \infty$. Hence from (49), we have

$$\lim_{n \to \infty} ||(I - \Phi^{B_i}_{\lambda_i})Lx^*|| = 0. \tag{51}$$

Therefore, $Lx^* = \Phi_{\lambda_i}^{B_i}Lx^*$, that is $Lx^* \in \bigcap_{i=1}^{N} B_i^{-1}(0)$. Further, we show that $x^* \in GMEP(F, g, \phi)$. From (40), we have $u_{n_j} \rightharpoonup x^*$. Since $u_n = K_{r_n}^F(x_n - r_n g(x_n))$, for any $y \in C$, we have

$$F(u_n, y) + \langle g(u_n), y - u_n \rangle + \phi(y) - \phi(u_n) + \frac{1}{r_n}\langle y - u_n, u_n - x_n \rangle \geq 0. \tag{52}$$

It follows from condition (C2) of the bifunction F, that

$$\langle g(u_n), y - u_n \rangle \phi(y) - \phi(u_n) + \frac{1}{r_n}\langle y - u_n, u_n - x_n \rangle \geq F(y, u_n).$$

Replacing n by n_j, we have

$$\langle g(u_{n_j}), y - u_{n_j} \rangle + \frac{1}{r_{n_j}}\langle y - u_{n_j}, u_{n_j} - x_{n_j} \rangle \geq F(y, u_{n_j}) + \phi(u_{n_j}) - \phi(y). \tag{53}$$

Let $y_t = ty + (1-t)x^*$ for all $t \in (0,1]$ and $y \in C$. Then we have $y_t \in C$. So from (53), we have

$$\begin{aligned}\langle g(y_t), y_t - u_{n_j} \rangle &\geq \langle y_t - u_{n_j}, g(y_t) \rangle - \langle y_t - u_{n_j}, g(x_{n_j}) \rangle - \left\langle y_t - u_{n_j}, \frac{u_{n_j} - x_{n_j}}{r_{n_j}} \right\rangle + F(y_t, u_{n_j}) + \phi(u_{n_j}) - \phi(y_t) \\ &= \langle y_t - u_{n_j}, g(y_t) - g(u_{n_j}) \rangle + \langle y_t - u_{n_j}, g(u_{n_j}) - g(x_{n_j}) \rangle - \left\langle y_t - u_{n_j}, \frac{u_{n_j} - x_{n_j}}{r_{n_j}} \right\rangle \\ &\quad + F(y_t, u_{n_j}) + \phi(u_{n_j}) - \phi(y_t).\end{aligned} \tag{54}$$

Since $\lim_{n \to \infty} \|u_n - x_n\| = 0$, we obtain $\|g(u_{n_j}) - g(x_{n_j})\| \to 0$ as $n \to \infty$. Moreover, since g is monotone, we have $\langle y_t - u_{n_j}, g(y_t) - g(u_{n_j}) \rangle \geq 0$. Therefore by (C4) of the bifunction F and the weak lower semicontinuity of ϕ, taking the limit of (54), we obtain

$$\langle y_t - x^*, g(y_t) \rangle \geq F(y_t, x^*) + \phi(x^*) - \phi(y_t). \tag{55}$$

Using (C1) of bifunction F and (55), we obtain

$$\begin{aligned}0 &= F(y_t, y_t) + \phi(y_t) - \phi(y_t) \\ &\leq tF(y_t, y) + (1-t)F(y_t, x^*) + t\phi(y) + (1-t)\phi(x^*) - \phi(y_t) \\ &= t[F(y_t, y) + \phi(y) - \phi(y_t)] + (1-t)[F(y_t, x^*) + \phi(x^*) - \phi(y_t)] \\ &\leq t[F(y_t, y) + \phi(y) - \phi(y_t)] + (1-t)\langle y_t - x^*, g(y_t) \rangle \\ &\leq t[F(y_t, y) + \phi(y) - \phi(y_t)] + (1-t)t\langle y - x^*, g(y_t) \rangle,\end{aligned}$$

this implies that

$$F(y_t, y) + (1-t)\langle y - x^*, g(y_t) \rangle + \phi(y) - \phi(y_t) \geq 0. \tag{56}$$

By letting $t \to 0$, we have

$$F(x^*, y) + \langle g(x^*), y - x^* \rangle + \phi(y) - \phi(x^*) \geq 0, \ y \in C, \tag{57}$$

which implies $x^* \in GMEP(F, g, \phi)$.

Finally we show that $x_n \to p = P_\Omega \nabla f(p)$. Let $\{x_{n_j}\}$ be subsequence of $\{x_n\}$, such that $x_{n_j} \rightharpoonup x^*$ and

$$\limsup_{n \to \infty} \frac{2}{1-c^2}\langle \nabla f(p) - p, x_{n+1} - p \rangle = \lim_{j \to \infty} \frac{2}{1-c^2}\langle \nabla f(p) - p, x_{n_j+1} - p \rangle, \tag{58}$$

since $||x_{n+1} - x_n|| \to 0$ as $n \to \infty$ and $x_{n_j} \rightharpoonup x^*$, it follows that $x_{n_j+1} \rightharpoonup x^*$. Consequently, we obtain by (12), that

$$\limsup_{n \to \infty} \frac{2}{1-c^2} \langle \nabla f(p) - p, x_{n+1} - p \rangle = \frac{2}{1-c^2} \langle \nabla f(p) - p, x^* - p \rangle \leq 0. \tag{59}$$

By using Lemma 4 in (28), we conlude that $||x_n - p|| \to 0$ as $n \to \infty$. Thus, $x_n \to p$ as $n \to \infty$, ditto for both $\{u_n\}$ and $\{z_n\}$.

Case 2: Let $\Gamma_n = ||x_n - p||$ be monotonically nondecreasing. Define $\tau : \mathbb{N} \to \mathbb{N}$ for all $n \geq n_0$ (for some n_0 large enough) by

$$\tau(n) := \max\{k \in \mathbb{N} : k \leq n, \Gamma_k \leq \Gamma_{k+1}\}.$$

Clearly, τ is nondecreasing, $\tau(n) \to \infty$ as $n \to \infty$ and

$$0 \leq \Gamma_{\tau(n)} \leq \Gamma_{\tau(n)+1}, \quad \forall \ n \geq n_0.$$

By using similar argument as in Case 1, we make the following conclusions

$$\lim_{n \to \infty} ||(I - \Phi^{B_i}_{\lambda_i})Lx_{\tau(n)}|| = 0,$$

$$\lim_{n \to \infty} ||L^*(I - \Phi^{B_i}_{\lambda_i})Lx_{\tau(n)}|| = 0,$$

$$\lim_{n \to \infty} ||u_{\tau(n)} - x_{\tau(n)}|| = 0,$$

$$\lim_{n \to \infty} ||x_{\tau(n)+1} - x_{\tau(n)}|| = 0$$

and

$$\limsup_{n \to \infty} \frac{2}{1-c^2} \langle \nabla f(p) - p, x_{\tau(n)+1} - p \rangle \leq 0. \tag{60}$$

Using the boundedness of $\{x_{\tau(n)}\}$, we can obtain a subsequence of $\{x_{\tau(n)}\}$ which converges weakly to $x^* \in \bigcap_{i=1}^{N} A_i^{-1}(0)$, $Lx^* \in \bigcap_{i=1}^{N} B_i^{-1}(0)$ and $x^* \in GMEP(F, \phi, g)$. Therefore, it follows from (28), that

$$||x_{\tau(n)+1} - p||^2 \leq [1 - \alpha_{\tau(n)}(1-c^2)]||x_{\tau(n)} - p||^2$$
$$+ \alpha_{\tau(n)}(1-c^2)\left(\frac{2}{1-c^2}\nabla f(p) - p, x_{\tau(n)+1} - p\right). \tag{61}$$

Since $\Gamma_{\tau(n)} \leq \Gamma_{\tau(n)+1}$, we obtain $||x_{\tau(n)} - x_{\tau(n)+1}|| \leq 0$. Thus, from (61), we obtain

$$\alpha_{\tau(n)}(1-c^2)||x_{\tau(n)} - p||^2 \leq \alpha_{\tau(n)}(1-c^2)\left(\frac{2}{1-c^2}\nabla f(p) - p, x_{\tau(n)+1} - p\right). \tag{62}$$

We note that $\alpha_{\tau(n)}(1-c^2) > 0$, then from (62), we get

$$\lim_{n \to \infty} ||x_{\tau(n)} - p||^2 \leq 0.$$

This implies

$$\lim_{n \to \infty} ||x_{\tau(n)} - p||^2 = 0,$$

hence
$$\lim ||x_{\tau(n)} - p|| = 0.$$

Using this and $\lim_{n\to\infty} ||x_{\tau(n)+1} - x_{\tau(n)}|| = 0$, we obtain

$$||x_{\tau(n)+1} - p|| \leq ||x_{\tau(n)+1} - x_{\tau(n)}|| + ||x_{\tau(n)} - p|| \to 0, \text{ as } n \to \infty.$$

Further, for $n \geq n_0$, we clearly observe that $\Gamma_{\tau(n)} \leq \Gamma_{\tau(n)+1}$ if $n \neq \tau(n)$, (i.e., $\tau(n) < n$). Since $\Gamma_j \geq \Gamma_{j+1}$ for $\tau(u) + 1 \leq j \leq n$. Consequently, for all $n \geq n_0$

$$0 \leq \Gamma_n \max\{\Gamma_{\tau(n)}, \Gamma_{\tau(n)+1}\} = \Gamma_{\tau(n)+1}. \tag{63}$$

Using (63), we conclude that $\lim_{n\to\infty} ||x_n - p|| = 0$, that is $x_n \to p$. □

The following are some consequences of our main theorem.
Let $u = \nabla f(z_n)$ in (21), we have the following corollary:

Corollary 1. *Let C and Q be nonempty, closed and convex subsets of real Hilbert spaces H_1 and H_2, respectively and $L : H_1 \to H_2$ be a bounded linear operator. Assume F is a real valued bifunction on $C \times C$ which admits condition C1-C4. Let $\phi : H_1 \to \mathbb{R} \cup \{+\infty\}$ be a proper, lower semicontinuous function, g be a β-inverse strongly monotone mapping. For $i = 1, 2 \cdots, N$, let $A_i : H_1 \to 2^{H_1}$ and $B_i : H_2 \to 2^{H_2}$ be finite families of monotone mappings. Assume $\Omega = GMEP(F, g, \phi) \cap \Gamma \neq \emptyset$, where $\Gamma = \{p \in H_1 : 0 \in \bigcap_{i=1}^{N} A_i(p)$ and $Lp \in H_2 : 0 \in \bigcap_{i=1}^{N} B_i(Lp)\}$. For an arbitrary $u, x_0 \in H_1$, let $\{x_n\} \subset H_1$ be a sequence defined iteratively by*

$$\begin{cases} F(u_n, y) + \langle g(u_n), y - u_n \rangle + \phi(y) - \phi(u_n) + \frac{1}{r_n} \langle y - u_n, u_n - x_n \rangle \geq 0, y \in H_1, \\ z_n = u_n - \gamma_n L^*(I - \Phi^{B_i}_{\lambda_{i,n}}) L u_n, \\ x_{n+1} = \alpha_n u + (1 - \alpha_n) \Phi^{A_i}_{\lambda_{i,n}} z_n, \end{cases} \tag{64}$$

where $\{r_n\}$ is a nonnegative sequence of real numbers, $\{\alpha_n\}$ and $\{\lambda_{i,n}\}$ are sequences in $(0,1)$, γ_n is a nonnegative sequence defined by (19), satisfying the following restrictions:

(i) $\sum_{n=1}^{\infty} \alpha_n = \infty$, $\lim_{n\to\infty} \alpha_n = 0$;
(ii) $0 < \lambda_i \leq \lambda_{i,n}$;
(iii) $0 < a \leq r_n \leq b < 2\beta$.

Then x_n converges strongly to $p \in \Omega$, where $p = P_\Omega \nabla f(p)$.

For $i = 1, 2$, we obtain the following result for approximation a common solution of a split null point for a sum of monotone operators and generalized mixed equilibrium problem.

Corollary 2. *Let C and Q be nonempty, closed and convex subsets of real Hilbert spaces H_1 and H_2, respectively and $L : H_1 \to H_2$ be a bounded linear operator. Assume F is a real valued bifunction on $C \times C$ which admits condition C1-C4. Let $\phi : H_1 \to \mathbb{R} \cup \{+\infty\}$ be a proper, lower semicontinuous function, g be a β-inverse strongly monotone mapping. For $i = 1, 2$, let $A_i : H_1 \to 2^{H_1}$ and $B_i : H_2 \to 2^{H_2}$ be finite families of monotone mappings. Assume $\Omega = GMEP(F, g, \phi) \cap \Gamma \neq \emptyset$, where $\Gamma = \{p \in H_1 : 0 \in \bigcap_{i=1}^{2} A_i(p)$ and $Lp \in H_2 : 0 \in \bigcap_{i=1}^{2} B_i(Lp)\}$. For an arbitrary $u, x_0 \in H_1$, let $\{x_n\} \subset H_1$ be a sequence defined iteratively by*

$$\begin{cases} F(u_n, y) + \langle g(u_n), y - u_n \rangle + \phi(y) - \phi(u_n) + \frac{1}{r_n} \langle y - u_n, u_n - x_n \rangle \geq 0, y \in H_1, \\ z_n = u_n - \gamma_n L^*(I - (J^{B_2}_{\lambda_{2,n}} \circ J^{B_1}_{\lambda_{1,n}})) L u_n, \\ x_{n+1} = \alpha_n u + (1 - \alpha_n)(J^{A_2}_{\lambda_{2,n}} \circ J^{A_1}_{\lambda_{1,n}}) z_n, \end{cases} \tag{65}$$

where $\{r_n\}$ is a nonnegative sequence of real numbers, $\{\alpha_n\}$ and $\{\lambda_{i,n}\}$ are sequences in $(0,1)$, γ_n is a nonnegative sequence defined by (19), satisfying the following restrictions:

(i) $\sum_{n=1}^{\infty} \alpha_n = \infty,\ \lim_{n\to\infty} \alpha_n = 0;$

(ii) $0 < \lambda_i \leq \lambda_{i,n};$

(iii) $0 < a \leq r_n \leq b < 2\beta.$

Then x_n converges strongly to $p \in \Omega$, where $p = P_\Omega \nabla f(p)$.

4. Numerical Example

In this section, we provide some numerical examples. The algorithm was coded in MATLAB 2019a on a Dell i7 Dual core 8.00 GB(7.78 GB usable) RAM laptop.

Example 1. Let $E_1 = E_2 = C = Q = \ell_2(\mathbb{R})$ be the linear spaces of 2-summable sequences $\{x_j\}_{j=1}^{\infty}$ of scalars in $\mathbb{R}$, that is

$$\ell_2(\mathbb{R}) := \left\{ x = (x_1, x_2, \cdots, x_j, \cdots),\ x_j \in \mathbb{R}\ \text{and}\ \sum_{j=1}^{\infty} |x_i|^2 < \infty \right\},$$

with the inner product $\langle \cdot, \cdot \rangle : \ell_2 \times \ell_2 \to \mathbb{R}$ defined by $\langle x, y \rangle := \sum_{j=1}^{\infty} x_j y_j$ and the norm $\|\cdot\| : \ell_2 \to \mathbb{R}$ by $\|x\| := \sqrt{\sum_{i=1}^{\infty} |x_j|^2}$, where $x = \{x_j\}_{j=1}^{\infty},\ y = \{y_j\}_{j=1}^{\infty}$. Let $L : \ell_2 \to \ell_2$ be given by $Lx = (x_1, x_2, \cdots, x_j, \cdots,)$ for all $x = \{x_i\}_{i=1}^{\infty} \in \ell_2$, then $L^*y = (y_1, y_2, \cdots, y_j, \cdots,)$ for each $y = \{y_i\}_\infty \in \ell_2$.

Let $f(x) = \frac{1}{2} x(s)^2,\ \forall x \in \ell_2$, it is easy to that f is differentiable with $\nabla f = x$. For each $i = 1, 2 \cdots N$, define $A_i(x) : \ell_2 \to \ell_2$ and $B_i(x) : \ell_2 \to \ell_2$ by $A_i(x) = ix$ and $B_i(x) = \frac{2}{3} ix$ respectively for all $x \in \ell_2$.

For each $u, v \in \ell_2$, define the bifunction $F : C \times C \to \mathbb{R}$ by $F(u,v) = uv + 15v - 15u - u^2$, the function $g : C \to H_1$ by $g(u) = u,\ \forall u \in H_1$ and $\phi : H_1 \to \mathbb{R} \cup \{+\infty\}$ by $\phi(u) = 0$, for each $u \in H_1$. For each $x \in C$, we have the following steps to get $\{u_n\}$: Find u such that

$$\begin{aligned}
0 &\leq F(u,v) + \langle g(u), v - u \rangle + \phi(v) - \phi(u) + \frac{1}{r}\langle v - u, u - x \rangle \\
&= uv + 15v - 15u - u^2 + v - u + \frac{1}{r}\langle v - u, u - x \rangle \\
&= uv + 16v - 16u - u^2 + \frac{1}{r}\langle v - u, u - x \rangle \\
&= (u + 16)(v - u) + \frac{1}{r}\langle v - u, u - x \rangle \\
&= (v - u)\left(u + 16 + \frac{1}{r}\langle v - u, u - x \rangle \right)
\end{aligned}$$

for all $v \in C$. Hence, by Lemma 3 (2), it follows that $u = \dfrac{x - 16r}{r+1}$. Therefore, $u_n = \dfrac{x_n - 16r_n}{r_n + 1}$.

For $i = 1, 2$, choose the sequences $\alpha_n = \dfrac{1}{n+1},\ r_n = \dfrac{1}{2n^2 - 1},\ \lambda_{i,n} = \dfrac{1}{in+2}$ and $\gamma = 0.25$. We obtain the graph of errors against the number of iterations for different values of x_0. The following cases are presented in Figure 1 below:

Case 1 $x_0 = (0.435, 0.896, 1.004, \cdots 0, \cdots),$
Case 2 $x_0 = (-0.987, 0.615, -2.804, \cdots 0, \cdots),$

Case 3 $x_0 = (3.45, 6.000, 1.53, \cdots 0, \cdots)$.

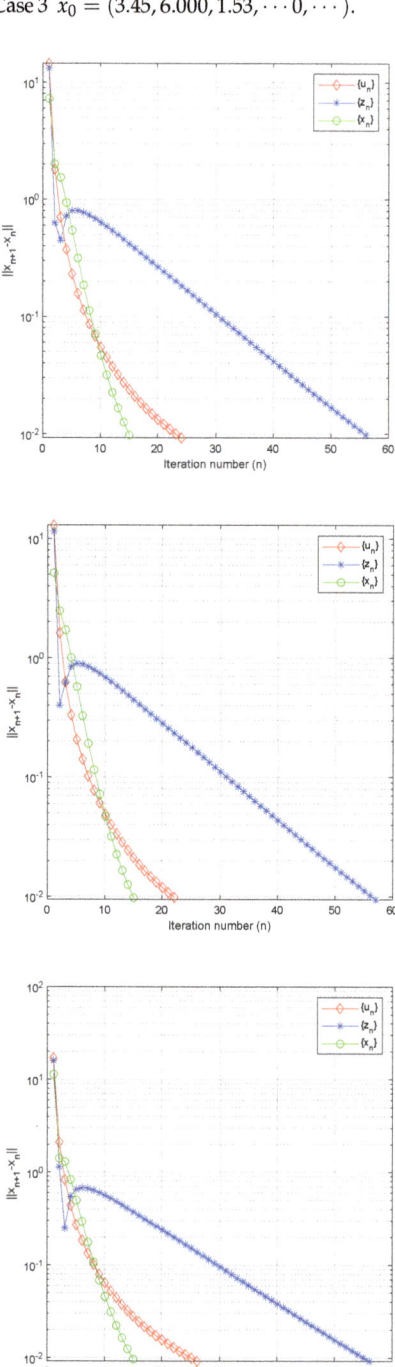

Figure 1. Case 1 (**top**); Case 2 (**middle**); Case 3 (**bottom**).

Example 2. Let $H_1 = H_2 = \mathbb{R}^2$ be endowed with an inner product $\langle x,y \rangle = x \cdot y = x_1y_1 + x_2y_2$, where $x = (x_1, x_2), y = (y_1, y_2)$ and the euclidean norm. Let $L : \mathbb{R}^2 \to \mathbb{R}^2$ be defined by $L(x) = (x_1 + x_2, 2x_1 + 2x_2)$, $x = (x_1, x_2)$ and $f(x) = \frac{1}{4}x^2$. For each $i = 1, 2 \cdots N$, define $A_i(x) : \mathbb{R}^2 \to \mathbb{R}^2$ and $B_i(x) : \mathbb{R}^2 \to \mathbb{R}^2$ by $A_i(x) = ix$ and $B_i(x) = \frac{2}{3}ix$ respectively, where $x = (x_1, x_2)$. Let $y = (y_1, y_2), z = (z_1, z_2) \in \mathbb{R}^2$. Define $F(z, y) = -3z^2 + 2zy + y^2$, $g(z) = z$ and $\phi(z) = z$. By simple calculation, we obtain that

$$u_n = \frac{x_n}{8r_n + 1}.$$

Choose the sequences $\alpha_n = \frac{1}{\sqrt{2n^2 + 3}}$, $r_n = \frac{n-1}{2n^2 - 1}$, $\lambda_{i,n} = \frac{1}{in + 2}$ and $\gamma = 0.25$. For $i = 1, 2$, (21) becomes

$$\begin{cases} F(u_n, y) + \langle g(u_n), y - u_n \rangle + \phi(y) - \phi(u_n) + \frac{1}{r_n}\langle y - u_n, u_n - x_n \rangle \geq 0, \ y \in H_1, \\ z_n = u_n - \gamma_n L^*(I - J_{\lambda_n}^{B_1} \circ J_{\lambda_n}^{B_2})Lu_n, \\ x_{n+1} = \frac{1}{\sqrt{2n^2 + 3}}\nabla f(z_n) + \left(1 - \frac{1}{\sqrt{2n^2 + 3}}\right) J_{\lambda_n}^{A_1} \circ J_{\lambda_n}^{A_2} z_n, \end{cases} \quad (66)$$

We make different choices of our initial value as follow:

Case 1, $x = (0.5, 1)$, Case 2, $x = (-0.05, 0.5)$, and Case 3, $x = (-1.5, 1.0)$.

We use $\|x_{n+1} - x_n\|^2 < 2 \times 10^{-3}$ as our stopping criterion and plot the graphs of errors against the number of iterations. See Figure 2.

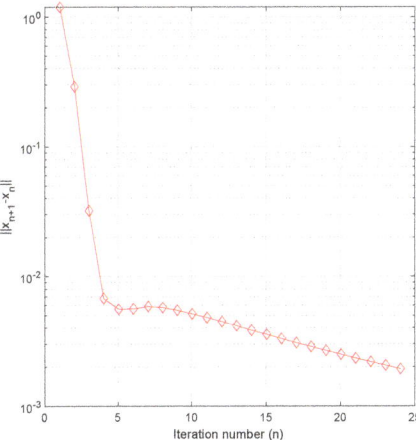

Figure 2. Cont.

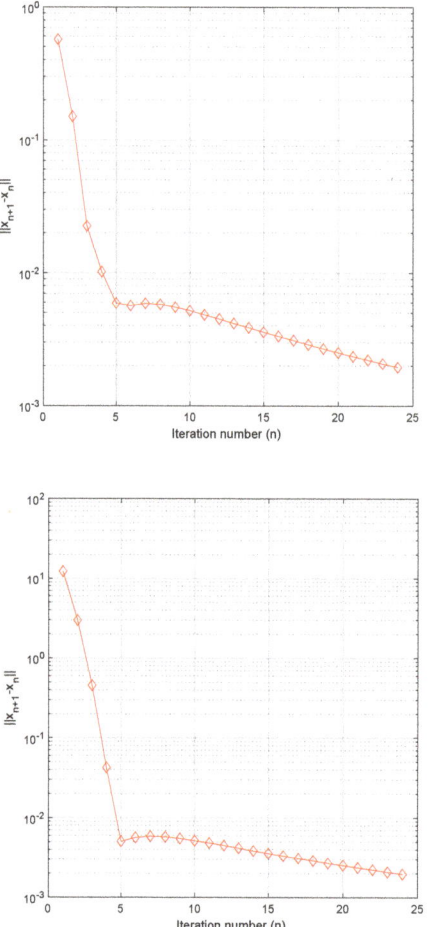

Figure 2. Case 1 (**top**); Case 2 (**middle**); Case 3 (**bottom**).

5. Conclusions

This paper considered the approximation of common solutions of a split null point problem for a finite family of maximal monotone operators and generalized mixed equilibrium problem in real Hilbert spaces. We proposed an iterative algorithm which does not depend on the prior knowledge of the operator norm as being used by many authors in the literature [39,42]. We proved a strong convergence of the proposed algorithm to a common solution of the two problems. We displayed some numerical examples to illustrate our method. Our result improves some existing results in the literature.

Author Contributions: Conceptualization of the article was given by O.K.O. and O.T.M., methodology by O.K.O., software by O.K.O., validation by O.T.M., formal analysis, investigation, data curation, and writing–original draft preparation by O.K.O. and O.T.M., resources by O.K.O. and O.T.M., writing–review and editing by O.K.O. and O.T.M., visualization by O.K.O. and O.T.M., project administration by O.T.M., Funding acquisition by O.T.M. All authors have read and agreed to the published version of the manuscript.

Funding: O.K.O is funded by Department of Science and Innovation and National Research Foundation, Republic of South Africa Center of Excellence in Mathematical and Statistical Sciences (DSI-NRF COE-MaSS) and O.T.M. is funded by National Research Foundation (NRF) of South Africa Incentive Funding for Rated Researchers (grant number 119903).

Acknowledgments: The authors sincerely thank the reviewers for their careful reading, constructive comments and fruitful suggestions that substantially improved the manuscript. The first author acknowledges with thanks the bursary and financial support from Department of Science and Innovation and National Research Foundation, Republic of South Africa Center of Excellence in Mathematical and Statistical Sciences (DST-NRF COE-MaSS) Doctoral Bursary. The second author is supported by the National Research Foundation (NRF) of South Africa Incentive Funding for Rated Researchers (Grant Number 119903). Opinions expressed and conclusions arrived are those of the authors and are not necessarily to be attributed to the CoE-MaSS and NRF.

Conflicts of Interest: The authors declare that they have no competing interests.

References

1. Blum, E.; Oettli, W. From optimization and variational inequalities to equilibrium problems. *Math. Stud.* **1994**, *63*, 123–145.
2. Chang, S.S.; Joseph, H.W.L.; Chan, C.K. A new method for solving equilibrium problem, fixed point problem and variational inequality problem with application to optimization. *Nonlinear Anal.* **2009**, *70*, 3307–3319. [CrossRef]
3. Nguyen, V.T. Golden Ratio Algorithms for Solving Equilibrium Problems in Hilbert Spaces. *arXiv* **2018**, arXiv:1804.01829
4. Oyewole, O.K.; Mewomo, O.T.; Jolaoso, L.O.; Khan, S.H. An extragradient algorithm for split generalized equilibrium problem and the set of fixed points of quasi-φ-nonexpansive mappings in Banach spaces. *Turkish J. Math.* **2020**, *44*, 1146–1170. [CrossRef]
5. Alakoya, T.O.; Taiwo, A.; Mewomo, O.T.; Cho, Y.J. An iterative algorithm for solving variational inequality, generalized mixed equilibrium, convex minimization and zeros problems for a class of nonexpansive-type mappings. *Ann. Univ. Ferrara Sez. VII Sci. Mat.* **2020**. [CrossRef]
6. Alakoya, T.O.; Jolaoso, L.O.; Mewomo, O.T. Modified inertia subgradient extragradient method with self adaptive stepsize for solving monotone variational inequality and fixed point problems. *Optimization* **2020**. [CrossRef]
7. Khan, S.H.; Alakoya, T.O.; Mewomo, O.T. Relaxed projection methods with self-adaptive step size for solving variational inequality and fixed point problems for an infinite family of multivalued relatively nonexpansive mappings in Banach spaces. *Math. Comput. Appl.* **2020**, *25*, 54. [CrossRef]
8. Qin, X.; Shang, M.; Su, Y. A general iterative method for equilibrium problem and fixed point problem in Hilbert spaces. *Nonlinear Anal.* **2008**, *69*, 3897–3909. [CrossRef]
9. Oyewole, O.K.; Abass, H.A.; Mewomo, O.T. Strong convergence algorithm for a fixed point constraint split null point problem. *Rend. Circ. Mat. Palermo II* **2020**. [CrossRef]
10. Jolaoso, L.O.; Alakoya, T.O.; Taiwo, A.; Mewomo, O.T. Inertial extragradient method via viscosity approximation approach for solving Equilibrium problem in Hilbert space. *Optimization* **2020**. [CrossRef]
11. Li, S.; Li, L.; Cao, L.; He, X.; Yue, X. Hybrid extragradient method for generalized mixed equilibrium problem and fixed point problems in Hilbert space. *Fixed Point Theory Appl.* **2013**, *2013*, 240. [CrossRef]
12. Alakoya, T.O.; Jolaoso, L.O.; Mewomo, O.T. A general iterative method for finding common fixed point of finite family of demicontractive mappings with accretive variational inequality problems in Banach spaces. *Nonlinear Stud.* **2020**, *27*, 1–24.
13. Aremu, K.O.; Izuchukwu, C.; Ogwo, G.N.; Mewomo, O.T. Multi-step Iterative algorithm for minimization and fixed point problems in p-uniformly convex metric spaces. *J. Ind. Manag. Optim.* **2020**. [CrossRef]
14. Ceng, L.-C.; Yao, J.-C. A hybrid iterative scheme for mixed equilibrium problems and fixed point problems. *J. Comput. Appl. Math.* **2008**, *214*, 186–201. [CrossRef]
15. Flam, S.D.; Antiprin, A.S. Equilibrium programming using proximal-like algorithm. *Math. Programm.* **1997**, *78*, 29–41. [CrossRef]
16. Combettes, P.L.; Histoaga, S.A. Equilibrium programming in Hilbert spaces. *J. Nonlinear Convex Anal.* **2005**, *6*, 117–136.
17. Izuchukwu, C.; Ogwo, G.N.; Mewomo, O.T. An Inertial Method for solving Generalized Split Feasibility Problems over the solution set of Monotone Variational Inclusions. *Optimization* **2020**. [CrossRef]
18. Oyewole, O.K.; Mewomo, O.T. A subgradient extragradient algorithm for solving split equilibrium and fixed point problems in reflexive Banach spaces. *J. Nonlinear Funct. Anal.* **2020**, *2020*, 19.
19. Taiwo, A.; Alakoya, T.O.; Mewomo, O.T. Halpern-type iterative process for solving split common fixed point and monotone variational inclusion problem between Banach spaces. *Numer. Algorithms* **2020**. [CrossRef]
20. Taiwo, A.; Jolaoso, L.O.; Mewomo, O.T.; Gibali, A. On generalized mixed equilibrium problem with α-β-μ bifunction and μ-τ monotone mapping. *J. Nonlinear Convex Anal.* **2020**, *21*, 1381–1401.
21. Taiwo, A.; Jolaoso, L.O.; Mewomo, O.T. Inertial-type algorithm for solving split common fixed-point problem in Banach spaces. *J. Sci. Comput.* **2014**, *2014*, 389689. [CrossRef]
22. Martinet, B. Regularisation d'inequations variationelles par approximations successives. *Rev. Franaise Inf. Rech. Oper.* **1970**, *4*, 154–158.

23. Alakoya, T.O.; Jolaoso, L.O.; Mewomo, O.T. A self adaptive inertial algorithm for solving split variational inclusion and fixed point problems with applications. *J. Ind. Manag. Optim.* **2020**. [CrossRef]
24. Alakoya, T.O.; Jolaoso, L.O.; Mewomo, O.T. Two modifications of the inertial Tseng extragradient method with self-adaptive step size for solving monotone variational inequality problems. *Demonstr. Math.* **2020**, *53*, 208–224. [CrossRef]
25. Cho, S.Y.; Dehaish, B.A.B.; Qin, X. Weak convergence on a splitting algorithm in Hilbert spaces. *J. Appl. Anal. Comput.* **2017**, *7*, 427–438.
26. Dehghan, H.; Izuchukwu, C.; Mewomo, O.T.; Taba, D.A.; Ugwunnadi, G.C. Iterative algorithm for a family of monotone inclusion problems in CAT(0) spaces. *Quaest. Math.* **2020**, *43*, 975–998. [CrossRef]
27. Gibali, A.; Jolaoso, L.O.; Mewomo, O.T.; Taiwo, A. Fast and simple Bregman projection methods for solving variational inequalities and related problems in Banach spaces. *Results Math.* **2020**, *75*, 36. [CrossRef]
28. Godwin, E.C.; Izuchukwu, C.; Mewomo, O.T. An inertial extrapolation method for solving generalized split feasibility problems in real Hilbert spaces. *Boll. Unione Mat. Ital.* **2020**. [CrossRef]
29. Izuchukwu, C.; Aremu, K.O.; Mebawondu, A.A.; Mewomo, O.T. A viscosity iterative technique for equilibrium and fixed point problems in a Hadamard space. *Appl. Gen. Topol.* **2019**, *20*, 193–210. [CrossRef]
30. Izuchukwu, C.; Mebawondu, A.A.; Aremu, K.O.; Abass, H.A.; Mewomo, O.T. Viscosity iterative techniques for approximating a common zero of monotone operators in an Hadamard space. *Rend. Circ. Mat. Palermo II* **2020**, *69*, 475–495. [CrossRef]
31. Izuchukwu, C.; Mebawondu, A.A.; Mewomo, O.T. A New Method for Solving Split Variational Inequality Problems without Co-coerciveness. *J. Fixed Point Theory Appl.* **2020**, *22*, 1–23. [CrossRef]
32. Jolaoso, L.O.; Taiwo, A.; Alakoya, T.O.; Mewomo, O.T. Strong convergence theorem for solving pseudo-monotone variational inequality problem using projection method in a reflexive Banach space. *J. Optim. Theory Appl.* **2020**, *185*, 744–766. [CrossRef]
33. Jolaoso, L.O.; Oyewole, O.K; Aremu, K.O.; Mewomo, O.T. A new efficient algorithm for finding common fixed points of multivalued demicontractive mappings and solutions of split generalized equilibrium problems in Hilbert spaces. *Int. J. Comput. Math.* **2020**. [CrossRef]
34. Jolaoso, L.O.; Taiwo, A.; Alakoya, T.O.; Mewomo, O.T. A unified algorithm for solving variational inequality and fixed point problems with application to the split equality problem. *Comput. Appl. Math.* **2020**, *39*, 38. [CrossRef]
35. Oyewole, O.K.; Jolaoso, L.O.; Izuchukwu, C.; Mewomo, O.T. On approximation of common solution of finite family of mixed equilibrium problems involving $\mu - \alpha$ relaxed monotone mapping in a Banach space. *Politehn. Univ. Bucharest Sci. Bull. Ser. A Appl. Math. Phys.* **2019**, *81*, 19–34.
36. Shehu, Y.; Iyiola, O.S. Iterative algorithms for solving fixed point problems and variational inequalities with uniformly continuous monotone operators. *Numer. Algorithms* **2018**, *79*, 529–553. [CrossRef]
37. Suwannaprapa, M.; Petrot, N. Finding a solution of split null point of the sum of monotone operators without prior knowledge of operator norms in Hilbert space. *J. Nonlinear Sci. Appl.* **2018**, *11*, 683–700. [CrossRef]
38. Censor, Y.; Elfving, T. A multiprojection algorithm using Bregman projections in product space. *Numer. Algorithms* **1994**, *8*, 221–239. [CrossRef]
39. Bryne, C. Iterative oblique projection onto convex sets and split feasibility problem. *Inverse Probl.* **2002**, *18*, 441–453. [CrossRef]
40. Censor, Y.; Bortfield, T.; Martin, B.; Trofimov, A. A unified approach for inversion problems in intensity-modulated radiation therapy. *Phys. Med. Biol.* **2006**, *51*, 2353–2365. [CrossRef]
41. Lopez, G.; Martin-Marquez, V.; Wang, F.; Xu, H.K. Solving the split feasibility problem without prior knowledge of matrix norms. *Inverse Probl.* **2012**, *28*, 085004. [CrossRef]
42. Bryne, C.; Censor, Y.; Gibali, A.; Reich, S. Weak and strong convergence of algorithms for the split common null point problem. *J. Nonlinear Convex Anal.* **2012**, *13*, 759–775.
43. Bauschke, H.H.; Combettes, P.L. *Convex Analysis and Monotone Operator Theory in Hilbert Spaces*; Springer: Berlin, Germany, 2011.
44. Boikanyo, O.A., The viscosity approximation forward-backward splitting method for zeros of the sum of monotone operators. *Abstr. Appl. Anal.* **2016**, *2016*, 10. [CrossRef]
45. Xu, H.K. Averaged mappings and the gradient-projection algorithm. *J. Optim. Theory. Appl.* **2011**, *150*, 360–378. [CrossRef]
46. Rockafellar, R.T. On the maximality of sums of nonlinear monotone operators. *Trans. Amer. Math. Soc.* **1970**, *149*, 75–88. [CrossRef]
47. Taiwo, A.; Owolabi, A.O.-E.; Jolaoso, L.O.; Mewomo, O.T.; Gibali, A. A new approximation scheme for solving various split inverse problems. *Afr. Mat.* **2020**. [CrossRef]
48. Taiwo, A.; Jolaoso, L.O.; Mewomo, O.T. Viscosity approximation method for solving the multiple-set split equality common fixed-point problems for quasi-pseudocontractive mappings in Hilbert Spaces. *J. Ind. Manag. Optim.* **2020**. [CrossRef]
49. Ogwo, G.N.; Izuchukwu, C.; Aremu, K.O.; Mewomo, O.T. A viscosity iterative algorithm for a family of monotone inclusion problems in an Hadamard space. *Bull. Belg. Math. Soc. Simon Stevin* **2020**, *27*, 127–152. [CrossRef]
50. Xu, H.K. Iterative algorithms for nonlinear operators. *J. Lond. Math. Soc.* **2002**, *66*, 240–256. [CrossRef]

Article

Approximation of the Fixed Point for Unified Three-Step Iterative Algorithm with Convergence Analysis in Busemann Spaces

Hassan Almusawa [1], Hasanen A. Hammad [2,*] and Nisha Sharma [3]

[1] Department of Mathematics, College of Sciences, Jazan Univesity, Jazan 45142, Saudi Arabia; almusawah@mymail.vcu.edu
[2] Department of Mathematics, Faculty of Science, Sohag University, Sohag 82524, Egypt
[3] Department of Mathematics, Pt. J.L.N. Govt. College, Faridabad 121002, India; nnishaa.bhardwaj@gmail.com
* Correspondence: hassanein_hamad@science.sohag.edu.eg

Abstract: In this manuscript, a new three-step iterative scheme to approximate fixed points in the setting of Busemann spaces is introduced. The proposed algorithms unify and extend most of the existing iterative schemes. Thereafter, by making consequent use of this method, strong and Δ-convergence results of mappings that satisfy the condition $(\mathcal{E}_\mu)$ in the framework of uniformly convex Busemann space are obtained. Our results generalize several existing results in the same direction.

Keywords: the condition $(\mathcal{E}_\mu)$; standard three-step iteration algorithm; fixed point; uniformly convex Busemann space

MSC: AMS Subject Classification: 47H09; 47H10; 54H25; 54E40

Citation: Almusawa, H.; Hammad, H.A.; Sharma, N. Approximation of the Fixed Point for Unified Three-Step Iterative Algorithm with Convergence Analysis in Busemann Spaces. *Axioms* **2021**, *10*, 26. https://doi.org/10.3390/axioms10010026

Received: 12 January 2021
Accepted: 23 February 2021
Published: 27 February 2021

Publisher's Note: MDPI stays neutral with regard to jurisdictional claims in published maps and institutional affiliations.

Copyright: © 2021 by the authors. Licensee MDPI, Basel, Switzerland. This article is an open access article distributed under the terms and conditions of the Creative Commons Attribution (CC BY) license (https://creativecommons.org/licenses/by/4.0/).

1. Introduction

Throughout this paper, $\mathcal{R}$, $\mathcal{R}^+$, and $f_\wp$ denote the set of all real numbers, positive real numbers, and fixed points of the mapping $\wp$, respectively.

The fixed point theory is considered one of the most powerful analytical techniques in mathematics, especially in nonlinear analysis, where it plays a prominent role in algorithm technology. The purpose of investing in algorithms is to obtain the best algorithms with a faster convergence rate, because the lower the convergence rate, the faster the speed of obtaining the solution. This is probably the drawback of using the iterative methods.

It should be noted that the Mann iteration converges faster than the Ishikawa iteration for the class of Zamfirescu operators [1], and hence the convergence behavior of proclaimed and empirically proven faster iterative schemes need not always be faster. There was extensive literature on proclaimed new and faster iteration schemes in ancient times. Some of the iteration schemes are undoubtedly better versions of previously existed iteration schemes, whereas a few are only the special cases. There are more than twenty iteration schemes in the present literature. Our analysis's focal objective is to unify the existing results in the framework of Busemann spaces (see [2] for the precise definitions and properties of Busemann spaces). This analysis has a special significance in terms of unification, and numerous researchers have intensively investigated various aspects of it.

Apart from Picard, Mann, and Ishikawa, many iterative schemes with better convergence rates are obtained; see, for example, [3–11]. In many cases, these algorithms cannot obtain strong convergence; therefore, it was necessary to investigate new effective algorithms. Recently, several authors were able to apply the strong convergence of algorithms, see [12–16].

Recall that a metric space $(\mathcal{B}, \partial)$ is called a *geodesic* path (or simply a *geodesic* [17]) in $\mathcal{B}$ if there is a path $\gamma : [a,b] \to \mathcal{B}$, such that γ is an isometry for $[a,b] \subset [0,\infty)$. A *geodesic ray* is an isometry $\gamma : \mathcal{R}^+ \to \mathcal{B}$, and a *geodesic* line is an isometry $\gamma : \mathcal{R} \to \mathcal{B}$. For more details about *geodesic* path in metric fixed point theory, see [18–24].

Definition 1 ([17]). *Let $(\mathcal{B}, \partial)$ be a metric space and $\jmath, \ell \in \mathcal{B}$. A geodesic path joining $\jmath$ to ℓ is a mapping $\gamma : [\alpha, \beta] \subseteq \mathcal{R}^+ \to \mathcal{B}$ such that $\gamma(\alpha) = \jmath, \gamma(\beta) = \ell$ and*

$$\partial(\gamma(t), \gamma(t')) = |t - t'|$$

for all $t, t' \in [\alpha, \beta]$. Particularly, γ is an isometry and $\partial(\jmath, \ell) = \beta - \alpha$.

A geodesic segment joining $\jmath$ and ℓ in $\mathcal{B}$ is the image of a geodesic path in $\mathcal{B}$. The space $\mathcal{B}$ is said to be a geodesic space, if every two points of $\jmath$ are joined by a geodesic.

Definition 2 ([17]). *A metric space $\mathcal{B}$ is said to be a geodesic space if given two arbitrary points of $\mathcal{B}$ there exists a geodesic path that joins them.*

Definition 3 ([17]). *The geodesic metric space $(\mathcal{B}, \partial)$ is said to be Busemann space, if for any two affinely reparametrized geodesices $\gamma : [\alpha, \beta] \to \mathcal{B}$ and $\gamma' : [\alpha', \beta'] \to \mathcal{B}$, the map $\mathcal{D}_{\gamma, \gamma'} : [\alpha, \beta] \times [\alpha', \beta'] \to \mathcal{R}$ defined by*

$$\mathcal{D}_{\gamma, \gamma'}(t, t') = \partial(\gamma(t), \gamma'(t'))$$

is a convex; that is, the metric of Busemann space is convex. In a Busemann space the geodesic joining any two points is unique.

Proposition 1 ([25]). *In such spaces, the hypotheses below hold:*
(1) $\partial(\varepsilon, (1-\alpha)\jmath \oplus \alpha\ell) \leq (1-\alpha)\partial(\varepsilon, \jmath) + \alpha\partial(\varepsilon, \ell)$,
(2) $\partial((1-\alpha)\jmath \oplus \alpha\ell, (1-\alpha')\jmath \oplus \alpha'\ell) = |\alpha - \alpha'|\partial(\jmath, \ell)$,
(3) $(1-\alpha)\jmath \oplus \alpha\ell = \alpha\ell \oplus (1-\alpha)\jmath$,
(4) $\partial((1-\alpha)\jmath \oplus \alpha\varepsilon, ((1-\alpha)\ell \oplus \alpha\omega)) \leq (1-\alpha)\partial(\jmath, \ell) + \alpha\partial(\varepsilon, \omega)$,
where $\jmath, \ell, \varepsilon, \omega \in \mathcal{B}$ and $\alpha, \alpha' \in [0,1]$.

Busemann spaces are also hyperbolic spaces, which were introduced by Kohlenbach [26]. Further, $\mathcal{B}$ is said to be uniquely geodesic [17] if there is exactly one *geodesic* joining $\jmath$ and ℓ for each $\jmath, \ell \in \mathcal{B}$.

Definition 4 ([17]). *Suppose that $\mathcal{B}$ is a uniquely geodesic space and $\gamma([\alpha, \beta])$ is a geodesic segment joining $\jmath$ and ℓ and $\alpha \in [0,1]$. Then,*

$$\varepsilon = \gamma((1-\alpha)\jmath + \alpha\ell)$$

will be a unique point in $\gamma([\alpha, \beta])$ satisfying

$$\partial(\varepsilon, \jmath) = \alpha\partial(\jmath, \ell)$$

and

$$\partial(\varepsilon, \ell) = (1-\alpha)\partial(\jmath, \ell).$$

In the sequel, the notation $[\jmath, \ell]$ is used for *geodesic* segment $\gamma([\alpha, \beta])$ and ε is denoted by $(1-\alpha)\jmath \oplus \alpha\ell$. A subset $\kappa \subseteq \mathcal{B}$ is said to be *geodesically* convex if κ includes every geodesic segment joining any two of its points. Let $\mathcal{B}$ be a *geodesic* metric space and $\wp : \mathcal{B} \to \mathcal{R}$. We say that $\wp$ is convex if for every *geodesic* path $\gamma : [\alpha, \beta] \to \mathcal{B}$, the map $\wp \circ \gamma : [\alpha, \beta] \to$ is a convex. It is known that if $\wp : \mathcal{B} \to \mathcal{R}$ is a convex function and $\wp' : \mathcal{B} \to \mathcal{R}$ is an increasing convex function, then $\wp' \circ \wp : \mathcal{B} \to \mathcal{R}$ is convex.

We now introduce our algorithm.

Let $\mathcal{B}$ be a complete Busemann space, $\mathcal{B}_s$ be a nonempty convex subset of $\mathcal{B}$ and $\wp : \mathcal{B}_s \to \mathcal{B}_s$ be a mapping. For any $v_0 \in \mathcal{B}_s$,

$$\begin{cases} \ell_\eta &= \tau_\eta^0 v_\eta \oplus \kappa_\eta^0 \wp v_\eta \oplus \iota_\eta^0 J_\eta \oplus \omega_\eta^0 \wp J_\eta, \\ J_\eta &= \tau_\eta^1 v_\eta \oplus \kappa_\eta^1 \wp v_\eta \oplus \iota_\eta^1 \ell_\eta \oplus \omega_\eta^1 \wp \ell_\eta, \\ v_{n+1} &= \tau_\eta^2 v_\eta \oplus \kappa_\eta^2 \wp v_\eta \oplus \iota_\eta^2 J_\eta \oplus \omega_\eta^2 \wp J_\eta \oplus \varepsilon_\eta \ell_\eta \oplus \sigma_\eta \wp \ell_\eta, \end{cases}$$

where $\{\varepsilon_\eta\}, \{\sigma_\eta\}, \{\tau_\eta^i\}, \{\kappa_\eta^i\}, \{\iota_\eta^i\}$ and $\{\omega_\eta^i\}$ for $i = 0, 1, 2$ are sequences in $[0, 1]$. Moreover, $\tau_\eta^0 + \kappa_\eta^0 + \iota_\eta^0 + \omega_\eta^0 = 1$, $\tau_\eta^1 + \kappa_\eta^1 + \iota_\eta^1 + \omega_\eta^1 = 1$ and $\tau_\eta^2 + \kappa_\eta^2 + \iota_\eta^2 + \varepsilon_\eta + \sigma_\eta = 1$.

Remark 1. *For distinct values of ε_η, $\{\sigma_\eta\}$, τ_η^i, κ_η^i, ι_η^i and ω_η^i for $i = 0, 1, 2$, we have well-known distinct iteration schemes as follows:*

$(\mathcal{R}_1)$ $\iota_\eta^0 = \omega_\eta^0 = \iota_\eta^1 = \kappa_\eta^1 = \iota_\eta^2 = \kappa_\eta^2 = \varepsilon_\eta = \sigma_\eta = 0$, $\tau_\eta^0 = (1 - \kappa_\eta^0)$, $\tau_\eta^1 = (1 - \omega_\eta^1)$, $\tau_\eta^1 = (1 - \omega_\eta^2)$ *in the standard three-step iteration scheme, we obtain the Noor iterative scheme* [27].

$(\mathcal{R}_2)$ $\iota_\eta^0 = \omega_\eta^0 = \iota_\eta^1 = \tau_\eta^1 = \kappa_\eta^1 = \iota_\eta^2 = \omega_\eta^2 = \varepsilon_\eta = \sigma_\eta = 0$, $\tau_\eta^0 = (1 - \kappa_\eta^0)$, $\iota_\eta^1 = (1 - \omega_\eta^1)$ *and* $\iota_\eta^2 = (1 - \omega_\eta^2)$ *in the standard three-step iteration scheme, we obtain the SP iterative scheme* [28].

$(\mathcal{R}_3)$ $\iota_\eta^0 = \omega_\eta^0 = \iota_\eta^1 = \tau_\eta^1 = \kappa_\eta^1 = \tau_\eta^2 = \iota_\eta^2 = \omega_\eta^2 = \varepsilon_\eta = \sigma_\eta = 0$, $\iota_\eta^0 = (1 - \kappa_\eta^0)$, $\kappa_\eta^1 = (1 - \omega_\eta^1)$ *and* $\kappa_\eta^2 = 1$ *in the standard three-step iteration scheme, we obtain the Picard-S iterative scheme* [29].

$(\mathcal{R}_4)$ $\tau_\eta^0 = \kappa_\eta^0 = \iota_\eta^0 = \omega_\eta^0 = \iota_\eta^1 = \omega_\eta^1 = \tau_\eta^2 = \kappa_\eta^2 = \varepsilon_\eta = \sigma_\eta = 0$ *and* $\tau_\eta^0 = (1 - \kappa_\eta^0)$, $\kappa_\eta^1 = (1 - \omega_\eta^1)$ *and* $\iota_\eta^2 = (1 - \omega_\eta^2)$ *in the standard three-step iteration scheme, we obtain the CR iterative scheme* [30].

$(\mathcal{R}_5)$ $\tau_\eta^0 = \kappa_\eta^0 = \iota_\eta^1 = \tau_\eta^1 = \tau_\eta^2 = \iota_\eta^2 = \varepsilon_\eta = \sigma_\eta = 0$, $\tau_\eta^1 = (1 - \kappa_\eta^1)$, $\kappa_\eta^1 = (1 - \omega_\eta^1)$ *in the standard three-step iteration scheme, we obtain the Abbas and Nazir iterative scheme* [31].

$(\mathcal{R}_6)$ $\iota_\eta^0 = \omega_\eta^0 = \tau_\eta^1 = \kappa_\eta^1 = \kappa_\eta^2 = \iota_\eta^2 = \omega_\eta^2 = \varepsilon_\eta = 0$, $\tau_\eta^0 = (1 - \kappa_\eta^0)$, $\iota_\eta^1 = (1 - \omega_\eta^1)$ *and* $\sigma_\eta^2 = (1 - \omega_\eta^2)$ *in the standard three-step iteration scheme, we obtain the P iterative scheme* [32].

$(\mathcal{R}_7)$ $\iota_\eta^0 = \omega_\eta^0 = \tau_\eta^1 = \iota_\eta^1 = \kappa_\eta^2 = \iota_\eta^2 = \tau_\eta^2 = \varepsilon_\eta = 0$, $\tau_\eta^0 = (1 - \kappa_\eta^0)$, $\kappa_\eta^1 = (1 - \omega_\eta^1)$ *and* $\sigma_\eta^2 = (1 - \omega_\eta^2)$ *in the standard three-step iteration scheme, we obtain the D iterative scheme* [33].

$(\mathcal{R}_8)$ $\iota_\eta^0 = \omega_\eta^0 = \tau_\eta^0 = \kappa_\eta^0 = \iota_\eta^1 = \omega_\eta^1 = \tau_\eta^1 = \kappa_\eta^1 = \omega_\eta^2 = \iota_\eta^2 = \varepsilon_\eta = 0$, $\tau_\eta^2 = (1 - \kappa_\eta^2)$ *in the standard three-step iteration scheme, we obtain the Mann iterative scheme* [34].

$(\mathcal{R}_9)$ $\iota_\eta^0 = \omega_\eta^0 = \tau_\eta^0 = \kappa_\eta^0 = \iota_\eta^1 = \omega_\eta^1 = \kappa_\eta^2 = \iota_\eta^2 = \varepsilon_\eta = \sigma_\eta = 0$, $\tau_\eta^1 = (1 - \kappa_\eta^1)$ *and* $\tau_\eta^2 = (1 - \omega_\eta^2)$ *in the standard three-step iteration scheme, we obtain the Ishikawa iterative scheme* [35].

2. Preliminaries

In this section, we present some relevant and essential definitions, lemmas, and theorems needed in the sequel.

Definition 5 ([36]). *The Busemann space $\mathcal{B}$ is called uniformly convex if for any $\zeta > 0$ and $\epsilon \in (0, 2]$, there exists a map δ such that for every three points $\alpha, J, \ell \in \mathcal{B}$, $\partial(J, \alpha) \leq \zeta$, $\partial(\ell, \alpha) \leq \zeta$ and $\partial(J, \ell) \geq \epsilon \zeta$ implies that*

$$\partial(m, \alpha) \leq (1 - \delta)\zeta,$$

where m denotes the midpoint of any geodesic segment $[J, \ell]$ (i.e., $m = \frac{1}{2}J \oplus \frac{1}{2}\ell$) and $\inf\{\delta : \zeta > 0\}$. A mapping $\wp : (0, \infty) \times (0, 2] \to (0, 1]$ is called a modulus of uniform convexity, for $\wp(\eta, \epsilon) := \delta$ and for a given $\eta > 0$, $\epsilon \in (0, 2]$.

Henceforth, the uniform convexity modulus with a decreasing modulus concerning η (for a fixed ϵ) is termed as the uniform convexity monotone modulus. The subsequent lemmas and geometric properties, which are instrumental throughout the discussion to learn about essential terms of Busemann spaces, are necessary to achieve our significant findings and are as follows:

Lemma 1. *If $\wp$ is a mapping satisfying condition $(\mathcal{E})$ and has a fixed point then it is a quasi-nonexpansive mapping.*

Let $\mathcal{B}_s$ be a nonempty closed convex subset of a Busemann space $\mathcal{B}$, and let $\{\jmath_\eta\}$ be a bounded sequence in $\mathcal{B}$. For $\jmath \in \mathcal{B}$, we set

$$\zeta(\jmath, \{\jmath_\eta\}) = \limsup_{\eta \to \infty} ||\jmath_\eta - \jmath||.$$

The asymptotic radius of $\zeta(\{\jmath_\eta\})$ is given by

$$\zeta(\mathcal{B}_s, \{\jmath_\eta\}) = \inf\{\zeta(\jmath, \{\jmath_\eta\}) : \jmath \in \mathcal{B}_s\}$$

and the asymptotic center $\mathcal{A}(\{\jmath_\eta\})$ of $\{\jmath_\eta\}$ relative to $\mathcal{B}_s$ is the set

$$\mathcal{A}(\mathcal{B}_s, \{\jmath_\eta\}) = \{\jmath \in \mathcal{B}_s : \zeta(\jmath, \jmath_\eta) = \zeta(\mathcal{B}_s, \{\jmath_\eta\})\}.$$

It is known that, in a Busemann space, $\mathcal{A}(\{\jmath_\eta\})$ consists of exactly one point [37].

Recall that a bounded sequence $\{\jmath_\eta\} \in \mathcal{B}$ is said to be regular [38], if $\zeta(\{\jmath_\eta\}) = \zeta(\{\jmath_{\eta_k}\})$ for every subsequence $\{\jmath_{\eta_k}\}$ of $\{\jmath_\eta\}$.

Lemma 2 ([4]). *Let $\mathcal{B}$ be a Busemann space and $\jmath \in \mathcal{B}$, $\{t_\eta\}$ a sequence in $[b,c]$, for some $b, c \in (0,1)$. If $\{\jmath_\eta\}$ and $\{\ell_\eta\}$ are sequences in $\mathcal{B}$ satisfying*

$$\limsup_{\eta \to \infty} \partial(\jmath_\eta, \jmath) \leq r$$

also,

$$\limsup_{\eta \to \infty} \partial(\ell_\eta, \ell) \leq r$$

and

$$\limsup_{\eta \to \infty} \partial(t_\eta \jmath_\eta \oplus (1 - t_\eta)\ell_\eta, \jmath) = r,$$

for some $r \geq 0$, then

$$\lim_{\eta \to \infty} \partial(\jmath_\eta, \ell_\eta) = 0.$$

Lemma 3 ([5]). *If $\mathcal{B}_s$ is a closed convex subset of a uniformly convex Busemann space $\mathcal{B}$ and $\{\jmath_\eta\}$ is a bounded sequence in $\mathcal{B}_s$, then the asymptotic center of $\{\jmath_\eta\}$ belongs to $\mathcal{B}_s$.*

Lemma 4 ([38]). *Let $\mathcal{B}$ be a Busemann space, $\{\jmath_\eta\}$ be a bounded sequence in $\mathcal{B}$ and $\mathcal{B}_s$ be a subset of $\mathcal{B}$. Then $\{\jmath_\eta\}$ has a subsequence, which is regular in $\mathcal{B}_s$.*

Definition 6 ([38]). *A sequence $\{\jmath_\eta\}$ in Busemann space $\mathcal{B}$ is said to be Δ − convergent if there exists some $\jmath \in \mathcal{B}$ such that $\jmath$ is the unique asymptotic center of $\jmath_{\eta_k}$ for every subsequence $\{\jmath_{\eta_k}\}$ of $\{\jmath_\eta\}$. In this case we write $\Delta - \lim_{\eta \to \infty} \jmath_\eta = \jmath$ and it is called the $\Delta - \lim$ of $\{\jmath_\eta\}$.*

Lemma 5 ([21]). *Every bounded sequence in a complete Busemann space always has a $\Delta -$ convergent subsequence.*

Lemma 6 ([21]). *Suppose that $\mathcal{B}_s$ is a closed convex subset of a Busemann space $\mathcal{B}$ and $\wp : \mathcal{B}_s \to \mathcal{B}$ satisfies the condition $(\mathcal{E})$. Then $\{\jmath_\eta\}, \Delta -$ converges to $\jmath$ and $\partial(\wp \jmath_\eta, \jmath_\eta) \to 0$, implying that $\jmath \in \mathcal{B}_s$*

and $\wp_J = J$.

Definition 7. *Assume that $\mathcal{B}_s \neq \emptyset$ is a subset of a Busemann space $\mathcal{B}$. For $J, \ell \in \mathcal{B}_s$, a mapping $\wp : \mathcal{B}_s \to \mathcal{B}_s$ is called:*

(i) *Contraction if there is $\mu \in (0,1)$ so that $\partial(\wp_J, \wp\ell) \leq \mu \partial(J, \ell)$,*
(ii) *Nonexpansive if $\partial(\wp_J, \wp\ell) \leq \partial(J, \ell)$,*
(iii) *Quasi-nonexpansive if $\partial(\wp_J, \varkappa) \leq \partial(J, \varkappa)$, $\varkappa \in F(\wp)$ and $F(\wp)$ denote the set $\{\varkappa \in \mathcal{B} : \varkappa = \wp\varkappa\}$.*
(iv) *Satisfy Condition $(\mathcal{E})$ if*

$$\frac{1}{2}\partial(J, \wp_J) \leq \partial(J, \ell) \Rightarrow \partial(\wp_J, \wp\ell) \leq \partial(J, \ell).$$

(v) *Suzuki generalized nonexpansive if it verifies Condition $(\mathcal{E})$.*

Garcia-Falset et al. [6] introduced the generalization for nonexpansive mappings known as condition $(\mathcal{E}_\mu)$.

Definition 8 ([6]). *Let $\mu \geq 1$. A mapping $\wp : \mathcal{B}_s \to \mathcal{B}_s$ is said to satisfy condition $(\mathcal{E}_\mu)$ if for all $J, \ell \in \mathcal{B}_s$, we have*

$$\partial(J, \wp\ell) \leq \mu \partial(J, \wp_J) + \partial(J, \ell).$$

We say that $\wp$ satisfies condition $(\mathcal{E})$, if $\wp$ satisfies condition $(\mathcal{E}_\mu)$ for some $\mu \geq 1$ [39].

Theorem 1. *Let $\mathcal{C}_s$ be a nonempty bounded, closed and convex subset of a complete $CAT(0)$ space $\mathcal{C}$. If $\wp : \mathcal{C}_s \to \mathcal{C}_s$ is a generalized nonexpansive mapping, then $\wp$ has a fixed point in $\mathcal{C}_s$. Moreover, $f_\wp$ is closed and convex.*

3. Main Results

We begin this section with the proof of the following lemmas:

Lemma 7. *Let $\mathcal{B}_s$ be a nonempty closed convex subset of a complete Busemann space $\mathcal{B}$, and let $\wp : \mathcal{B}_s \to \mathcal{B}_s$ be a mapping satisfying condition $(\mathcal{E}_\mu)$. For an arbitrary chosen $v_0 \in \mathcal{B}_s$, let the sequence $\{v_\eta\}$ be generated by a standard three-step iteration algorithm with the condition*

$$((\tau^1_\eta + \kappa^1_\eta) + (\iota^1_\eta + \omega^1_\eta)(\tau^0_\eta + \kappa^0_\eta))(1 - (\iota^1_\eta + \omega^1_\eta)(\iota^0_\eta + \omega^0_\eta))^{-1} \leq 1$$

and

$$(\tau^2_\eta + \kappa^2_\eta + \iota^2_\eta + \omega^2_\eta + (\varepsilon_\eta + \sigma_\eta)(\tau^0_\eta + \kappa^0_\eta + \iota^0_\eta + \omega^0_\eta)) \leq 1.$$

Then, $\lim_{n \to \infty} \partial(v_\eta, v_*)$ exists for all $v_* \in f_\wp$.

Proof. Let $v_* \in f_\wp$ and $z \in \mathcal{B}_s$. Since $\wp$ satisfies condition $(\mathcal{E}_\mu)$, and hence

$$\partial(v_*, \wp_J) \leq \mu \partial(v_*, \wp v_*) + \partial(v_*, J).$$

From standard three-step iteration algorithm, we have

$$\begin{aligned}
\partial(\ell_\eta, v_*) &= \partial(\tau^0_\eta v_\eta \oplus \kappa^0_\eta \wp v_\eta \oplus \iota^0_\eta J_\eta \oplus \omega^0_\eta \wp J_\eta, v_*) \\
&\leq \tau^0_\eta \partial(v_\eta, v_*) + \kappa^0_\eta(\mu \partial(v_*, \wp v_*) + \partial(v_*, v_n)) + \iota^0_\eta \partial(J_\eta, v_*) + \omega^0_\eta(\mu \partial(v_*, \wp v_*) + \partial(v_*, J_n)) \\
&\leq \tau^0_\eta \partial(v_\eta, v_*) + \kappa^0_\eta \partial(v_\eta, v_*) + \iota^0_\eta \partial(J_\eta, v_*) + \omega^0_\eta \partial(J_\eta, v_*) \\
&= (\tau^0_\eta + \kappa^0_\eta) \partial(v_\eta, v_*) + (\iota^0_\eta + \omega^0_\eta) \partial(J_\eta, v_*).
\end{aligned}$$

Also,

$$\partial(j_\eta, v_*) = \partial(\tau_\eta^1 v_\eta \oplus \kappa_\eta^1 \wp v_\eta \oplus \iota_\eta^1 \ell_\eta \oplus \omega_\eta^1 \wp \ell_\eta, v_*)$$
$$\leq \tau_\eta^1 \partial(v_\eta, v_*) + \kappa_\eta^1 \partial(\wp v_\eta, v_*) + \iota_\eta^1 \partial(\ell_\eta, v_*) + \omega_\eta^1 \partial(\wp \ell_\eta, v_*)$$
$$\leq \tau_\eta^1 \partial(v_\eta, v_*) + \kappa_\eta^1(\mu \partial(v_*, \wp v_*) + \partial(v_*, v_n)) + \iota_\eta^1 \partial(\ell_\eta, v_*) + \omega_\eta^1(\mu \partial(v_*, \wp v_*) + \partial(v_*, \ell_n))$$
$$= (\tau_\eta^1 + \kappa_\eta^1)\partial(v_\eta, v_*) + (\iota_\eta^1 + \omega_\eta^1)\partial(\ell_\eta, v_*).$$

Using the value of $\partial(\ell_\eta, v_*)$, we have

$$\partial(j_\eta, v_*) \leq (\tau_\eta^1 + \kappa_\eta^1)\partial(v_\eta, v_*) + (\iota_\eta^1 + \omega_\eta^1)((\tau_\eta^0 + \kappa_\eta^0)\partial(v_\eta, v_*) + (\iota_\eta^0 + \omega_\eta^0)\partial(j_\eta, v_*))$$
$$\leq ((\tau_\eta^1 + \kappa_\eta^1) + (\iota_\eta^1 + \omega_\eta^1)(\tau_\eta^0 + \kappa_\eta^0))\partial(v_\eta, v_*) + (\iota_\eta^1 + \omega_\eta^1)(\iota_\eta^0 + \omega_\eta^0)\partial(j_\eta, v_*)$$
$$\partial(j_\eta, v_*) \leq \left(\frac{(\tau_\eta^1 + \kappa_\eta^1) + (\iota_\eta^1 + \omega_\eta^1)(\tau_\eta^0 + \kappa_\eta^0)}{1 - (\iota_\eta^1 + \omega_\eta^1)(\iota_\eta^0 + \omega_\eta^0)}\right)\partial(v_\eta, v_*).$$

Since $((\tau_\eta^1 + \kappa_\eta^1) + (\iota_\eta^1 + \omega_\eta^1)(\tau_\eta^0 + \kappa_\eta^0))(1 - (\iota_\eta^1 + \omega_\eta^1)(\iota_\eta^0 + \omega_\eta^0))^{-1} \leq 1$, we have

$$\partial(j_\eta, v_*) \leq \partial(v_\eta, v_*).$$

Now,

$$\partial(v_{n+1}, v_*) \leq \partial(\tau_\eta^2 v_\eta \oplus \kappa_\eta^2 \wp v_\eta \oplus \iota_\eta^2 j_\eta \oplus \omega_\eta^2 \wp j_\eta \oplus \varepsilon_\eta \ell_\eta \oplus \sigma_\eta \wp \ell_\eta, v_*)$$
$$\leq \tau_\eta^2 \partial(v_\eta, v_*) + \kappa_\eta^2 \partial(\wp v_\eta, v_*) + \iota_\eta^2 \partial(j_\eta, v_*) + \omega_\eta^2 \partial(\wp j_\eta, v_*) + \varepsilon_\eta \partial(\ell_\eta, v_*) + \sigma_\eta \partial(\wp \ell_\eta, v_*)$$
$$\leq \tau_\eta^2 \partial(v_\eta, v_*) + \kappa_\eta^2(\mu \partial(\wp v_*, v_*) + \partial(v_\eta, v_*)) + \iota_\eta^2 \partial(j_\eta, v_*) + \omega_\eta^2(\mu \partial(\wp v_*, v_*) + \partial(j_\eta, v_\eta))$$
$$+ \varepsilon_\eta \partial(\ell_\eta, v_*) + \sigma_\eta(\mu \partial(\wp v_*, v_*) + \partial(\ell_\eta, v_*))$$
$$\leq (\tau_\eta^2 + \kappa_\eta^2)\partial(v_\eta, v_*) + (\iota_\eta^2 + \omega_\eta^2)\partial(j_\eta, v_*) + (\varepsilon_\eta + \sigma_\eta)\partial(\ell_\eta, v_*).$$

Since

$$\partial(j_\eta, v_*) \leq \partial(v_\eta, v_*),$$

we have

$$\partial(v_{n+1}, v_*) \leq (\tau_\eta^2 + \kappa_\eta^2 + \iota_\eta^2 + \omega_\eta^2)\partial(v_\eta, v_*) + (\varepsilon_\eta + \sigma_\eta)\partial(\ell_\eta, v_*).$$

On substituting

$$\partial(\ell_\eta, v_*) = (\tau_\eta^0 + \kappa_\eta^0)\partial(v_\eta, v_*) + (\iota_\eta^0 + \omega_\eta^0)\partial(j_\eta, v_*),$$

we have

$$\partial(v_{n+1}, v_*) \leq (\tau_\eta^2 + \kappa_\eta^2 + \iota_\eta^2 + \omega_\eta^2)\partial(v_\eta, v_*) + ((\varepsilon_\eta + \sigma_\eta) \times (\tau_\eta^0 + \kappa_\eta^0)\partial(v_\eta, v_*) + (\iota_\eta^0 + \omega_\eta^0)\partial(j_\eta, v_*))$$
$$\leq (\tau_\eta^2 + \kappa_\eta^2 + \iota_\eta^2 + \omega_\eta^2)\partial(v_\eta, v_*) + (\varepsilon_\eta + \sigma_\eta)(\tau_\eta^0 + \kappa_\eta^0 + \iota_\eta^0 + \omega_\eta^0)\partial(v_\eta, v_*)$$
$$= (\tau_\eta^2 + \kappa_\eta^2 + \iota_\eta^2 + \omega_\eta^2 + (\varepsilon_\eta + \sigma_\eta)(\tau_\eta^0 + \kappa_\eta^0 + \iota_\eta^0 + \omega_\eta^0))\partial(v_\eta, v_*).$$

Also, it is given that

$$(\tau_\eta^2 + \kappa_\eta^2 + \iota_\eta^2 + \omega_\eta^2 + (\varepsilon_\eta + \sigma_\eta)(\tau_\eta^0 + \kappa_\eta^0 + \iota_\eta^0 + \omega_\eta^0)) \leq 1,$$

we have

$$\partial(v_{n+1}, v_*) \leq \partial(v_\eta, v_*).$$

This implies that $\{\partial(v_\eta, v_*)\}$ is bounded and non-increasing for all $v_* \in f_\wp$. Hence, $\lim_{n \to \infty} \partial(v_\eta, v_*)$ exists, as required. □

Lemma 8. *Let $\mathcal{B}_s$ be a nonempty closed convex subset of complete Busemann space $\mathcal{B}$, and $\wp : \mathcal{B}_s \to \mathcal{B}_s$ be a mapping satisfying condition $(\mathcal{E}_\mu)$. For an arbitrary chosen $v_0 \in \mathcal{B}_s$, let the sequence $\{v_\eta\}$ be generated by a standard three-step iteration algorithm. Then, $f_\wp$ is nonempty if and only if $\{v_\eta\}$ is bounded and $\lim_{n\to\infty} \partial(\wp v_\eta, v_\eta) = 0$ for a unique asymptotic center.*

Proof. Since $f_\wp \neq \emptyset$, let $v_* \in f_\wp$ and $z \in \mathcal{B}_s$. Using Lemma 7, there is an existence of $\lim_{n\to\infty} \partial(v_\eta, v_*)$, which confirms the boundedness of $\{v_\eta\}$. Assuming

$$\lim_{n\to\infty} \partial(v_\eta, v_*) = r,$$

on combining this result with the values of $\partial(\jmath_\eta, v_*)$ and $\partial(\ell_\eta, v_*)$ of Lemma 7

$$\limsup_{n\to\infty} \partial(\ell_\eta, v_*) \leq \limsup_{n\to\infty} \partial(v_\eta, v_*) = r. \quad (2.1)$$

Also,

$$\limsup_{n\to\infty} \partial(\wp v_\eta, v_*) = \limsup_{n\to\infty} (\mu \partial(\wp v_*, v_*) + \partial(v_\eta, v_*))$$
$$\leq \limsup_{n\to\infty} \partial(v_\eta, v_*)$$
$$= r.$$

On the other hand, by using the value of $\partial(\ell_\eta, v_*)$ of Lemma 7, we have

$$\partial(v_{n+1}, v_*) \leq \partial(\tau_\eta^2 v_\eta \oplus \kappa_\eta^2 \wp v_\eta \oplus \iota_\eta^2 \jmath_\eta \oplus \omega_\eta^2 \wp \jmath_\eta \oplus \varepsilon_\eta \ell_\eta \oplus \sigma_\eta \ell_\eta, v_*)$$
$$\leq \tau_\eta^2 \partial(v_\eta, v_*) + \kappa_\eta^2 \partial(\wp v_\eta, v_*) + \iota_\eta^2 \partial(\jmath_\eta, v_*) + \omega_\eta^2 \partial(\wp \jmath_\eta, v_*)$$
$$+ \varepsilon_\eta \partial(\ell_\eta, v_*) + \sigma_\eta \partial(\wp \ell_\eta, v_*)$$
$$\leq \tau_\eta^2 \partial(v_\eta, v_*) + \kappa_\eta^2 (\mu \partial(\wp v_*, v_*) + \partial(v_\eta, v_*)) + \iota_\eta^2 \partial(\jmath_\eta, v_*) + \omega_\eta^2 (\mu \partial(\wp v_*, v_*)$$
$$+ \partial(\jmath_\eta, v_*)) + \varepsilon_\eta \partial(\ell_\eta, v_*) + \sigma_\eta (\mu \partial(\wp v_*, v_*) + \partial(\ell_n, v_*))$$
$$\leq (\tau_\eta^2 + \kappa_\eta^2 + \iota_\eta^2 + \omega_\eta^2) \partial(v_\eta, v_*) + (\varepsilon_\eta + \sigma_\eta) \partial(\ell_\eta, v_*)$$
$$\leq (\tau_\eta^2 + \kappa_\eta^2 + \iota_\eta^2 + \omega_\eta^2) \partial(v_\eta, v_*) + (\varepsilon_\eta + \sigma_\eta) \partial(\ell_\eta, v_*),$$

by the above-mentioned standard three-step iteration algorithm,

$$\partial(v_{n+1}, v_*) \leq (1 - (\varepsilon_\eta + \sigma_\eta)) \partial(v_\eta, v_*) + (\varepsilon_\eta + \sigma_\eta) \partial(\ell_\eta, v_*)$$
$$\leq \partial(v_\eta, v_*) - (\varepsilon_\eta + \sigma_\eta) \partial(v_\eta, v_*) + (\varepsilon_\eta + \sigma_\eta) \partial(\ell_\eta, v_*).$$

This implies that,

$$\partial(v_{n+1}, v_*) \leq \partial(v_\eta, v_*) - (\varepsilon_\eta + \sigma_\eta) \partial(v_\eta, v_*) + (\varepsilon_\eta + \sigma_\eta) \partial(\ell_\eta, v_*).$$
$$\leq \partial(v_\eta, v_*) + (\varepsilon_\eta + \sigma_\eta)(\partial(\ell_\eta, v_*) - \partial(v_\eta, v_*)).$$
$$\partial(v_{n+1}, v_*) - \partial(v_\eta, v_*) \leq (\varepsilon_\eta + \sigma_\eta)(\partial(\ell_\eta, v_*) - \partial(v_\eta, v_*)).$$
$$\frac{\partial(v_{n+1}, v_*) - \partial(v_\eta, v_*)}{(\varepsilon_\eta + \sigma_\eta)} \leq (\partial(\ell_\eta, v_*) - \partial(v_\eta, v_*)).$$

This implies that,

$$\partial(v_{n+1}, v_*) - \partial(v_\eta, v_*) \leq \frac{\partial(v_{n+1}, v_*) - \partial(v_\eta, v_*)}{(\varepsilon_\eta + \sigma_\eta)} \leq (\partial(\ell_\eta, v_*) - \partial(v_\eta, v_*))$$

and hence, we have

$$\partial(v_{n+1}, v_*) \leq \partial(\ell_\eta, v_*).$$

Therefore,
$$r \leq \lim_{n \to \infty} \partial(\ell_\eta, v_*). \tag{2.2}$$

By using Equations (2.1) and (2.2), we have

$$\begin{aligned}
r &= \lim_{n \to \infty} \partial(\ell_\eta, v_*) \\
&= \lim_{n \to \infty} \partial(\tau_\eta^0 v_\eta \oplus \kappa_\eta^0 \wp v_\eta \oplus \iota_\eta^0 \jmath_\eta \oplus \omega_\eta^0 \wp \jmath_\eta, v_*) \\
&= \lim_{n \to \infty} (\partial(\tau_\eta^0 v_\eta + \kappa_\eta^0 \wp v_\eta, v_*) + (\iota_\eta^0 + \omega_\eta^0)\partial(\jmath_\eta, v_*)) \\
&= \lim_{n \to \infty} ((\tau_\eta^0 + \iota_\eta^0 + \omega_\eta^0)\partial(v_\eta, v_*) + \kappa_\eta^0 \partial(\wp v_\eta, v_*)) \\
&= \lim_{n \to \infty} ((\tau_\eta^0 + \iota_\eta^0 + \omega_\eta^0)\partial(v_\eta, v_*) + \kappa_\eta^0 (\mu \partial(\wp v_*, v_*) + \partial(v_\eta, v_*))) \\
&= \lim_{n \to \infty} ((1 - \kappa_\eta^0)\partial(v_\eta, v_*) + \kappa_\eta^0 \partial(v_\eta, v_*)). \\
&= \lim_{n \to \infty} \partial(v_\eta, v_*). \tag{2.3}
\end{aligned}$$

Using Equations (2.1) and (2.3) and the above-mentioned inequalities, we have
$$\limsup_{n \to \infty} \partial(\ell_\eta, v_*) \leq \limsup_{n \to \infty} \partial(v_\eta, v_*) = r,$$

and hence, by Lemma 2, we have
$$\lim_{n \to \infty} \partial(\wp v_\eta, v_\eta) = 0.$$

Conversely, suppose that $\{v_\eta\}$ is bounded and $\lim_{n \to \infty} \partial(\wp v_\eta, v_\eta) = 0$. Then, by Lemma 4 $\{v_\eta\}$ has a subsequence that is regular with respect to $\mathcal{B}_s$. Let $v_{\{\eta_\kappa\}}$ be a subsequence of $\{v_\eta\}$ in such a way that $A(\mathcal{B}_s, \{v_\eta\}) = v$. Hence, we have

$$\limsup_{n \to \infty} \partial(v_\eta, \wp v_*) \leq \limsup_{n \to \infty} (\mu \partial(v_\eta, \wp v_\eta) + \partial(v_\eta, v))$$
$$\leq \limsup_{n \to \infty} \partial(v_\eta, v)$$

As a consequence, the uniqueness of the asymptotic center ensures that v is a fixed point of $\wp$ so this concludes the proof. $\square$

Now, we state and prove our main theorems in this section.

Theorem 2. *Let $\mathcal{B}_s$ be a nonempty closed convex subset of a compete Busemann space $\mathcal{B}$ and $\wp : \mathcal{B}_s \to \mathcal{B}_s$ be a mapping satisfying condition $(\mathcal{E}_\mu)$. For an arbitrary chosen $v_0 \in \mathcal{B}_s$, assume that $\{v_\eta\}$ is a sequence generated by a standard three-step iteration algorithm. Then $f_\wp \neq \emptyset$ and $\{v_\eta\} \Delta$-converges to a fixed point of $\wp$.*

Proof. Since $f_\wp \neq \emptyset$, so by Lemma 8, we have bounded $\{v_\eta\}$ and
$$\lim_{n \to \infty} \partial(\wp v_\eta, v_\eta) = 0.$$

Also, let
$$\omega_\omega\{v_\eta\} := \bigcup A(v_{\eta_\kappa})$$
where the union is taken over all subsequences $\{v_{\eta_\kappa}\}$ of $\{v_\eta\}$. We claim that $\omega_\omega\{v_\eta\} \subset f_\wp$. Considering $v_* \in \omega_\omega\{v_\eta\}$, then there is an existence of subsequence $\{v_{\eta_\kappa}\}$ of $\{v_\eta\}$ in such a way that $A(\{v_\eta\}) = \{v_*\}$. Using Lemmas 3 and 5 there is an existence of subsequence $\{v'_{\eta_\kappa}\}$ of $\{v_{\eta_\kappa}\}$ in such a way that $\Delta - \lim_{\eta \to \infty} \{v'_{\eta_\kappa}\} = v'_* \in \mathcal{B}_s$. Since $\lim_{\eta \to \infty} \partial(v'_{\eta_\kappa}, \wp v'_{\eta_\kappa}) = 0$, then by Lemma 6 $v' \in f_\wp$. We claim that $v_* = v'_*$. In contrast,

since $\wp$ is a mapping satisfying condition $(\mathcal{E})$ and $v_* \in f_\wp$, then by Lemma 7 there is an existence of $\lim_{\eta \to \infty} \partial(v_\eta, v_*)$. Using the uniqueness of asymptotic centers, we have

$$\limsup_{n \to \infty} \partial(v_\eta, v'_*) < \limsup_{n \to \infty} \partial(v'_{\eta_\kappa}, v_*)$$
$$\leq \lim_{n \to \infty} \partial(v_{\eta_\kappa}, v_*)$$
$$< \lim_{n \to \infty} \partial(v'_{\eta_\kappa}, v'_*)$$
$$= \lim_{n \to \infty} \partial(v_\eta, v'_*)$$
$$= \lim_{n \to \infty} \partial(v'_{\eta_\kappa}, v'_*)$$

which is a contradiction. So $v'_{\eta_\kappa} = v_{\eta_\kappa} \in f_\wp$. To prove that $\{v_\eta\}$ Δ-converges to a fixed point of $\wp$, it is sufficient to show that $\omega_\omega\{v_\eta\}$ consists of exactly one point. Considering a subsequence $\{v_{\eta_\kappa}\}$ of $\{v_\eta\}$. By Lemmas 3 and 5 there is existence of subsequence $\{v'_{\eta_\kappa}\}$ of $\{v_{\eta_\kappa}\}$, which is how $\Delta - \lim_{\eta \to \infty}\{v'_{\eta_\kappa}\} = v'_* \in \mathcal{B}_s$. Let $\mathcal{A}(\{v_{\eta_\kappa}\}) = \{v_*\}$ and $\mathcal{A}(\{v_\eta\}) = v_{**}$. We can conclude the explanation by proving that $v_{**} = v_*$. On the contrary, since $\partial(v_\eta, v'_{\eta_\kappa})$ is convergent, then by the uniqueness of the asymptotic centers, we have

$$\limsup_{n \to \infty} \partial(v_\eta, v'_*) < \limsup_{n \to \infty} \partial(v'_{\eta_\kappa}, v_{**})$$
$$\leq \lim_{n \to \infty} \partial(v_\eta, v_{**})$$
$$< \lim_{n \to \infty} \partial(v_\eta, v'_*)$$
$$= \lim_{n \to \infty} \partial(v'_{\eta_\kappa}, v'_*)$$

which is a contradiction. Hence, $f_\wp \neq \emptyset$ and $\{v_\eta\}$ Δ-converges to a fixed point of $\wp$. □

Theorem 3. *Let $\mathcal{B}_s$ be a nonempty closed convex and complete Busemann space $\mathcal{B}$, and $\wp : \mathcal{B}_s \to \mathcal{B}_s$ be a mapping verification condition $(\mathcal{E}_\mu)$. For an arbitrary chosen $v_0 \in \mathcal{B}_s$, assume that $\{v_\eta\}$ is a sequence generated by a standard three-step iteration algorithm. Then $\{v_\eta\}$ converges strongly to a fixed point of $\wp$.*

Proof. By Lemmas 7, 8 and Theorem 2, we have $f_\wp \neq \emptyset$ so by Lemma 8 $\{v_\eta\}$ is bounded and Δ-converges to $v \in f_\wp$. Suppose on the contrary that $\{v_\eta\}$ does not converge strongly to v. By the compactness assumption, passing to subsequences if necessary, we may assume that there exists $v' \in \mathcal{B}_s$ with $v' \neq v$ such that $\{v_\eta\}$ converges strongly to v'. Therefore,

$$\lim_{n \to \infty} \partial(\wp v_\eta, v') = 0 \leq \lim_{n \to \infty} \partial(\wp v_\eta, v).$$

Since v is the unique asymptotic center of $\{v_\eta\}$, it follows that $v' = v$, which is a contradiction. Hence, $\{v_\eta\}$ converges strongly to a fixed point of $\wp$. □

4. Conclusions

The extension of the linear version of fixed point results to nonlinear domains has its own significance. To achieve the objective of replacing a linear domain with a nonlinear one, Takahashi [40] introduced the notion of a convex metric space and studied fixed point results of nonexpansive mappings in this direction. Since the standard three-step iteration scheme unifies various existing iteration schemes for different values of $\varepsilon_\eta, \sigma_\eta, \tau^i_\eta, \kappa^i_\eta, \omega^i_\eta$, and ι^i_η for $i = 0, 1, 2$, existing results of the standard three-step iteration scheme including strong and $\Delta - convergence$ results in the setting of Busemann spaces satisfying condition $\mathcal{E}$ are generalized.

Author Contributions: All authors have equally contributed to this work. All authors have read and agreed to the published version of the manuscript.

Funding: This work received no external funding.

Acknowledgments: The authors are very grateful to the anonymous reviewers for their constructive comments leading to the substantial improvement of the paper.

Conflicts of Interest: The authors declare no conflict of interest.

References

1. Babu, G.V.R.; Prasad, K. Mann iteration converges faster than Ishikawa iteration for the class of zamfirescu operators. *Fixed Point Theory Appl.* **2006**, *2006*, 49615. [CrossRef]
2. Busemann, H. Spaces with nonpositive curvature. *Acta Math.* **1948**, *80*, 259–310. [CrossRef]
3. Sosov, E.N. On analogues of weak convergence in a special metric space. *Izvestiya Vysshikh Uchebnykh Zavedenii. Matematika* **2004**, *5*, 84–89.
4. Lawaong, W.; Panyanak, B. Approximating fixed points of nonexpansive nonself-mappings in CAT(0) spaces. *Fixed Point Theory Appl.* **2010**, *2010*, 367274. [CrossRef]
5. Dhompongsa, S.; Inthakon, W.; Kaewkhao, A. Edelstien's method and Fixed point theorems for some generalized nonexpansive mappings. *J. Math. Anal. Appl.* **2009**, *350*, 12–17. [CrossRef]
6. Garcia-Falset, J.; Llorens-Fuster, E.; Suzuki, T. Fixed point theory for a class of generalized nonexpansive mappings. *J. Math. Anal. Appl.* **2011**, *375*, 185–195. [CrossRef]
7. Đukić, D.; Paunović, L.; Radenović, S. (). Convergence of iterates with errors of uniformly quasi-Lipschitzian mappings in cone metric spaces. *Kragujev. J. Math.* **2011**, *35*, 399–410.
8. Todorčević, V. *Harmonic Quasiconformal Mappings and Hyperbolic Type Metrics*; Springer Nature: Cham, Switzerland, 2019.
9. Xu, H.-K.; Altwaijry, N.; Chebbi, S. Strong convergence of mann's iteration process in Banach Spaces. *Mathematics* **2020**, *8*, 954. [CrossRef]
10. Ivanov, S.I. General local convergence theorems about the Picard iteration in arbitrary normed fields with applications to Super-Halley method for multiple polynomial zeros. *Mathematics* **2020**, *8*, 1599. [CrossRef]
11. Berinde, V. Iterative Approximation of Fixed Points. In *Lecture Notes in Mathematics*; Springer: Berlin/Heidelberg, Germany, 2007.
12. Ansari, Q.H.; Islam, M.; Yao, J.C. Nonsmooth variational inequalities on Hadamard manifolds. *Appl. Anal.* **2020**, *99*, 340–358. [CrossRef]
13. Hammad, H.A.; ur Rehman, H.; De la Sen, M. Advanced Algorithms and common solutions to variational inequalities. *Symmetry* **2020**, *12*, 1198. [CrossRef]
14. Dang, Y.; Sun, J.; Xu, H. Inertial accelerated algorithms for solving a split feasibility problem. *J. Ind. Manag. Optim.* **2017**, *13*, 1383–1394. [CrossRef]
15. Hammad, H.A.; ur Rehman, H.; De la Sen, M. Shrinking projection methods for accelerating relaxed inertial Tseng-type algorithm with applications. *Math. Probl. Eng.* **2020**, *2020*, 7487383. [CrossRef]
16. Tuyen, T.M.; Hammad, H.A. Effect of Shrinking Projection and CQ-Methods on Two Inertial Forward-Backward Algorithms for Solving Variational Inclusion Problems. In *Rendiconti del Circolo Matematico di Palermo Series 2*; Available online: https://doi.org/10.1007/s12215-020-00581-8 (accessed on 27 February 2021).
17. Papadopoulos, A. Metric Spaces, Convexity and Nonpositive Curvature. In *IRMA Lectures in Mathematics and Theoretical Physics*; European Mathematical Society: Strasbourg, France, 2005.
18. Sharma, N.; Mishra, L.N.; Mishra, V.N.; Almusawa, H. End point approximation of standard three-step multivalued iteration algorithm for nonexpansive mappings. *Appl. Math. Inf. Sci.* **2021**, *15*, 73–81.
19. Khamsi, M.A.; Kirk, W.A. *An Introduction to Metric Spaces and Fixed Point Theory*; John Wiley: New York, NY, USA, 2011.
20. Bridson, M.; Haefliger, A. *Metric Spaces of Non-Positive Curvature*; Springer: Berlin, Germany, 1999.
21. Kirk, W.A.; Panyanak, B. A concept of convergence in geodesic spaces. *Nonlinear Anal.* **2008**, *68*, 3689–3696. [CrossRef]
22. Kirk, W.A. Geodesic geometry and fixed point theory. In *Seminar of Mathematical Analysis*; Colecc. Abierta: Malaga, Spain, 2003.
23. Foertsch, T.; Lytchak, A.; Schroeder, V. Non-positive curvature and the Ptolemy inequality. *Int. Math. Res. Not.* **2007**, *2007*, rnm100.
24. Kirk, W.A. Geodesic geometry and fixed point theory II. In *International Conference on Fixed Point Theory and Applications*; Yokohama Publ.: Yokohama, Japan, 2004; pp. 113–142.
25. Bagherboum, M. Approximating fixed points of mappings satisfying condition (E) in Busemann space. *Numer. Algorithms* **2016**, *71*, 25–39. [CrossRef]
26. Kohlenbach, U. Some Logical theorems with applications in functional analysis. *Tran. Am. Math. Soc.* **2005**, *357*, 89–128. [CrossRef]
27. Noor, M.A. New approximation schemes for general variational inequalities. *J. Math. Anal. Appl.* **2000**, *251*, 217–229. [CrossRef]
28. Phuengrattana, W.; Suantai, S. On the rate of convergence of Mann, Ishikawa, Noor and SP-iterations for continuous functions on an arbitrary interval. *J. Comput. Appl. Math.* **2011**, *235*, 3006–3014. [CrossRef]

29. Gursoy, F.; Karakaya, V. A Picard-S hybrid type iteration method for solving a differential equation with retarded argument. *arXiv* **2014**, arXiv:1403.25-46.
30. Chugh, R.; Kumar, V.; Kumar, S. Strong Convergence of a new three step iterative scheme in Banach spaces. *Am. J. Comput. Math.* **2012**, *2*, 345–357. [CrossRef]
31. Abbas, M.; Nazir, T. A new faster iteration process applied to constrained minimization and feasibility problems. *Mat. Vesn.* **2014**, *66*, 223–234.
32. Sainuan, P. Rate of convergence of P-iteration and S-iteration for continuous functions on closed intervals. *Thai J. Math.* **2015**, *13*, 449–457.
33. Daengsaen, J.; Khemphet, A. On the Rate of Convergence of P-Iteration, SP-Iteration, and D-Iteration Methods for Continuous Non-decreasing Functions on Closed Intervals. *Abstr. Appl. Anal.* **2018**, *2018*, 7345401. [CrossRef]
34. Mann, W.R. Mean value methods in iteration. *Proc. Am. Math. Soc.* **1953**, *4*, 506–510. [CrossRef]
35. Ishikawa, S. Fixed points by a new iteration method. *Proc. Am. Math. Soc.* **1974**, *44*, 147–150. [CrossRef]
36. Goebel, K.; Reich, S. *Uniform Convexity, Hyperbolic Geometry, and Nonexpansive Mappings*; Marcel Dekker: New York, NY, USA; Basel, Switzerland, 1984.
37. Dhompongsa, S.; Kirk, W.A.; Sims, B. Fixed points of uniformly lipschitzian mappings. *Nonlinear Anal.* **2006**, *65*, 762–772. [CrossRef]
38. Kirk, W.; Shahzad, N. *Fixed Point Theory in Distance Spaces*; Springer: Cham, Switzerland; Heidelberg, Germany; New York, NY, USA; Dordrecht, The Netherlands; London, UK, 2014.
39. Ullah, K.; Khan, B.A.; Ozer, O.; Nisar, Z. Some Convergence Results Using K_* Iteration Process in Busemann Spaces. *Malays. J. Math. Sci.* **2019**, *13*, 231–249.
40. Takahashi, T. A convexity in metric spaces and nonexpansive mappings. *Kodai Math. Semin. Rep.* **1970**, *22*, 142–149. [CrossRef]

Article

Note on Common Fixed Point Theorems in Convex Metric Spaces

Anil Kumar [1,†] and Aysegul Tas [2,*,†]

1. Department of Mathematics, Government College for Women Sampla, Rohtak 124501, Haryana, India; anilkshk84@gmail.com
2. Department of Management, Cankaya University, 06790 Ankara, Turkey
* Correspondence: aysegul@cankaya.edu.tr
† These authors contributed equally to this work.

Abstract: In the present paper, we pointed out that there is a gap in the proof of the main result of Rouzkard et al. (The Bulletin of the Belgian Mathematical Society 2012). Then after, utilizing the concept of (E.A.) property in convex metric space, we obtained an alternative and correct version of this result. Finally, it is clarified that in the theory of common fixed point, the notion of (E.A.) property in the set up of convex metric space develops some new dimensions in comparison to the hypothesis that a range set of one map is contained in the range set of another map.

Keywords: compatible maps; common fixed points; convex metric spaces; q-starshaped

1. Introduction

A point on which a self-map remains invariant is called a fixed point for that map. Fixed point theory plays an important role in solving different kinds of problems of nonlinear analysis and so it has applications in engineering, medical science, physical science, computer science, etc. In 1922, Banach [1] proved that a contraction map in a complete metric space has a fixed point and this result is known as the Banach contraction principle. Due to the simplicity and usefulness of this result, fixed point theory became a more aggressive area of research. Many researchers so far have worked in this field, extending this contraction principle in several possible ways [2–4].

In 1976, Jungck [5] extended the Banach contraction principle for the pair of commuting self-maps by ensuring the existence of a common fixed point for this pair. Sessa [6] relaxed the condition of commutativity and introduced the class of weak commuting maps. Again, Jungck [7] gave the weaker version of the commutativity condition by introducing the class of compatible maps and proved that weak commuting maps are compatible but the converse is not true in general. After that, many authors obtained more comprehensive common fixed point theorems under some given hypothesis [8,9].

On the other side, Takahashi [10] defined the notion of a convex structure in a metric space and called such a space a convex metric space. Further, he studied several properties of this space and ensure the existence of a fixed point for nonexpansive maps in the setup of convex metric space. In the last forty years, many fixed point and common fixed point theorems in the context of convex metric space have been established; for example, see [11–15].

2. Preliminaries

In the present section we recall some standard notations, basic definitions and auxiliary results, which are required in the sequel.

In 2014, inspired by the idea of Aamri and Moutawakil [16], the concept of (E.A.) property in the context of convex metric space was introduced by Kumar and Rathee [17].

In the present article, we shall show that there is a gap in the proof of the Theorem 1, which is one of the main results of Rouzkard et al. [18]. Then, we obtain a correct version of Theorem 1 by utilizing the concept of (E.A.) property in a convex metric space defined by Kumar and Rathee [17].

Finally, we clarify the importance of the notion (E.A.) property in a convex metric space in comparison to the hypothesis that the range set of one map is contained in the range set of another map.

Definition 1. *[10] Let (S, ρ) be a metric space. A continuous mapping $W : S \times S \times [0,1] \to S$ is called a convex structure on S, if for all $x, y \in S$ and $\lambda \in [0,1]$, we have*

$$d(u, W(x, y, \lambda)) \leq \lambda d(u, x) + (1 - \lambda) d(u, y)$$

for all $u \in S$. A metric space (S, ρ) equipped with a convex structure is called a convex metric space.

Let M be a subset of a convex metric space (S, ρ). The set M is said to be

(i) convex if $W(x, y, \lambda) \in M$ for all $x, y \in M$ and $\lambda \in [0,1]$;
(ii) q-starshaped if there exists $q \in M$ such that $W(x, q, \lambda) \in M$ for all $x \in M$ and $\lambda \in [0,1]$.

In addition, the map $I : M \to M$ is said to be

(i) affine if M is convex and $I(W(x, y, \lambda)) = W(Ix, Iy, \lambda)$ for all $x, y \in M$ and $\lambda \in [0,1]$;
(ii) q-affine if M is q-starshaped and $I(W(x, q, \lambda)) = W(Ix, q, \lambda)$ for all $x \in M$ and $\lambda \in [0,1]$.

Clearly, each convex set M is q-starshaped for any $q \in M$ but the converse assertion is not necessarily true (see Example 7 of [19]).

Definition 2. *[10] A convex metric space (S, ρ) is said to satisfy the Property (I), if for all $x, y, z \in S$ and $\lambda \in [0,1]$, we have $\rho(W(x, z, \lambda), W(y, z, \lambda)) \leq \lambda \rho(x, y)$.*

Notice that Property (I) is always satisfied in a normed linear space and each of its convex subsets.

Definition 3. *[19] Let $T, I : S \to S$ be mappings on a metric space (S, ρ). The pair (T, I) is said to be compatible if*

$$\rho(TIx_n, ITx_n) \to 0$$

whenever $\{x_n\}$ is a sequence in S such that

$$Tx_n, Ix_n \to t \in S$$

Definition 4. *[16] Let $T, I : S \to S$ be mappings on a metric space (S, ρ). The pair (T, I) is said to satisfy (E.A.) property if there is a sequence $\{x_n\} \in S$ such that*

$$Tx_n, Ix_n \to t \in S$$

Definition 5. *Let (S, ρ) be a metric space and $T, I : S \to S$. Then the pair (T, I) is said to be reciprocally continuous if*

$$\lim_{n \to +\infty} TIx_n = Tt \quad \text{and} \quad \lim_{n \to +\infty} ITx_n = It$$

whenever $\{x_n\}$ is a sequence in S such that $\lim_{n \to +\infty} Tx_n = \lim_{n \to +\infty} Ix_n = t$ for some $t \in X$.

It is easy to see that if T and I are continuous, then the pair (T, I) is reciprocally continuous but the converse is not true in general (see Example 2.3 of [20]).

Moreover, in the setting of common fixed point theorems for compatible pairs of self-mappings satisfying some contractive conditions, continuity of one of the mappings implies their reciprocal continuity.

Definition 6. *A pair (T, I) of self-maps of a metric space (S, ρ) is said to be sub-compatible if there exists a sequence $\{x_n\}$ such that*

$$\lim_{n \to +\infty} Tx_n = \lim_{n \to +\infty} Ix_n = t \text{ for some } t \in X \text{ and } \lim_{n \to +\infty} \rho(TIx_n, ITx_n) = 0.$$

Recently, Rouzkard et al. [18] proved the following common fixed point theorem for the pair of compatible maps in a convex metric space.

Theorem 1. *Let C be a nonempty closed convex subset of a convex metric space (X, ρ) satisfying the Property (I). Denote $[x, q] = \{W(x, q, k) : 0 \leq k \leq 1\}$ where W is a convex structure on the metric space.*

If T and I are compatible self-maps defined on C such that $I(C) = C$, I is q-affine and nonexpansive, which satisfy the inequality

$$\rho(Tx, Ty) \leq \rho(Ix, Iy) + \frac{(1-k)}{k} \max\{\rho(Ix, [Tx, q]), \rho(Iy, [Ty, q])\} \tag{1}$$

for all $x, y \in C$, where $1/2 < k < 1$, then T and I have a common fixed point provided $cl(T(C))$ is compact and T is continuous.

3. Results

3.1. Compatibility in Proof of Theorem 1.

Let us recall the lines of the proof given in Rouzkard et al. [18]. First of all, for each $n \in \mathbb{N}$, the authors define $T_n : C \to C$ by

$$T_n x = W(Tx, q, k_n) \text{ for all } x \in C, \tag{2}$$

where k_n is a sequence in $(\frac{1}{2}, 1)$ such that $k_n \to 1$. Afterward, to accomplish the compatibility of the maps T_n and I for each $n \in \mathbb{N}$, the authors choose an arbitrary sequence $\{x_m\}$ in C such that

$$\lim_{m \to +\infty} Ix_m = \lim_{m \to +\infty} T_n x_m = t \in C \tag{3}$$

Using the definition of T_n, it has been written that

$$\begin{aligned}
\rho(Tx_m, T_n x_m) &= \rho(Tx_m, W(Tx_m, q, k_n)) \\
&\leq k_n \rho(Tx_m, Tx_m) + (1 - k_n)\rho(Tx_m, q) \\
&= (1 - k_n)\rho(Tx_m, q).
\end{aligned}$$

Then by taking $m \to +\infty$ and using (3), the authors get

$$\rho(\lim_{m \to +\infty} Tx_m, t) \leq (1 - k_n)\rho(\lim_{m \to +\infty} Tx_m, q). \tag{4}$$

Again, on making $n \to +\infty$ in (4), the authors wrote the following (see [18], page 323, line 20–21)

$$\rho(\lim_{m \to +\infty} Tx_m, t) \leq 0. \tag{5}$$

Then, by using this expression, the authors claim the compatibility of the maps T_n and I for each $n \in \mathbb{N}$.

Here, it is pertinent to mention that the compatibility of the maps T_n and I is to be shown for each $n \in \mathbb{N}$ and so the compatibility of T_n and I is to be shown for arbitrarily fixed natural number n. If n is fixed, then it is superfluous to approach $n \to +\infty$, therefore (5) is not valid because this is obtained by taking $n \to \infty$ in (4). So the compatibility of the maps T_n and I for each $n \in \mathbb{N}$ proved by this way is totally wrong. The same mistake occurred when the authors tried to prove the reciprocal continuity of T_n and I for each $n \in N$ (see [18], page 324, line 3–15).

3.2. Modified Version of Theorem 1

The following definition given by Kumar and Rathee [17] is required to prove the modified version of Theorem 1.

Definition 7. Let M be a q-starshaped subset of a convex metric space (S, ρ) and let $T, I : M \to M$ with $q \in F(I)$. The pair (T, I) is said to satisfy (E.A.) property with respect to q if there exists a sequence $\{x_n\}$ in M such that for all $\lambda \in [0, 1]$.

$$\lim_{n \to +\infty} Ix_n = \lim_{n \to +\infty} T_\lambda x_n = t \text{ for some } t \in M, \tag{6}$$

where $T_\lambda x = W(Tx, q, \lambda)$.

The following lemma is a direct consequence of Theorem 3.2 of Rouzkard et al. [18].

Lemma 1. Let T and I be self-maps of a metric space (S, ρ). If the pair (T, I) is sub-compatible, reciprocally continuous and satisfies the inequality

$$\rho(Tx, Ty) \le a\, \rho(Ix, Iy) + (1-a) \max\{\rho(Ix, Tx), \rho(Iy, Ty)\} \tag{7}$$

for all $x, y \in X$, where $0 < \alpha < 1$. Then T and I have a unique common fixed point in X.

Now we modify Theorem 1 by replacing the condition $I(M) = M \supseteq T(M)$ with the assumption that the pair (T, I) satisfies (E.A.) property with respect to some $q \in M$.

Theorem 2. Let M be a nonempty q-starshaped subset of a convex metric space (X, ρ) with Property (I) and let T and I be continuous self-maps of M such that the pair (T, I) satisfies (E.A.) property with respect to q. Assume that I is q-affine, $cl(T(M))$ is compact. If T and I are compatible and satisfy the inequality

$$\rho(Tx, Ty) \le \rho(Ix, Iy) + \frac{1-k}{k} \max\{\rho(Ix, [Tx, q]), \rho(Iy, [Ty, q])\} \tag{8}$$

for all $x, y \in M$, where $\frac{1}{2} < k < 1$, then T and I have a common fixed point in M.

Proof. For each $n \in \mathbb{N}$, we define $T_n : M \to M$ by

$$T_n(x) = W(Tx, q, k_n) \text{ for all } x \in M, \tag{9}$$

where k_n is a sequence in $(\frac{1}{2}, 1)$ such that $k_n \to 1$.

Now, we have to show that for each $n \in \mathbb{N}$, the pair (T_n, I) is sub-compatible. Since T and I satisfy (E.A.) property with respect to q, there exists a sequence $\{x_m\}$ in M such that for all $\lambda \in [0, 1]$

$$\lim_{m \to +\infty} Ix_m = \lim_{m \to +\infty} T_\lambda x_m = t \in M, \tag{10}$$

where $T_\lambda x_m = W(Tx_m, q, \lambda)$.

Since $k_n \in (0,1)$, in light of (9) and (10), for each $n \in N$, we have

$$\lim_{m \to +\infty} T_n x_m = \lim_{m \to +\infty} W(Tx_m, q, k_n)$$
$$= \lim_{m \to \infty} T_{k_n} x_m = t \in M.$$

Thus, we have

$$\lim_{m \to +\infty} Ix_m = \lim_{m \to +\infty} T_n x_m = t \in M. \tag{11}$$

Now using the fact that I is q-affine and Property (I) is satisfied, we get

$$\rho(T_n Ix_m, IT_n x_m) = \rho(W(TIx_m, q, k_n), I(W(Tx_m, q, k_n)))$$
$$= \rho(W(TIx_m, q, k_n), W(ITx_m, q, k_n))$$
$$\leq k_n \rho(TIx_m, ITx_m). \tag{12}$$

Since (T, I) satisfies (E.A.) property with T and I are compatible, in view of (10) we have

$$\lim_{m \to +\infty} \rho(TIx_m, ITx_m) = 0.$$

Now, letting $m \to \infty$ in (12), we obtain

$$\lim_{m \to +\infty} \rho(T_n Ix_m, IT_n x_m) = 0. \tag{13}$$

Hence, on account of (11) and (13), it follows that the pair (T_n, I) is sub-compatible for each $n \in \mathbb{N}$. Since T and I are continuous, for each $n \in \mathbb{N}$, the pair (T_n, I) is reciprocally continuous. Furthermore, by (8),

$$\rho(T_n x, T_n y) = \rho(W(Tx, q, k_n), W(Ty, q, k_n))$$
$$\leq k_n \rho(Tx, Ty)$$
$$\leq k_n[\rho(Ix, Iy) + \frac{1-k_n}{k_n} \max\{dist(Ix, [Tx, q]), dist(Iy, [Ty, q])\}]$$
$$\leq k_n \rho(Ix, Iy) + (1-k_n) \max\{\rho(Ix, T_n x), \rho(Iy, T_n y)\} \tag{14}$$

for each $x, y \in M$ and $\frac{1}{2} < k_n < 1$. By Lemma 1, for each $n \in \mathbb{N}$, there exists $x_n \in M$ such that $x_n = Ix_n = T_n x_n$.

Now the compactness of $cl(T(M))$ implies that there exists a sub-sequence $\{Tx_m\}$ of $\{Tx_n\}$ such that $Tx_m \to z$ as $m \to +\infty$. Further, it follows that

$$x_m = T_m x_m = W(Tx_m, q, k_m) \to z \text{ as } m \to +\infty.$$

Then, by the continuity of T and I, we obtain $Iz = z = Tz$ and so z is a common fixed point of T and I. □

The following remark clarifies that in the context of a convex metric space, the notion of (E.A.) property introduced by Kumar and Rathee [17] for proving the common fixed point theorems has importance in comparison to the hypothesis that a range set of one map is contained in the range set of another map.

Remark 1.

(a) In 2011, Haghi et al. [21] showed that several common fixed point generalizations in the theory of fixed point are not a real generalization because they can be obtained from the corresponding fixed point theorems. After the critical analysis of this paper, we reached the conclusion that the claim of Haghi et al. [21] is true only in the case if we make the assumption that the range set of one map is contained in the range set of another map.

So, keeping this in view, we replaced the condition $I(M) = M \supseteq T(M)$ of Theorem 1 with the assumption that the pair (T, I) satisfies (E.A.) property with respect to some $q \in M$ and due to this we have been able to obtain the modified and correct version of Theorem 1 in the form of Theorem 2.

(b) (see Example 17 of [16]) Let $S = \mathbb{R}$ with usual metric and $M = [0, 1]$. Define $T, I : M \to M$ by

$$T(x) = \begin{cases} \frac{1}{2} & \text{if } 0 \leq x \leq \frac{1}{2} \\ \frac{x}{2} + \frac{1}{4} & \text{if } \frac{1}{2} \leq x \leq 1. \end{cases} \quad \text{and} \quad I(x) = \begin{cases} \frac{1}{2} & \text{if } 0 \leq x \leq \frac{1}{2} \\ 1 - x & \text{if } \frac{1}{2} \leq x \leq 1. \end{cases}$$

Then (S, ρ) is a convex metric space with $W(x, y, \lambda) = \lambda x + (1 - \lambda)y$. It is easy to verify that the pair (T, I) satisfies (E.A.) property with respect to $q = \frac{1}{2}$, but the pair violates the condition that the range set of one map is contained in the range set of another map since $T(M) = [\frac{1}{2}, \frac{3}{4}] \nsubseteq [0, \frac{1}{2}] = I(M)$ and $I(M) = [0, \frac{1}{2}] \nsubseteq [\frac{1}{2}, \frac{3}{4}] = T(M)$.

In this way, we can say that there are certain pairs of self-maps, namely T and I, defined on a set (say M), which satisfies (E.A.) property in the set up of convex metric space but violates the condition $T(M) \subseteq I(M)$. Thus, the common fixed point theorems in which the pair of maps satisfy (E.A.) property with some other hypotheses will ensure the existence of a common fixed point for such maps.

Remark 2. *As an application of Theorem 1, the authors in [18] obtained two more theorems (see Theorems 4.1 and 4.2 of [18]). Since we have quoted a gap in the proof of Theorem 1, Theorems 4.1 and 4.2 of [18] are no longer valid. Thus, these theorems can also be modified by using the notion of (E.A.) property in the set up of a convex metric space.*

4. Conclusions

In this work, a gap in the proof of the main result of Rouzkard et al. (The Bulletin of the Belgian Mathematical Society 2012) is detected. Then after, utilizing the concept of (E.A.) property in convex metric space, we obtained an alternative and correct version of this result.

In the set up of a convex metric space, the notion of (E.A.) property introduced by Kumar and Rathee [17] for proving the common fixed point theorems is more important than the hypothesis that a range set of one map is contained in the range set of another map and it develops some new extensions.

Author Contributions: Authors contributed equally to this work. All authors have read and agreed to the published version of the manuscript.

Funding: This research received no external funding.

Institutional Review Board Statement: Not applicable.

Informed Consent Statement: Not applicable.

Data Availability Statement: This research did not report any data.

Acknowledgments: Thanks to Nasser Shahzad, King Abdulaziz University, Saudi Arabia, for providing us the preprint copy of Haghi et al. [21].

Conflicts of Interest: The authors declare no conflict of interest.

References

1. Banach, S. Sur les operations dans les ensembles abstraits et leur application aux equations integrals. *Fundam. Math.* **1922**, *3*, 133–181. [CrossRef]
2. Aleksic, S.; Kadelburg, Z.; Mitrovic, Z.D.; Radenovic, S. A new survey: Cone metric spaces. *J. Int. Math. Virtual Inst.* **2019**, *9*, 93–121.

3. Paunovic, L.J. *Teorija Apstraktnih Metrickih Prostora-Neki Novi Rezultati*; University of Priština: Leposavic, Serbia, 2017.
4. Ciric, L. *Some Recent Results in Metrical fixed Point Theory*; University of Belgrade: Beograd, Serbia, 2003.
5. Jungck, G. Commuting mappings and fixed points. *Am. Math. Mon.* **1976**, *83*, 261–263. [CrossRef]
6. Sessa, S. *On a Weak Commutativity Condition of Mappings in Fixed Point Consideration*; Publications de l'Institute, Mathématique: Belgrade, Serbia, 1982; Volume 32, pp. 149–153.
7. Jungck, G. Compatible mappings and common fixed points. *Int. J. Math. Math. Sci.* **1986**, *9*, 771–779. [CrossRef]
8. Mustafa, Z.; Aydi, H.; Karapınar, E. On common fixed points in G-metric spaces using (EA) property. *Comput. Math. Appl.* **2012**, *64*, 1944–1956. [CrossRef]
9. Karapınar, E.; Shahi, P.; Tas, K. Generalized $\alpha - \psi$ -contractive type mappings of integral type and related fixed point theorems. *J. Inequalities Appl.* **2014**, *160*, 1–18. [CrossRef]
10. Takahashi, W.A. A convexity in metric spaces and nonexpansive mapping. *Kōdai Math. Semin. Rep.* **1970**, *22*, 142–149. [CrossRef]
11. Guay, M.D.; Singh, K.L.; Whitfield, J.H.M. *Fixed Point Theorems for Nonexpansive Mappings in Convex Metric Spaces*; Lecture Notes in Pure and Applied Mathematics; Dekker: New York, NY, USA, 1982; Volume 80, pp. 179–189.
12. Huang, N.J.; Li, H.X. Fixed point theorems of compatible mappings in convex metric spaces. *Soochow J. Math.* **1996**, *22*, 439–447.
13. Fu, J.Y.; Huang, N.J. Common fixed point theorems for weakly commuting mappings in convex metric spaces. *J. Jiangxi Univ.* **1991**, *3*, 39–43.
14. Beg, I.; Shahzad, N.; Iqbal, M. Fixed point theorems and best approximation in convex metric space. *Approx. Theory Its Appl.* **1992**, *8*, 97–105.
15. Ding, D.P. Iteration processes for nonlinear mappings in convex metric spaces. *J. Math. Anal. Appl.* **1998**, *132*, 114–122. [CrossRef]
16. Aamri, M.; Moutawakil, D.E. Some new common fixed point theorems under strict contractive conditions. *J. Math. Anal. Appl.* **2002**, *270*, 181–188. [CrossRef]
17. Kumar, A.; Rathee, S. Some common fixed point and invariant approximation results for nonexpansive mappings in convex metric space. *Fixed Point Theory Appl.* **2014**, *2014*, 182. [CrossRef]
18. Rouzkard, F.; Imdad, M.; Nashine, H.K. New common fixed point theorems and invariant approximation in convex metric space. *Bull. Belg. Math. Soc. Simon Stevin* **2012**, *19*, 311–328. [CrossRef]
19. Rathee, S.; Kumar, A. Some common fixed point results for modified subcompatible maps and related invariant approximation results. *Abstr. Appl. Anal.* **2014**, *24*, 505067. [CrossRef]
20. Imdad, M.; Ali, J.; Tanveer, M. Remarks on some recent metrical common fixed point theorems. *Appl. Math. Lett.* **2011**, *24*, 1165–1169. [CrossRef]
21. Haghi, R.H.; Rezapour, S.; Shahzad, N. Some fixed point generalizations are not real generalizations. *Nonlinear Anal. Theory Method Appl.* **2011**, *74*, 1799–1803. [CrossRef]

Article

Fixed-Point Study of Generalized Rational Type Multivalued Contractive Mappings on Metric Spaces with a Graph

Binayak S. Choudhury [1], Nikhilesh Metiya [2], Debashis Khatua [1] and Manuel de la Sen [3,*]

[1] Department of Mathematics, Indian Institute of Engineering Science and Technology, Shibpur, Howrah 711103, West Bengal, India; binayak12@yahoo.co.in (B.S.C.); debashiskhatua@yahoo.com (D.K.)

[2] Department of Mathematics, Sovarani Memorial College, Jagatballavpur, Howrah 711408, West Bengal, India; metiya.nikhilesh@gmail.com

[3] Institute of Research and Development of Processes, Faculty of Science and Technology, Campus of Leioa, University of the Basque Country, Bizkaia, 48940 Leioa, Spain

* Correspondence: manuel.delasen@ehu.eus

Abstract: The main result of this paper is a fixed-point theorem for multivalued contractions obtained through an inequality with rational terms. The contraction is an F-type contraction. The results are obtained in a metric space endowed with a graph. The main theorem is supported by illustrative examples. Several results as special cases are obtained by specific choices of the control functions involved in the inequality. The study is broadly in the domain of setvalued analysis. The methodology of the paper is a blending of both graph theoretic and analytic methods.

Keywords: fixed-point; multivalued maps; F-contraction; directed graph; metric space

MSC: 47H10; 54H10; 54H25

1. Introduction and Mathematical Preliminaries

Let (X, d) be a metric space. The following standard notations and definitions will be used. $N(X)$ is the family of all nonempty subsets of X, $B(X)$ is the family of all nonempty bounded subsets of X, $CB(X)$ is the family of all nonempty closed and bounded subsets of X, $K(X)$ is the family of all nonempty compact subsets of X and

$$D(x, B) = \inf\{d(x, y) : y \in B\}, \text{ where } x \in X \text{ and } B \in B(X),$$

$$H(A, B) = \max\{\sup_{x \in A} D(x, B), \sup_{y \in B} D(y, A)\}, \text{ where } A, B \in CB(X).$$

H is known as the Hausdorff metric induced by the metric d on $CB(X)$ [1]. Furthermore, if (X, d) is complete then $(CB(X), H)$ is also complete.

Let X be a nonempty set and $\mho = \{(x,x) : x \in X\}$. Consider a directed graph G such that the set $V(G)$ of its vertices coincides with X and the set $E(G)$ of its edges contains all loops, i.e., $\mho \subseteq E(G)$. Assume that G has no parallel edges. By G^{-1} we denote the graph obtained from G by reversing the directions of the edges. Thus, $V(G^{-1}) = V(G)$ and $E(G^{-1}) = \{(x,y) \in X \times X : (y,x) \in E(G)\}$. By $\widetilde{G}$ we denote the undirected graph obtained from G by ignoring the direction of edges. Actually, it will be more convenient for us to treat $\widetilde{G}$ as a directed graph for which the $V(\widetilde{G}) = V(G)$ and $E(\widetilde{G}) = E(G) \cup E(G^{-1})$. A nonempty set X is said to be endowed with a directed graph $G(V,E)$ if $V(G) = X$ and $\mho \subseteq E(G)$.

Let $F : (0, \infty) \to \mathbb{R}$ be a function with the following properties:

(F1) F is strictly increasing, i.e., $x < y \implies F(x) < F(y)$;

(F2) For each sequence $\{\alpha_n\}_{n=1}^{\infty}$ in $(0, \infty)$, $\lim_{n \to \infty} \alpha_n = 0$ if and only if $\lim_{n \to \infty} F(\alpha_n) = -\infty$;

(F3) There exists $k \in (0,1)$ such that $\lim_{\alpha \to 0^+} \alpha^k F(\alpha) = 0$;

(F4) $F(\inf A) = \inf F(A)$ for all $A \subset (0, \infty)$ with $\inf A > 0$.

We denote the set of all functions F satisfying (F1 − F3) by $\Im$ and the set of all functions F satisfying (F1 − F4) by $\Im_*$.

Wardowski [2] introduced the notion of F-contraction and established a new type of generalization of the Banach's contraction mapping principle.

Definition 1 ([2]). *Let (X, d) be a metric space. A mapping $T : X \to X$ is said to be an F-contraction if there exist $F \in \Im$ and $\tau > 0$ such that*

$$\tau + F(d(Tx, Ty)) \leq F(d(x, y))$$

holds for any $x, y \in X$ with $d(Tx, Ty) > 0$.

Theorem 1 ([2]). *Let (X, d) be a complete metric space and $T : X \to X$ be an F-contraction. Then T has a unique fixed point ξ in X.*

Definition 2 ([3]). *Let (X, d) be a metric space endowed with a directed graph $G(V, E)$. A mapping $T : X \to X$ is graph-preserving if*

$$(x, y) \in E(G), \text{ for } x, y \in X \implies (Tx, Ty) \in E(G).$$

Definition 3 ([4]). *Let (X, d) be a metric space endowed with a directed graph $G(V, E)$. A mapping $T : X \to X$ is said to be an GF-contraction if T is graph-preserving and there exist $F \in \Im$ and $\tau > 0$ such that*

$$\tau + F(d(Tx, Ty)) \leq F(d(x, y))$$

holds for any $x, y \in X$ with $(x, y) \in E(G)$ and $d(Tx, Ty) > 0$.

Definition 4 ([5]). *Let (X, d) be a metric space endowed with a directed graph $G(V, E)$. A multivalued mapping $T : X \to CB(X)$ is graph-preserving if*

$$(x, y) \in E(G), \text{ for } x, y \in X \implies (u, v) \in E(G), \text{ whenever } u \in Tx \text{ and } v \in Ty.$$

Lemma 1 ([5]). *Let (X, d) be a metric space and $T : X \to N(X)$ be an upper semi-continuous mapping such that Tx is closed for all $x \in X$. If $x_n \to x_0$, $y_n \to y_0$ and $y_n \in Tx_n$, then $y_0 \in Tx_0$.*

Definition 5 ([5]). *Let (X, d) be a metric space endowed with a directed graph $G(V, E)$. A multivalued mapping $T : X \to CB(X)$ is weakly graph-preserving if $(x, y) \in E(G)$ where $x \in X$ and $y \in Tx$, implies that $(y, z) \in E(G)$ for all $z \in Ty$.*

Let X be a nonempty set and $T : X \to N(X)$ be a multivalued mapping. We define

$$P_T = \{x \in X : x \in Tx\}, \quad T_G = \{(x, y) \in E(G) : H(Tx, Ty) > 0\} \text{ and}$$

$$X_T = \{x \in X : (x, y) \in E(G) \text{ for some } y \in Tx\}.$$

The following class of functions will be used in our results in the next section.

Let $\psi \colon [0, \infty)^5 \to [0, \infty)$ be such that (i) ψ is continuous and monotone nondecreasing in each coordinate, (ii) $\psi(t, t, t, t, t) \leq t$ for all $t \geq 0$. We denote the collection of such functions ψ by the symbol Ψ.

Let $\phi \colon [0, \infty)^4 \to [0, \infty)$ be such that (i) ϕ is continuous and monotone nondecreasing in each coordinate, (ii) $\phi(x_1, x_2, x_3, x_4) = 0$ if $x_1 x_2 x_3 x_4 = 0$. We denote the collection of such functions ϕ by the symbol Φ.

Using the above mathematical notions in this paper we establish an F-contraction type multivalued fixed-point result in a metric space with a graph. Fixed-point theory on metric spaces with the additional structure of a graph is a recent development. Some works from this line of research can be found in works such as [3,5–10]. We make specific choices of a particular function used in the metric inequality to discuss special cases of the main theorem. This demonstrates the generality of our result. It may be further mentioned that F-contractions are new concepts in metric fixed-point theory which have been extended in various ways in works such as [2,4–6,11,12]. Essentially our results are in the domain of setvalued analysis to which the Banach contraction mapping principle was extended by Nadler [1]. In his result Nadler used the Hausdorff distance. The work was followed by several other works such as [5,6,13–15]. The contractive inequality which we use in our problem involves some rational terms. Dass and Gupta [16] generalized the Banach's contraction mapping principle by using a contractive condition of rational type. Fixed-point theorems for contractive type conditions satisfying rational inequalities in metric spaces have been developed in several works [17–20]. Finally, we support our main theorem with illustrative examples.

2. Main Result

Theorem 2. *Let (X, d) be a complete metric space endowed with a directed graph G and $T : X \to K(X)$ be a multivalued map. Suppose that (i) T is upper semi-continuous and weakly graph-preserving, (ii) X_T is nonempty, (iii) there exist $\tau > 0$, $F \in \Im$, $\psi \in \Psi$ and $\phi \in \Phi$ such that for $x, y \in X$ with $(x, y) \in T_G$,*

$$\tau + F(H(Tx, Ty)) \leq F(M(x,y) + N(x,y)),$$

where $N(x,y) = \phi\Big(D(x,Tx), D(y,Ty), D(x,Ty), D(y,Tx)\Big)$ and

$$M(x,y) = \psi\Big(d(x,y), D(x,Tx), D(y,Ty), \frac{D(x,Tx)D(y,Ty) + D(x,Ty)D(y,Tx)}{1 + d(x,y)},$$
$$\frac{D(x,Tx)D(y,Ty) + D(x,Ty)D(y,Tx)}{1 + H(Tx,Ty)}\Big).$$

Then P_T is nonempty.

Proof. Let us assume T has no fixed point. Then $D(x, Tx) > 0$ for all $x \in X$. Let $x_0 \in X_T$. Then there exists $x_1 \in Tx_0$ such that $(x_0, x_1) \in E(G)$. Now $0 < D(x_1, Tx_1) \leq H(Tx_0, Tx_1)$, which implies that $(x_0, x_1) \in T_G$. Using the assumption (iii) and a property of F, we have

$$F(D(x_1, Tx_1)) \leq F(H(Tx_0, Tx_1)) \leq F(M(x_0, x_1) + N(x_0, x_1)) - \tau, \qquad (1)$$

where

$$M(x_0, x_1) = \psi\Big(d(x_0, x_1), D(x_0, Tx_0), D(x_1, Tx_1),$$
$$\frac{D(x_0, Tx_0)D(x_1, Tx_1) + D(x_0, Tx_1)D(x_1, Tx_0)}{1 + d(x_0, x_1)},$$
$$\frac{D(x_0, Tx_0)D(x_1, Tx_1) + D(x_0, Tx_1)D(x_1, Tx_0)}{1 + H(Tx_0, Tx_1)}\Big)$$
$$\leq \psi\Big(d(x_0, x_1), d(x_0, x_1), D(x_1, Tx_1), \frac{D(x_0, Tx_0)D(x_1, Tx_1)}{1 + d(x_0, x_1)},$$
$$\frac{D(x_0, Tx_0)D(x_1, Tx_1)}{1 + H(Tx_0, Tx_1)}\Big)$$
$$\leq \psi\Big(d(x_0, x_1), d(x_0, x_1), D(x_1, Tx_1), \frac{D(x_0, Tx_0)D(x_1, Tx_1)}{1 + d(x_0, x_1)},$$
$$\frac{d(x_0, x_1)D(x_1, Tx_1)}{1 + D(x_1, Tx_1)}\Big)$$
$$\leq \psi\Big(d(x_0, x_1), d(x_0, x_1), D(x_1, Tx_1), D(x_1, Tx_1), d(x_0, x_1)\Big) \qquad (2)$$

and

$$0 \leq N(x_0, x_1) = \phi(D(x_0, Tx_0), D(x_1, Tx_1), D(x_0, Tx_1), D(x_1, Tx_0))$$
$$\leq \phi\Big(d(x_0, x_1), D(x_1, Tx_1), D(x_0, Tx_1), d(x_1, x_1)\Big) = 0,$$

that is, $N(x_0, x_1) = 0$. □

If possible, suppose that $d(x_0, x_1) \leq D(x_1, Tx_1)$. Then from (2), using the properties of ψ, we have

$$M(x_0, x_1) \leq \psi\Big(D(x_1, Tx_1), D(x_1, Tx_1), D(x_1, Tx_1), D(x_1, Tx_1), D(x_1, Tx_1)\Big)$$
$$\leq D(x_1, Tx_1).$$

Using (1) and a property of F, we have

$$F(D(x_1, Tx_1)) \leq F(D(x_1, Tx_1)) - \tau,$$

which is a contradiction. Thus, $D(x_1, Tx_1) < d(x_0, x_1)$. Using (2) and the properties of ψ, we have

$$M(x_0, x_1) \leq \psi\Big(d(x_0, x_1), d(x_0, x_1), d(x_0, x_1), d(x_0, x_1), d(x_0, x_1)\Big)$$
$$\leq d(x_0, x_1).$$

By (1) and a property of F, we have

$$F(D(x_1, Tx_1)) \leq F(d(x_0, x_1)) - \tau. \tag{3}$$

Since Tx_1 is compact, there exists $x_2 \in Tx_1$ such that $d(x_1, x_2) = D(x_1, Tx_1)$. Hence from (3), we have

$$F(d(x_1, x_2)) \leq F(d(x_0, x_1)) - \tau. \tag{4}$$

As T is weakly graph-preserving, $(x_0, x_1) \in E(G)$, $x_1 \in Tx_0$ and $x_2 \in Tx_1$, we have $(x_1, x_2) \in E(G)$. Now, $0 < D(x_2, Tx_2) \leq H(Tx_1, Tx_2)$, which implies that $(x_1, x_2) \in T_G$. By the assumption (iii) and a property of F, we have

$$F(D(x_2, Tx_2)) \leq F(H(Tx_1, Tx_2)) \leq F(M(x_1, x_2) + N(x_1, x_2)) - \tau. \tag{5}$$

Arguing similarly as before, we have

$$F(D(x_2, Tx_2)) \leq F(d(x_1, x_2)) - \tau. \tag{6}$$

Since Tx_2 is compact, there exists $x_3 \in Tx_2$ such that $d(x_2, x_3) = D(x_2, Tx_2)$. From (6), we have

$$F(d(x_2, x_3)) \leq F(d(x_1, x_2)) - \tau. \tag{7}$$

Continuing this process, we construct a sequence $\{x_n\}$ such that for all $n \geq 0$,

$$x_{n+1} \in Tx_n, \ (x_n, x_{n+1}) \in T_G \tag{8}$$

and

$$F(d(x_n, x_{n+1})) \leq F(d(x_{n-1}, x_n)) - \tau. \tag{9}$$

Let $\gamma_n = d(x_n, x_{n+1})$ for all $n \geq 0$.

From (9), we have

$$F(\gamma_n) \leq F(\gamma_{n-1}) - \tau \leq F(\gamma_{n-2}) - 2\tau \leq \ldots \leq F(\gamma_0) - n\tau. \tag{10}$$

Taking limit as $n \to \infty$ in the above inequality, we get $\lim_{n \to \infty} F(\gamma_n) = -\infty$, which by property (F_2) of F, implies that $\lim_{n \to \infty} \gamma_n = 0$. Then by property (F_3) of F, there exists $k \in (0,1)$ such that $\lim_{n \to \infty} \gamma_n^k F(\gamma_n) = 0$. Now, using (10), we have

$$\gamma_n^k F(\gamma_n) - \gamma_n^k F(\gamma_0) \leq -\gamma_n^k n\tau \leq 0.$$

Letting $n \to \infty$ in the above inequality, we obtain

$$\lim_{n \to \infty} n\gamma_n^k = 0.$$

Then there exists $n_1 \in N$ such that $n\gamma_n^k \leq 1$ for all $n \geq n_1$, which implies that $\gamma_n \leq \frac{1}{n^{\frac{1}{k}}}$ for all $n \geq n_1$. Then we have

$$\sum_{n=n_1}^{\infty} d(x_n, x_{n+1}) = \sum_{n=n_1}^{\infty} \gamma_n \leq \sum_{n=n_1}^{\infty} \frac{1}{n^{\frac{1}{k}}}.$$

As $0 < k < 1$, $\sum_{n=n_1}^{\infty} \frac{1}{n^{\frac{1}{k}}}$ is convergent. Then it follows that $\sum d(x_n, x_{n+1})$ is convergent. This implies that $\{x_n\}$ is a Cauchy sequence. As X is complete, there exists $z \in X$ such that $\lim_{n \to \infty} x_n = z$. Since T is upper semi-continuous, by Lemma 1, we have $z \in Tz$, which contradicts the assumption that T has no fixed point. Hence T has a fixed point, i.e., P_T is nonempty.

Remark 1. *Varying the functions ψ and ϕ in the assumption (iii) of Theorem 2, we have different form of F-contractions for which Theorems 2 hold. For some examples, choosing*
(a) $\psi(t_1, t_2, t_3, t_4, t_5) = t_1$ and $\phi(x_1, x_2, x_3, x_4) = 0$,
(b) $\psi(t_1, t_2, t_3, t_4, t_5) = \max\{t_2, t_3\}$ and $\phi(x_1, x_2, x_3, x_4) = 0$,
(c) $\psi(t_1, t_2, t_3, t_4, t_5) = \max\{t_4, t_5\}$ and $\phi(x_1, x_2, x_3, x_4) = 0$,
(d) $\psi(t_1, t_2, t_3, t_4, t_5) = \max\{t_1, t_2, t_3, t_4, t_5\}$ and $\phi(x_1, x_2, x_3, x_4) = 0$,
respectively, we have the following form of F-contractions respectively

$1_{(a)}: \tau + F(H(Tx, Ty)) \leq F(d(x,y))$,

$2_{(b)}: \tau + F(H(Tx, Ty)) \leq F\Big(\max\{D(x, Tx), D(y, Ty)\}\Big)$,

$3_{(c)}: \tau + F(H(Tx, Ty)) \leq F\Big(\max\Big\{\dfrac{D(x, Tx)D(y, Ty) + D(x, Ty)D(y, Tx)}{1 + d(x,y)},$

$$\dfrac{D(x, Tx)D(y, Ty) + D(x, Ty)D(y, Tx)}{1 + H(Tx, Ty)}\Big\}\Big),$$

$4_{(d)}: \tau + F(H(Tx, Ty)) \leq F(M(x,y))$,

where $M(x,y) = \max\Big\{d(x,y), D(x, Tx), D(y, Ty), \dfrac{D(x, Tx)D(y, Ty) + D(x, Ty)D(y, Tx)}{1 + d(x,y)},$

$$\dfrac{D(x, Tx)D(y, Ty) + D(x, Ty)D(y, Tx)}{1 + H(Tx, Ty)}\Big\}.$$

Remark 2. *Theorem 2 is a generalization of Theorem 2 in [6].*

Remark 3. Theorem 2 is true for the class of functions $T : X \to CB(X)$ under the consideration of the class of function $\Im_*$ instead of $\Im$. Arguing similarly as in the proof of Theorem 2 and taking into account the condition (F4) of F, we get

$$F(D(x_1, Tx_1)) = F(\inf \{d(x_1, z) : z \in Tx_1\})$$
$$= \inf(F(\{d(x_1, z) : z \in Tx_1\})).$$

From (3), we have

$$\inf(F(\{d(x_1, z) : z \in Tx_1\})) \leq F(d(x_0, x_1)) - \tau < F(d(x_0, x_1)) - \frac{\tau}{2}.$$

Then there exists $x_2 \in Tx_1$ such that

$$F(d(x_1, x_2)) \leq F(d(x_0, x_1)) - \frac{\tau}{2}.$$

Arguing similarly as in the proof of Theorem 2, it can be proved that P_T is nonempty.

Example 1. Take the metric space $X = [0, \infty)$ with usual metric d. Assume that G is a directed graph with $V(G) = X$ and $E(G) = \{(x, y) : \text{if } x, y \in [0, 1]\} \cup \{(x, x) : x > 1\}$. Define a multivalued mapping $T : X \to K(X)$ as $Tx = \begin{cases} [0, \frac{e^{-\tau}}{5}x] & \text{if } x \in [0, 1], \\ \{\frac{e^{-\tau}}{5}\} & \text{if } x > 1. \end{cases}$
Let $F(x) = \ln(x)$, $\psi(x_1, x_2, x_3, x_4, x_5) = x_1$, $\phi(x_1, x_2, x_3, x_4) = 0$ and $\tau > 0$. Then T is upper semi-continuous and weakly graph-preserving. Let $x, y \in X$ with $(x, y) \in E(G)$ and $H(Tx, Ty) > 0$. Then $x, y \in [0, 1]$ with $x \neq y$. Without loss of generality, assume that $y < x$. Then

$$H(Tx, Ty) = e^{-\tau}\frac{1}{5}|x - y| \leq e^{-\tau}|x - y| = e^{-\tau}d(x, y).$$

Taking 'ln' on both sides of the above equation, we get

$$F(H(Tx, Ty)) \leq -\tau + F(d(x, y)),$$
$$\tau + F(H(Tx, Ty)) \leq F(d(x, y)),$$
$$\tau + F(H(Tx, Ty)) \leq F(M(x, y) + N(x, y)),$$

where $N(x, y) = \phi\Big(D(x, Tx), D(y, Ty), D(x, Ty), D(y, Tx)\Big)$ and

$$M(x, y) = \psi\Big(d(x, y), D(x, Tx), D(y, Ty), \frac{D(x, Tx)D(y, Ty) + D(x, Ty)D(y, Tx)}{1 + d(x, y)},$$
$$\frac{D(x, Tx)D(y, Ty) + D(x, Ty)D(y, Tx)}{1 + H(Tx, Ty)}\Big).$$

Thus, all the conditions of Theorem 2 are satisfied and here $P_T = \{0\}$ is the fixed-point set of T.

Example 2. Let $X = \{0, 1, 2, 3, 4, 5, 6, 7, 8\}$ and G be a directed graph with $V(G) = X$ and $E(G) = \{(0, 0), (0, 1), (0, 4), (0, 5), (1, 1), (1, 0), (1, 2), (1, 3), (2, 2), (2, 3), (3, 2), (3, 3), (4, 4), (4, 5), (5, 4), (5, 5), (6, 6), (6, 7), (7, 1), (7, 7), (8, 7), (8, 8)\}$. Let d be a metric defined on X as
$d(x, y) = \begin{cases} 0 & \text{if } x = y, \\ x + y & \text{if } x \neq y. \end{cases}$
Let $T : X \to K(X)$ be defined as

$$T(x) = \begin{cases} \{4, 5\}, & \text{if } x \in \{0, 4, 5\}, \\ \{2, 3\}, & \text{if } x \in \{1, 2, 3\}, \\ \{7\}, & \text{if } x \in \{6, 8\}, \\ \{1\}, & \text{if } x = 7. \end{cases}$$

Let $F(x) = \ln(x)$, $\psi(x_1, x_2, x_3, x_4, x_5) = max(x_1, x_2, x_3, x_4, x_5)$, $\phi(x_1, x_2, x_3, x_4) = x_1 x_2 x_3 x_4$ and $\tau = 0.2$. Then all the conditions of Theorem 2 are satisfied and here $P_T = \{2, 3, 4, 5\}$ is the fixed-point set of T.

Remark 4. Take $x = 0$ and $y = 1$. Then $H(T0, T1)= 7$, $d(0, 1) = 1$, $D(0, T0) = 4$, $D(1, T1) = 3$, $D(0, T1) = 2$, $D(1, T0) = 5$. It is easy to verify that the inequality (3.1) of Theorem 2 in [6] is not satisfied when $x = 0$ and $y = 1$. Therefore, the above example is not applicable in case of Theorem 2 in [6]. Hence Theorem 2 is an actual extension of Theorem 2 in [6].

3. Conclusions

In this paper, we combine several concepts which have featured prominently in the recent literature of fixed-point theory. Fixed-point theory has many applications as, for instances, those in [10,21]. It is our perception that the structure of graph on the metric space allows us to obtain fixed-point results with more flexibility and for making some new applications. These problems are supposed to be taken up in our future works.

Author Contributions: Individual contributions by the authors are the following: Conceptualization, writing, review and editing, N.M. and D.K.; formal analysis and methodology, B.S.C., N.M. and D.K.; validation, B.S.C. and M.d.l.S.; Funding, Project Administration and Supervision, M.d.l.S. All authors have read and agreed to the published version of the manuscript.

Funding: This paper has been supported by the Basque Government though Grant T1207-19.

Institutional Review Board Statement: Not applicable.

Informed Consent Statement: Not applicable.

Data Availability Statement: Not applicable.

Acknowledgments: The suggestions of the learned referee are gratefully acknowledged. The authors are grateful to the Spanish Government for Grant RTI2018-094336-B-I00 (MCIU/AEI/FEDER, UE) and to the Basque Government for Grant IT1207-19.

Conflicts of Interest: The authors declare no conflict of interest.

References

1. Nadler, S.B. Multi-valued contraction mappings. *Pac. J. Math.* **1969**, *30*, 475–488. [CrossRef]
2. Wardowski, D. Fixed points of a new type of contractive mappings in complete metric space. *Fixed Point Theory Appl.* **2012**, *94*, 11. [CrossRef]
3. Jachymski, J. The contraction principal for mapping on a complete metric space with a graph. *Proc. Am. Math. Soc.* **2008**, *136*, 1359–1373. [CrossRef]
4. Younus, A.; Azam, M.U.; Asif, M. Fixed point theorems for self and non-self F-contractions in metric spaces endowed with a graph. *J. Egypt. Math. Soc.* **2020**, *28*, 44. [CrossRef]
5. Acar, O.; Altun, I. Multivalued F-contractive mapping with a graph and some fixed point results. *Publ. Math. Debr.* **2016**, *88*, 305–317. [CrossRef]
6. Acar, Ö. Rational type multivalued F_G-contractive mapping with a graph. *Results Math.* **2018**, *73*, 52. [CrossRef]
7. Beg, I.; Butt, A.; Radojević, S. The contraction principal for set valued mapping on a metric space with a graph. *Comput. Math. Appl.* **2010**, *60*, 1214–1219. [CrossRef]
8. Cholamjiak, W.; Suantai, S.; Suparatulatorn, R.; Kesornprom, S.; Cholamjiak, P. Viscosity approximation methods for fixed point problems in Hilbert spaces endowed with graphs. *J. Appl. Numer. Optim.* **2019**, *1*, 25–38.
9. Ma, Z.; Li, X.; Ameer, E.; Arshad, M.; Khan, S.U. Fixed points of (Y, Λ)-graph contractive mappings in metric spaces endowed with a directed graph. *J. Nonlinear Funct. Anal.* **2019**, *2019*, 20.
10. Sultana, A.; Vetrivel, V. Fixed points of Mizoguchi-Takahashi contraction on a metric space with a graph and application. *J. Math. Anal. Appl.* **2014**, *417*, 336–344. [CrossRef]
11. Vujaković, J.; Mitrović, S.; Mitrović, Z.D.; Radenović, S. On $\mathcal{F}$-contractions for weak α-admissible mappings in metric-like spaces. *Mathematics* **2020**, *8*, 1629. [CrossRef]
12. Wardowski, D.; Dung, N.V. Fixed points of F-weak contractions on complete metric spaces. *Demonstr. Math.* **2014**, *XLVII*, 146–155. [CrossRef]
13. Choudhury, B.S.; Metiya, N.; Kundu, S. Existence, data-dependence and stability of coupled fixed point sets of some multivalued operators. *Chaos Solitons Fractals* **2020**, *133*, 109678. [CrossRef]
14. Ćirić, L.B. Multi-valued nonlinear contraction mappings. *Nonlinear Anal.* **2009**, *71*, 2716–2723. [CrossRef]

15. Mizoguchi, N.; Takahashi, W. Fixed point theorems for multivalued mappings on complete metric spaces. *J. Math. Anal. Appl.* **1989**, *141*, 177–188. [CrossRef]
16. Dass, B.K.; Gupta, S. An extension of Banach contraction principle through rational expressions. *Inidan J. Pure Appl. Math.* **1975**, *6*, 1455–1458.
17. Chandok, S.; Kim, J.K. Fixed point theorem in ordered metric spaces for generalized contractions mappings satisfying rational type expressions. *J. Nonlinear Funct. Anal. Appl.* **2012**, *17*, 301–306.
18. Harjani, J.; López, B.; Sadarangani, K. A fixed point theorem for mappings satisfying a contractive condition of rational type on a partially ordered metric space. *Abstr. Appl. Anal.* **2010**, *2010*, 190701. [CrossRef]
19. Jaggi, D.S.; Das, B.K. An extension of Banach's fixed point theorem through rational expression. *Bull. Calcutta Math. Soc.* **1980**, *72*, 261–264.
20. Luong, N.V.; Thuan, N.X. Fixed point theorem for generalized weak contractions satisfying rational expressions in ordered metric spaces. *Fixed Point Theory Appl.* **2011**, *2011*, 46. [CrossRef]
21. Nguyen, L.V.; Tram, N.T.N. Fixed point results with applications to involution mappings. *J. Nonlinear Var. Anal.* **2020**, *4*, 415–426.

Article
Existence of Coupled Best Proximity Points of p-Cyclic Contractions

Miroslav Hristov [1], Atanas Ilchev [2], Diana Nedelcheva [3] and Boyan Zlatanov [2,*]

[1] Department of Mathematical Analysis, Faculty of Mathematics and Informatics,
Konstantin Preslavsky University of Shumen, 115, Universitetska St, 9700 Shumen, Bulgaria;
miroslav.hristov@shu.bg

[2] Department of Real Analysis, Faculty of Mathematics and Informatics,
University of Plovdiv Paisii Hilendarski, 24 Tzar Assen Str., 4000 Plovdiv, Bulgaria;
atanasilchev@uni-plovdiv.bg

[3] Department of Mathematics, Technical University of Varna, 1, Studentska Str., 9000 Varna, Bulgaria;
diana.nedelcheva@tu-varna.bg

* Correspondence: bobbyz@uni-plovdiv.bg; Tel.: +359-89-847-7827

Abstract: We generalize the notion of coupled fixed (or best proximity) points for cyclic ordered pairs of maps to p-cyclic ordered pairs of maps. We find sufficient conditions for the existence and uniqueness of the coupled fixed (or best proximity) points. We illustrate the results with an example that covers a wide class of maps.

Keywords: coupled fixed points; cyclic maps; uniformly convex Banach space; error estimate

MSC: Primary 47H10; 58E30; 54H25

1. Introduction

Banach's fixed point theorem has proven to be a powerful tool in pure and applied mathematics. Coupled fixed points were initiated in [1] more than 30 year ago. It turns out that the last 10 years there is a great interest on coupled fixed points, both in fundamental results and their applications [2–5]. We would like to mention a new kind of applications in the theory of equilibrium in duopoly markets [6,7].

A notion that generalizes fixed point results for non-self maps is that of cyclic maps [8] i.e., $T : A \to B$, $T : B \to A$. Since a cyclic map T does not necessarily have a fixed point, one can alter the problem $x = Tx$ to a problem to find an element x which is in some sense closest to Tx. Best proximity points were introduced for cyclic maps in [9] (x is called a best proximity points of T in A if $\|x - Tx\| = \text{dist}(A, B) = \inf\{\|a - b\| : a \in A, b \in B\}$) and they are relevant in this perspective. The notion of best proximity points [9] actually generalizes the notion of cyclic maps from [8], as far as if $A \cap B \neq \emptyset$, then any best proximity point is a fixed point, too. It turns out that best proximity points are interesting not only as a pure mathematical results, but also as a possibility for a new approach in solving of different types of problems [2–7].

We would like to mention just a few very recent results about coupled best proximity points, that can be applied in solving of different types of problems. The authors have investigated a generalization of GKT cyclic Φ-contraction mapping in [10] and a non trivial application for solving of initial value problem is presented. The existence of coupled best proximity point for a class of cyclic (or noncyclic) condensing operators are studied in [11] an the main result applied for finding of an optimal solution for a system of differential equations. A new class of mappings called fuzzy proximally compatible mappings are considered in [12], where coupled best proximity point results are obtained and further applied in finding the fuzzy distance between two subsets of a fuzzy metric space.

Unfortunately all of the mentions above results are for 2–cyclic maps. It is not easy to generalize the results about 2–cyclic maps to p-cyclic maps. The first breakthrough was obtained in [13], where authors succeed to show that for wide classes of maps the distances between the successive sets are equal. The technique from [13] was later widely used [14–17].

We have tried to unify the techniques from [1,13] to get results for the existence and uniqueness of coupled fixed (or best proximity points) for p-cyclic maps.

The first results related to finding the error estimate for best proximity points is made in [18]. In [19], results for the existence and uniqueness of coupled best proximity points are obtained, as well as an error estimate is obtained. In this article, p-cyclic operators are considered, and the results obtained include as a special case the results obtained in [13,19].

2. Preliminaries

We will summarize the notions and the results that we will need.

If A and B are nonempty subsets of the metric space (X, d), then a distance between the sets A and B will be the number $\text{dist}(A, B) = \inf\{d(x, y) : x \in A, y \in B\}$.

Let $\{A_i\}_{i=1}^p$ be nonempty subsets of X. Just to simplify some of the formulas we will assume the convention that $A_{kp+i} = A_i$ for $i = 1, 2, \ldots, p$ and $k \in \mathbb{N}$.

Following [13], if $\{A_i\}_{i=1}^p$ be nonempty subsets of a metric space (X, d), then the map $T : \bigcup_{i=1}^p A_i \to \bigcup_{i=1}^p A_i$ is called a p-cyclic map if it is satisfied that $T(A_i) \subseteq T(A_{i+1})$ for every $i = 1, 2, \ldots, p$. A point $\xi \in A_i$ is called a best proximity point of the cyclic map T in A_i if $d(\xi, T\xi) = \text{dist}(A_i, A_{i+1})$.

The next two lemmas are fundamental to the best proximity points theory.

Lemma 1. ([9]) *Let A be a nonempty closed, convex subset, and B be a nonempty closed subset of a uniformly convex Banach space $(X, \|\cdot\|)$. Let $\{x_n\}_{n=1}^\infty$ and $\{z_n\}_{n=1}^\infty$ be two sequences in A and $\{y_n\}_{n=1}^\infty$ be a sequence in B so that:*
1) $\lim_{n\to\infty} \|x_n - y_n\| = \text{dist}(A, B)$;
2) $\lim_{n\to\infty} \|z_n - y_n\| = \text{dist}(A, B)$;
then $\lim_{n\to\infty} \|x_n - z_n\| = 0$.

Lemma 2. ([9]) *Let A be a nonempty closed, convex subset, and B be a nonempty closed subset of a uniformly convex Banach space $(X, \|\cdot\|)$. Let $\{x_n\}_{n=1}^\infty$ and $\{z_n\}_{n=1}^\infty$ be sequences in A and $\{y_n\}_{n=1}^\infty$ be a sequence in B satisfying:*
(1) $\lim_{n\to\infty} \|z_n - y_n\| = \text{dist}(A, B)$;
(2) for every $\varepsilon > 0$ there is a number $N_0 \in \mathbb{N}$, such that for any $m > n \geq N_0$, $\|x_n - y_n\| \leq \text{dist}(A, B) + \varepsilon$,
then for every $\varepsilon > 0$, there is a number $N_1 \in \mathbb{N}$, so that for all $m > n > N_1$, holds the inequality $\|x_m - z_n\| \leq \varepsilon$.

The geometric structure of the underlying space X plays a key role. When we consider the Banach space $(X, \|\cdot\|)$ we will always assume that the distance between the elements is generated by the norm $\|\cdot\|$ i.e., $d(x, y) = \|x - y\|$.

Definition 1. *[20] Let $(X, \|\cdot\|)$ be a Banach space. For every $\varepsilon \in (0, 2]$ we define the modulus of convexity of $\|\cdot\|$ by*

$$\delta_{\|\cdot\|}(\varepsilon) = \inf\left\{1 - \left\|\frac{x+y}{2}\right\| : x, y \in B_X, \|x - y\| \geq \varepsilon\right\}.$$

The norm is called uniformly convex if $\delta_X(\varepsilon) > 0$ for all $\varepsilon \in (0, 2]$. The space $(X, \|\cdot\|)$ is then called a uniformly convex space.

For any uniformly convex Banach space X there holds the inequality [9]

$$\left\| \frac{x+y}{2} - z \right\| \leq \left(1 - \delta_X\left(\frac{r}{R}\right)\right) R \qquad (1)$$

for any $x, y, z \in X$, such that $\|x - z\| \leq R$, $\|y - z\| \leq R$ and $\|x - y\| \geq r$, provided that R, r be real numbers and $R > 0$, $r \in [0, 2R]$.

For any uniformly convex Banach space $(X, \|\cdot\|)$ its modulus of convexity δ_X is strictly increasing function and thus its inverse function δ^{-1} exists. If there are constants $C > 0$ and $q > 0$, so that the inequality $\delta_{\|\cdot\|}(\varepsilon) \geq C\varepsilon^q$ holds for any $\varepsilon \in (0, 2]$ we say that the modulus of convexity is of power type q with a constant C.

An extensive study of the Geometry of Banach spaces can be found in [21–23].

3. Auxiliary Results

The iterated sequence $\{(x_n, y_n)\}_{n=0}^{\infty}$ (defined in ([1] in the statement of Theorem 1 for coupled fixed points and in [24] in the statement of Lemma 3.8 for coupled best proximity points) will play a crucial role in the proofs of the results, as far as the ordered pair (x, y) of coupled fixed (or best proximity) points is obtained as its limit.

Definition 2. ([1,24]) *Let $\{A_i\}_{i=1}^{p}$ be nonempty subsets of a metric space X and $T : \bigcup_{i=1}^{p} A_i \times A_i \to A_{i+1}$. For any $(x_0, y_0) \in A_i \times A_i$ the sequence $\{(x_n, y_n)\}_{n=0}^{\infty}$ is define inductively by $(x_1, y_1) = (T(x_0, y_0), T(y_0, x_0))$ and if (x_n, y_n) has been already defined then $(x_{n+1}, y_{n+1}) = (T(x_n, y_n), T(y_n, x_n))$.*

When we consider a sequence $\{(x_n, y_n)\}_{n=0}^{\infty}$ we will always assume that it is the iterated sequence defined in Definition 2. Sometimes we will consider a subsequence $\{(x_{n_k}, y_{n_k})\}_{k=1}^{\infty}$ of $\{(x_n, y_n)\}_{n=0}^{\infty}$.

The notion of a coupled best proximity point for cyclic maps was defined in [24] and the notion of best proximity point for p-cyclic maps was introduced in [13]. We will combine both definitions to define a coupled best proximity point for a p-cyclic maps.

Definition 3. *Let A_i, $i = 1, 2, \ldots, p$ be nonempty subsets of a metric space (X, d) and $T : A_i \times A_i \to A_{i+1}$ for $i = 1, 2, \ldots, p$. A point $(x, y) \in A_i \times A_i$ is said to be a best proximity point of T in $A_i \times A_i$, if $d(x, T(x, y)) = d(y, T(y, x)) = d(A_i, A_{i+1})$.*

Following [13] we will define a p-cyclic contractive condition for $T : \bigcup_p^{i=1} A_i \times A_i \to A_{i+1}$.

Definition 4. *Let $\{A_i\}_{i=1}^{p}$ be nonempty subsets of a metric space (X, d). The map T is called p-cyclic contraction, if it satisfies the following condition:*

- $T : A_i \times A_i \to A_{i+1}$
- *There exists $\alpha, \beta \geq 0$, $\alpha + \beta \in (0, 1)$, such that the inequality*

$$d(T(x,y), T(u,v)) \leq \alpha d(x, u) + \beta d(y, v) + (1 - (\alpha + \beta)) d(A_i, A_{i+1}) \qquad (2)$$

holds for every $(x, y) \in A_i \times A_i$, $(u, v) \in A_{i+1} \times A_{i+1}$, $1 \leq i \leq p$.

Lemma 3. *Let $\{A_i\}_{i=1}^{p}$ be nonempty subsets of a metric space (X, d) and T be a p-cyclic contraction map. Then $\mathrm{dist}(A_i, A_{i+1}) = \mathrm{dist}(A_{i+1}, A_{i+2})$ for $i = 1, 2, \ldots, p$.*

Proof. Let us put $d_{i+1} = \mathrm{dist}(A_{p-i}, A_{p-1-i})$ for $i = 0, 1, \ldots p - 1$ (where we use the convention $d_p = \mathrm{dist}(A_1, A_0) = \mathrm{dist}(A_1, A_p)$).

Let us suppose the contrary, that there are two indexes $k, j \in \{1, 2, \ldots, p\}$, such that $\max\limits_{i \in \{1,2,\ldots,p\}} \{d_i\} = d_k > d_j$. Without loss of generality we may assume, that $k = p$. There exists $s \in (0, 1]$, such that
$$d_j = (1-s)d_p. \tag{3}$$

Let $(x_0, y_0) \in A_p$, then $x_{pn+m}, y_{pn+m} \in A_m$, $x_{pn}, y_{pn} \in A_p$, $x_{pn+1}, y_{pn+1} \in A_1$ and from (2) we get

$$\begin{aligned} d(x_{np+1}, x_{np}) &= d(T(x_{np}, y_{np}), T(x_{np-1}, y_{np-1})) \\ &\leq \alpha d(x_{np}, x_{np-1}) + \beta d(y_{np}, y_{np-1}) + (1-\alpha-\beta)d_1) \end{aligned} \tag{4}$$

and

$$\begin{aligned} d(y_{np+1}, y_{np}) &= d(T(y_{np}, x_{np}), T(y_{np-1}, x_{np-1})) \\ &\leq \alpha(d(y_{np}, y_{np-1}) + \beta d(x_{np}, x_{np-1}) + (1-\alpha-\beta)d_1). \end{aligned} \tag{5}$$

Let us, for what follows, to use the notation $\gamma = \alpha + \beta$. From (4) and (5) we can write the chain of inequalities

$$\begin{aligned} S_1 &= d(x_{np+1}, x_{np}) + d(y_{np+1}, y_{np}) \\ &\leq \gamma(d(x_{np}, x_{np-1}) + d(y_{np} - y_{np-1})) + 2(1-\gamma)d_1 \\ &\leq \gamma[\gamma(d(x_{np-1}, x_{np-2}) + d(y_{np-1}, x_{np-2}) + 2(1-\gamma)d_2)] + 2(1-\gamma)d_1 \\ &= \gamma^2(d(x_{np-1}, x_{np-2}) + d(y_{np-1}, x_{np-2})) + 2(1-\gamma)(\gamma d_2 + d_1) \\ &\leq \gamma^3(d(x_{np-2}, x_{np-3}) + d(y_{np-2}, x_{np-3})) + 2(1-\gamma)(\gamma^2 d_3 + \gamma d_2 + d_1) \\ &\quad \cdots\cdots\cdots\cdots\cdots\cdots\cdots\cdots\cdots\cdots\cdots\cdots\cdots \\ &\leq \gamma^p(d(x_{n(p-1)+1}, x_{n(p-1)}) + d(y_{n(p-1)+1}, x_{n(p-1)})) + 2(1-\gamma)\sum_{i=0}^{p-1}\gamma^i d_{i+1} \\ &\quad \cdots\cdots\cdots\cdots\cdots\cdots\cdots\cdots\cdots\cdots\cdots\cdots\cdots \\ &\leq \gamma^{np}(d(x_1, x_0) + d(y_1, x_0)) + 2(1-2\alpha)\sum_{i=0}^{p-1}\gamma^i \sum_{k=0}^{n-1}\gamma^{kp} d_{i+1} \end{aligned} \tag{6}$$

and thus we get

$$\sum_{i=0}^{p-1}\gamma^i \sum_{k=0}^{n-1}\gamma^{kp} d_{i+1} = \sum_{j=0}^{p-1}\sum_{k=0}^{n-1}\gamma^{kp+j}d_{j+1} \leq \frac{1}{1-\gamma^p}\sum_{k=0}^{p-1}\gamma^k d_{k+1}.$$

There exists $N \in \mathbb{N}$, so that for any $n \geq N$ there holds the inequality

$$\gamma^{np}(d(x_1, x_0) + d(y_1, y_0)) \leq \frac{s}{2}\frac{(1-\gamma)\gamma^j}{1-\gamma^p}d_p,$$

where j and s are the index and the constant from (3), respectively. Therefore using the assumption that $d_j = (1-s)d_p = d_p - sd_p$ and that for any $k \neq p$ there holds $d_k \leq d_p$ we get

$$\begin{aligned} 2d_p &\leq d(x_{np+1}, x_{np}) + d(y_{np+1}, y_{np}) \\ &\leq \gamma^{np}(d(x_1 + x_0) + d(y_1 + y_0)) + \frac{2(1-\gamma)}{1-\gamma^p}\sum_{k=0}^{p-1}\gamma^k d_{k+1} \\ &\leq \gamma^{np}(d(x_1 + x_0) + d(y_1 + y_0)) - s\frac{(1-\gamma)\gamma^j}{1-\gamma^p}d_p + \frac{2(1-\gamma)}{1-\gamma^p}\sum_{k=0}^{p-1}\gamma^k d_p \\ &= \gamma^{np}(d(x_1 + x_0) + d(y_1 + y_0)) - s\frac{(1-\gamma)\gamma^j}{1-\gamma^p}d_p + 2d_p \\ &< -\frac{s}{2}\frac{(1-\gamma)\gamma^j}{1-\gamma^p}d_p + 2d_p < 2d_p, \end{aligned} \tag{7}$$

which is a contradiction and consequently the assumption that there exists j so that $d_j <$ $\max\{d_i : i = 1, 2, \ldots, p\}$ could not holds. □

We have just proven in Lemma 3 that for maps, which satisfy Definition 4, there holds $\text{dist}(A_1, A_2) = \text{dist}(A_2, A_3) = \cdots = \text{dist}(A_{p-1}, A_p) = \text{dist}(A_p, A_1)$ and thus we can denote in the rest of the article the distance between the consecutive sets by $d = \text{dist}(A_i, A_{i+1}), i = 1, 2, \ldots, p$.

An easier to apply inequality, which is a consequence from (2) is the inequality

$$d(T(x,y), T(u,v)) + d(T(y,x), T(v,u)) - 2d \leq \gamma(d(x,u) + d(y,v) - 2d) \qquad (8)$$

for every $(x,y) \in A_i \times A_i$, $(u,v) \in A_{i+1} \times A_{i+1}$, $1 \leq i \leq p$.

Lemma 4. *Let $\{A_i\}_{i=1}^p$ be nonempty closed subsets of a metric space (X,d) and T be a p-cyclic contraction. Then for every $(x_0, y_0) \in A_i \times A_i$ there hold $\lim_{n \to \infty} d(x_{pn}, x_{pn+1}) = d$, $\lim_{n \to \infty} d(y_{pn}, y_{pn+1}) = d$, $\lim_{n \to \infty} d(x_{pn \pm p}, x_{pn+1}) = d$ and $\lim_{n \to \infty} d(y_{pn \pm p}, y_{pn+1}) = d$.*

Proof. By Lemma 3 we have that $d(A_i, A_{i+1}) = d(A_{i+1}, A_{i+2})$ for $i = 1, 2, \ldots p - 1$. Let us put $(A_1, A_2) = d$. Therefore there holds the chain of inequalities

$$\begin{aligned}
0 &\leq d(x_{pn+1}, x_{pn}) + d(y_{pn+1}, y_{pn}) - 2d \\
&\leq \gamma(d(x_{pn}, y_{pn-1}) + d(y_{pn}, y_{pn-1}) - 2d) \\
&\leq \gamma^2(d(x_{pn-1}, x_{pn-2}) + d(y_{pn-1}, y_{pn-2}) - 2d) \qquad (9) \\
&\cdots\cdots\cdots\cdots\cdots\cdots\cdots\cdots\cdots \\
&\leq \gamma^{pn}(d(x_1, x_0) + d(y_1, y_0)) - 2d).
\end{aligned}$$

Consequently after taking a limit in (9) when $n \to \infty$ we get $\lim_{n \to \infty} (d(x_{pn}, x_{pn-1}) + d(y_{pn}, y_{pn-1})) = 2d$. From the inequalities $d \leq d(x_{pn}, x_{pn-1})$ and $d \leq d(y_{pn}, y_{pn-1})$ it follows that $\lim_{n \to \infty} d(x_{pn}, x_{pn-1}) = \lim_{n \to \infty} d(y_{pn}, y_{pn-1}) = d$.

The proofs of the other two (actually four, because of $\pm$) limits can be done in a similar fashion. □

Lemma 5. *If $(X, \|\cdot\|)$ be a uniformly convex Banach space, $\{A_i\}_{i=1}^p$ be nonempty and convex subsets of X. T be a p-cyclic contraction. Then for every $(x_0, y_0) \in A_i \times A_i$ there hold $\lim_{n \to \infty} \|x_{pn} - x_{pn+p}\| = 0$, $\lim_{n \to \infty} \|y_{pn} - y_{pn+p}\| = 0$, $\lim_{n \to \infty} \|x_{pn+1} - x_{pn \pm p+1}\| = 0$ and $\lim_{n \to \infty} \|y_{pn+1} - y_{pn \pm p+1}\| = 0$.*

Proof. By Lemma 4 we have that $\lim_{n \to \infty} \|x_{pn} - x_{pn+1}\| = \lim_{n \to \infty} \|x_{pn+p} - x_{pn+1}\| = d$. According to Lemma 3 it follows that $\lim_{n \to \infty} \|x_{pn} - x_{pn+p}\| = 0$. □

Lemma 6. *Let $\{A_i\}_{i=1}^p$ be nonempty closed subsets of a metric space (X,d) and T be a p-cyclic contraction. Let $(x_0, y_0) \in A_i \times A_i$ and the sequence $\{(x_{pn}, y_{pn})\}_{n=0}^\infty$ has a convergent (say to $(\xi, \eta) \in A_i \times A_i$) subsequence $\{(x_{pn_j}, y_{pn_j})\}_{j=1}^\infty$, then (ξ, η) is a best proximity point of T in $A_i \times A_i$.*

Proof. By the inequality $d \leq d(x_{pn_j-1}, \xi) \leq d(x_{pn_j-1}, x_{pn_j}) + d(x_{pn_j}, \xi)$, the assumption that $\lim_{j \to \infty} d(x_{pn_j}, \xi) = 0$ and $\lim_{j \to \infty} d(x_{pn_j-1}, x_{pn_j}) = d$ we get $\lim_{j \to \infty} d(x_{pn_j-1}, \xi) = d$.

By similar arguments it follows that $\lim_{j\to\infty} d(y_{pn_j-1}, \eta) = d$. Using the continuity of the metric function $d(\cdot,\cdot)$ and Lemma 4, we can write the chain of inequalities

$$\begin{aligned}
0 &< d(\xi, T(\xi, \eta)) + d(T(\eta, \xi), \eta) - 2d \\
&= \lim_{j\to\infty} (d(x_{pn_j}, T(\xi, \eta)) + d(T(\xi, \eta), y_{pn_j})) - 2d) \\
&= \lim_{j\to\infty} (d(T(x_{pn_j-1}, y_{pn_j-1}), T(\xi, \eta)) + d(T(\eta, \xi), T(y_{pn_j-1}, x_{pn_j-1})) - 2d) \\
&\leq \gamma \lim_{j\to\infty} (d(x_{pn_j-1}, \xi) + d(y_{pn_j-1}, \eta) - 2d) \\
&= \gamma \left(\lim_{j\to\infty} (d(x_{pn_j-1}, x_{pn_j}) + d(y_{pn_j-1}, y_{pn_j})) - 2d \right) = 0.
\end{aligned}$$

Consequently $d(\xi, T(\xi, \eta)) + d(T(\eta, \xi), \eta) = 2d$ and from the inequalities $d(\xi, T(\xi, \eta)) \geq d$ and $d(T(\eta, \xi), \eta) \geq d$ it follows that $d(\xi, T(\xi, \eta)) = d(T(\eta, \xi), \eta) = d$. □

For an arbitrary chosen $(z, v) \in A_i \times A_i$, let us denote $T^2(z, v) = T(T(z, v), T(v, z))$ and $T^2(v, z) = T(T(v, z), T(z, v))$ and if we have already defined $(T^{p-1}(z, v), T^{p-1}(v, z))$, then put

$$T^p(z, v) = T(T^{p-1}(z, v), T^{p-1}(v, z)) \text{ and } T^p(v, z) = T(T^{p-1}(v, z), T^{p-1}(z, v)).$$

Lemma 7. *Let $\{A_i\}_{i=1}^p$ be nonempty closed subsets of a metric space (X, d) and T be a p-cyclic contraction. If there exists a coupled best proximity point (z, v) of T in $A_i \times A_i$, then $(T^n(z, v), T^n(v, z))$ is a coupled best proximity point of T in $A_{i+n} \times A_{i+n}$. If (z, v) is a limit of the sequence $\{(x_{pn}, y_{pn})\}_{n=0}^\infty$, then the ordered pair (z, v) is a p–periodic point of T, i.e., $z = T^{pn}(z, v)$ and $v = T^{pn}(v, z)$ for $n \in \mathbb{N}$ and any sequence $\{(\xi_{pn}, \eta_{pn})\}_{n=0}^\infty$ converges to (z, v).*

Proof. Let (z, v) be any ordered pair, which is a coupled best proximity points of T in $A_i \times A_i$. From the inequality

$$\begin{aligned}
S_2 &= \|T(z, v) - T^2(z, v)\| + \|T(v, z) - T^2(v, z)\| - 2d \\
&\leq \gamma(\|z - T(z, v)\| + \|v - T(v, z)\| - 2d) = 0
\end{aligned}$$

it follows that $(T(z, v), T(v, z))$ is an ordered pair, which is a coupled best proximity points of T in $A_{i+1} \times A_{i+1}$. From

$$\begin{aligned}
S_3 &= \|T^2(z, v) - T^3(z, v)\| + \|T^2(v, z) - T^3(v, z)\| - 2d \\
&\leq \gamma^2(\|z - T(z, v)\| + \|v - T(v, z)\| - 2d) = 0
\end{aligned}$$

it follows that $(T^2(z, v), T^2(v, z))$ is a coupled best proximity points of T in $A_{i+2} \times A_{i+2}$. By induction we can prove that $(T^n(z, v), T^n(v, z))$ is a coupled best proximity points of T in $A_{i+n} \times A_{i+n}$.

Therefore we have

$$\begin{aligned}
0 &\leq \|z - T^{p+1}(z, v)\| + \|v - T^{p+1}(v, z)\| - 2d \\
&= \lim_{n\to\infty} \left(\|z_{pn} - T^{p+1}(z, v)\| + \|v_{pn} - T^{p+1}(v, z)\| - 2d \right) \\
&= \lim_{n\to\infty} \left(\|T(z_{pn-1}, v_{pn-1}) - T^{p+1}(z, v)\| + \|T(v_{pn-1}, z_{pn-1}) - T^{p+1}(v, z)\| - 2d \right) \\
&\leq \gamma^p \lim_{n\to\infty} \left(\|z_{p(n-1)} - T(z, v)\| + \|v_{p(n-1)} - T(v, z)\| - 2d \right) \\
&= \gamma^p(\|z - T(z, v)\| + \|v - T(v, z)\| - 2d) = 0.
\end{aligned}$$

Thus $\|z - T^{p+1}(z, v)\| = \|v - T^{p+1}(v, z)\| = d$. From $\|z - T(z, v)\| = \|v - T(v, z)\| = d$ and Lemma 2 it follows that $T(z, v) = T^{p+1}(z, v)$ and $T(v, z) = T^{p+1}(v, z)$. From

$$\|z - T(z, v)\| + \|v - T(v, z)\| - 2d = 0,$$

$$\begin{aligned}
S_4 &= \|T^p(z,v) - T(z,v)\| + \|T^p(v,z) - T(v,z)\| - 2d \\
&= \|T^p(z,v) - T^{p+1}(z,v)\| + \|T^p(v,z) - T^{p+1}(v,z)\| - 2d \\
&\leq \gamma^p(\|z - T(z,v)\| + \|v - T(v,z)\| - 2d) = 0,
\end{aligned}$$

Lemma 2 it follows that $z = T^p(z,v)$ and $v = T^p(v,z)$. Now, by a similar calculations we can obtain that $T^{2p}(z,v) = T^p(z,v) = z$, $T^{2p}(v,z) = T^p(v,z) = v$ and by induction, that $T^{np}(z,v) = z$ and $T^{np}(v,z) = v$.

Let there exists $(\xi, \eta) \in A_i \times A_i$, which is a coupled best proximity points of T in $A_i \times A_i$, i.e., $\|\xi - T(\xi, \eta)\| = \|\eta - T(\eta, \xi)\| = d$, that is different from (z,v) and obtained as a limit of a sequence $\{(\xi_{pn}, \eta_{pn})\}_{n=0}^{\infty}$. Using the continuity of the norm function, the equality $T(z,v) = T^{p+1}(z,v)$, $T(v,z) = T^{p+1}(v,z)$ we get the inequality

$$\begin{aligned}
S_5 &= \|\xi - T(z,v)\| + \|\eta - T(v,z)\| - 2d \\
&= \lim_{n\to\infty} \left(\|\xi_{pn} - T^{p+1}(z,v)\| + \|\eta_{pn} - T^{p+1}(v,z)\| - 2d \right) \\
&\leq \gamma^p \lim_{n\to\infty} (\|\xi_{pn-p} - T(z,v)\| + \|\eta_{pn-p} - T(v,z)\| - 2d) \\
&\leq \gamma^p (\|\xi - T(z,v)\| + \|\eta - T(v,z)\| - 2d),
\end{aligned}$$

and by the assumption $\gamma \in (0,1)$ we get the inequality

$$\|\xi - T(z,v)\| + \|\eta - T(v,z)\| - 2d < \|\xi - T(z,v)\| + \|\eta - T(v,z)\| - 2d = 0.$$

Consequently $\|\xi - T(z,v)\| + \|\eta - T(v,z)\| = 2d$. Therefore $\|\xi - T(z,v)\| = \|\eta - T(v,z)\| = d$ and from $\|z - T(z,v)\| = \|v - T(v,z)\| = d$ and by Lemma 2 it follows that $(z,v) = (\xi, \eta)$. □

4. Main Results

Theorem 1. *Let $\{A_i\}_{i=1}^p$ be nonempty, closed and convex subsets of a complete metric space (X,d). Let $T: \bigcup_{i=1}^p A_i \times A_i \to A_i \times A_i$ be a p-cyclic map, so that exist $\alpha, \beta \geq 0$, $\alpha + \beta \in (0,1)$, such that the inequality*

$$d(T(x,y), T(u,v)) \leq \alpha d(x,u) + \beta d(y,v) \tag{10}$$

holds for every $(x,y) \in A_i \times A_i$, $(u,v) \in A_{i+1} \times A_{i+1}$, $1 \leq i \leq p$.

Then there exists an order pair $(z,v) \in \cap_{i=1}^p (A_i \times A_i)$, such that, if $(x_0, y_0) \in A_i \times A_i$ be an arbitrary point of $A_i \times A_i$, the sequence $\{(x_n, y_n)\}\}_{n=0}^{\infty}$ converges to (z,v) and the order pair (z,v) is a unique coupled fixed point of T. Moreover, there hold

- *the a priori estimate $\max\{\rho(x_n, z), \rho(y_n, v)\} \leq \dfrac{\gamma^n}{1-\gamma}(\rho(x_1, x_0) + \rho(y_1, y_0))$*
- *the a posteriori estimate $\max\{\rho(x_n, z), \rho(y_n, v)\} \leq \dfrac{\gamma}{1-\gamma}(\rho(x_{n-1}, x_n) + \rho(y_{n-1}, y_n))$*
- *the rate of convergence $\rho(x_n, z) + \rho(y, v) \leq \gamma(\rho(x_{n-1}, z) + \rho(y_{n-1}, v))$,*

where $\gamma = \alpha + \beta$.

Proof. Let $(x_0, y_0) \in \bigcup_{i=1}^p (A_i \times A_i)$ be arbitrary chosen. Let us consider the iterated sequence $\{(x_n, y_n)\}_{n=1}^{\infty}$. Then there hold the inequalities

$$d(x_{n+1}, x_n) = d(T(x_n, y_n), T(x_{n-1}, y_{n-1})) \leq \alpha d(x_n, x_{n-1}) + \beta d(y_n, y_{n-1})$$

and

$$d(y_{n+1}, y_n) = d(T(y_n, x_n), T(y_{n-1}, x_{n-1})) \leq \alpha d(y_n, y_{n-1}) + \beta d(x_n, x_{n-1}).$$

After summing up the above two inequalities we get

$$d(x_{n+1}, x_n) + d(y_{n+1}, y_n) \leq \gamma(\rho(x_n, x_{n-1}) + \rho(y_n, y_{n-1})). \tag{11}$$

From (11) we get that there holds true

$$\max\{d(x_{n+1}, x_n), d(y_{n+1}, y_n)\} \leq d(x_{n+1}, x_n) + d(y_{n+1}, y_n) \leq \gamma^n(d(x_1, x_0) + d(y_1, y_0)).$$

Thus

$$d(x_n, x_{n+p}) \leq \sum_{k=0}^{p-1} d(x_{n+k}, x_{n+k+1}) \leq \sum_{k=0}^{p-1} \gamma^{n+k}(d(x_1, x_0) + d(y_1, y_0))$$
$$= \gamma^n \frac{1-\gamma^p}{1-\gamma}(d(x_1, x_0) + d(y_1, y_0)). \tag{12}$$

Therefore $\{x_n\}_{n=1}^{\infty}$ is a Cauchy sequence in $\cup_{i=1}^{p} A_i$. From the assumption that A_i are closed subsets of the complete metric space (X, d) it follows that $\{x_n\}_{n=1}^{\infty}$ is convergent to some point $z \in \cup_{i=1}^{p} A_i$. The sequence $\{x_n\}_{n=1}^{\infty}$ is an iterated sequence defined by the p-cyclic map T and thus it has infinite number of terms that belong to each A_i, $i = 1, 2, \ldots, p$. Consequently $z \in \cap_{i=1}^{p} A_i$.

By literary the same arguments we get that $\lim_{n \to \infty} y_n = v \in \cap_{i=1}^{p} A_i$.

We will show that (z, v) is a coupled fixed point of T. Indeed from

$$\begin{aligned} S_6 &= d(z, T(z, v)) + d(v, T(v, z)) \\ &= \lim_{n \to \infty}(d(x_n, T(z, v)) + d(y_n, T(v, z))) \\ &= \lim_{n \to \infty}(d(T(x_{n-1}, y_{n-1}), T(z, v)) + d(T(y_{n-1}, x_{n-1}), T(v, z))) \\ &\leq \gamma \lim_{n \to \infty}(d(x_{n-1}, z) + d(y_{n-1}, v)) = 0 \end{aligned}$$

it follows that $d(z, T(z, v)) = d(v, T(v, z)) = 0$, i.e., (z, v) is a coupled fixed point of T.

We will proof that (z, v) is a unique coupled fixed point by assuming the contrary. Let (x, y) be a coupled fixed point of T, different from (z, v). If $x \in A_i$, then by the definition of a coupled fixed point it follows that $y \in A_i$, too. From the assumption that T is a p-cyclic map it follows that $(x, y) = (T(x, y), T(y, x)) \in A_{i+1} \times A_{i+1}$ and therefore $(x, y) \in \cap_{i=1}^{p}(A_i \times A_i)$. From the inequality

$$d(x, z) + d(y, v) = d(T(x, y), T(z, v)) + d(T(y, x), T(v, z)) \leq \gamma(d(x, z) + d(y, v))$$

and the assumption that $\gamma \in (0, 1)$ it follows that $d(x, z) = d(y, v) = 0$, i.e., the coupled fixed point (z, v) of T is unique.

After taking a limit in (12) we get

$$d(x_n, z) = \lim_{p \to \infty} d(x_n, x_{n+p}) \leq \frac{\gamma^n}{1-\gamma}(d(x_1, x_0) + d(y_1, y_0)).$$

and

$$d(y_n, z) = \lim_{p \to \infty} d(y_n, y_{n+p}) \leq \frac{\gamma^n}{1-\gamma}(d(x_1, x_0) + d(y_1, y_0)).$$

Consequently there holds the a priori estimate

$$\max\{d(x_n, z), d(y_n, z)\} \leq \frac{\gamma^n}{1-\gamma}(d(x_1, x_0) + d(y_1, y_0)).$$

From the chain of inequalities

$$d(x_n, x_{n+p}) \leq \sum_{k=0}^{p-1} d(x_{n+k}, x_{n+k+1}) \leq \sum_{k=1}^{p} \gamma^k (d(x_{n-1}, x_n) + d(y_{n-1}, y_n))$$
$$= \gamma \frac{1-\gamma^p}{1-\gamma}(d(x_{n-1}, x_n) + d(y_{n-1}, y_n)).$$

After taking a limit, when $p \to \infty$, in the above inequality, we get

$$d(x_n, z) = \lim_{p \to \infty} d(x_n, x_{n+p}) \leq \frac{\gamma}{1-\gamma}(d(x_{n-1}, x_n) + d(y_{n-1}, y_n)).$$

Consequently, after using the same arguments for $d(y_n, v)$, there holds the a posteriori estimate
$$\max\{d(x_n, z), d(y_n, v)\} \leq \frac{\gamma}{1-\gamma}(d(x_{n-1}, x_n) + d(y_{n-1}, y_n)).$$

From the inequality
$$\begin{aligned} d(x_n, z) + d(y, v) &= d(T(x_{n-1}, y_{n-1}), T(z, v)) + d(T(y_{n-1}, x_{n-1}), T(v, z)) \\ &\leq \gamma(d(x_{n-1}, z) + d(y_{n-1}, v)) \end{aligned}$$

we get the estimate the rate of convergence. □

We will use the notations $P_{n,m} = \|x_n - x_m\| + \|y_n - y_m\|$ and $W_{n,m} = \|x_n - x_m\| + \|y_n - y_m\| - 2d$, where $\{x_n\}_{n=0}^{\infty}$ and $\{y_n\}_{n=0}^{\infty}$ be the sequences from Definition 2, when the text field is too short.

We have proven in Lemma 3, that for any p-cyclic contraction the distances between the consecutive sets are equal. Therefore in the next theorem we will denote $d = \text{dist}(A_i, A_{i+1}), i = 1, 2, \ldots, p$.

Theorem 2. *Let $\{A_i\}_{i=1}^{p}$ be nonempty, closed and convex subsets of a uniformly convex Banach space $(X, \|\cdot\|)$. Let $T : \bigcup_{i=1}^{p} A_i \times A_i \to A_{i+1}$ be a p-cyclic contraction. Then there exists a unique ordered pair $(z_i, v_i) \in A_i \times A_i$ ($1 \leq i \leq p$), which is a limit of the subsequence $\{(x_{pn}, y_{pn})\}_{n=0}^{\infty} \subset \{(x_n, y_n)\}_{n=0}^{\infty}$ for any initial guess $(x_0, y_0) \in A_i \times A_i$ and it is a coupled best proximity point of T in $A_i \times A_i$. Moreover, $(T(z_i, v_i), T(v_i, z_i))$ is a coupled best proximity point of T in $A_{i+1} \times A_{i+1}$ and (z_i, v_i) is a p–periodic point of T.*

- *If $d > 0$ and $(X, \|\cdot\|)$ be with a modulus of convexity of power type q with a constant C, then there hold the a priori error estimate*
$$\max\{\|x_{pn} - z_i\|, \|y_{pn} - v_i\|\} \leq P_{0,1} \sqrt[q]{\frac{W_{0,1}}{Cd}} \cdot \frac{\left(\sqrt[q]{\gamma}\right)^{pm}}{1 - \sqrt[q]{\gamma^p}}.$$

and the a posteriori error estimate
$$\max\{\|x_{pn} - z_i\|, \|y_{pn} - v_i\|\} \leq P_{pn, pn-1} \sqrt[q]{\frac{W_{pn, pn-1}}{Cd}} \frac{\sqrt[q]{\gamma}}{1 - \sqrt[q]{\gamma^p}},$$

where $\gamma = \alpha + \beta$, α and β be the constants form Definition 4.
- *If $d = 0$, then there hold the error estimates of Theorem 1.*

If $p = 2$, we get as a particular case the results from [19].

Proof. If $\text{dist}(A_i, A_{i+1}) = 0$ for some i, then by Lemma 3 it follows that $\text{dist}(A_i, A_{i+1}) = 0$ for all i. Then the contractive condition induced on T is equivalent to (10). Thus by Theorem 1, T has a unique coupled fixed point and the error estimates from Theorem 1 holds.

Let us assume that $d > 0$. Let $(x_0, y_0) \in A_i \times A_i$. Then $x_{np} \in A_i$ and $x_{np+1} \in A_{i+1}$ for all n. By Lemma 4, $\lim_{n \to \infty} \|x_{np} - x_{np+1}\| = d$. If, for any arbitrary chosen $\varepsilon > 0$, there exists an $n_0 \in \mathbb{N}$, such that for all $m > n > n_0$ the inequality to hold
$$\|x_{pm} - x_{pn+1}\| + \|y_{pm} - y_{pn+1}\| \leq 2d + \varepsilon, \tag{13}$$

by the inequalities $\|x_{pm} - x_{pn+1}\| \geq d$ and $\|y_{pm} - y_{pm+1}\| \geq d$ it follows the inequality $\max\{\|x_{pm} - x_{pn+1}\|, \|y_{pm} - y_{pn+1}\|\} \leq d + \varepsilon$ holds for all $m > n > n_0$. Then by Lemma 1, for any $\varepsilon_1 > 0$, there exists $n_1 \in \mathbb{N}$, such that for $m > n > n_1$ the inequality $\max\{\|x_{mp} - x_{np}\|, \|y_{mp} - y_{np}\|\} \leq \varepsilon_1$ holds, i.e., $\{x_{np}\}_{n=1}^{\infty}$ and $\{y_{np}\}_{n=1}^{\infty}$ are Cauchy sequences and thus converges to some $(z, v) \in A_i \times A_i$. By Lemma 6 (z, v) will be a best proximity point of T in $A_i \times A_i$.

Let us assume contrary of (13). Then, there exists an $\varepsilon_0 > 0$ such that, for every $k \in \mathbb{N}$, there exists $m_k > n_k \geq k$ such that,

$$\|x_{pm_k} - x_{pn_k+1}\| + \|y_{pm_k} - y_{pn_k+1}\| \geq 2d + \varepsilon_0. \tag{14}$$

Let m_k be the smallest integer greater than n_k, to satisfy the above inequality. Now

$$\begin{aligned}S_7 = 2d + \varepsilon_0 &\leq \|x_{pm_k} - x_{pn_k+1}\| + \|y_{pm_k} - y_{pn_k+1}\| \\ &\leq \|x_{pm_k} - x_{pm_k-p}\| + \|x_{pm_k-p} - x_{pn_k+1}\| + \|y_{pm_k} - y_{pm_k-p}\| + \|y_{pm_k-p} - y_{pn_k+1}\|.\end{aligned}$$

By Lemma 4 we have $\lim_{k\to\infty} \|x_{pm_k} - x_{pm_k+p}\| = 0$ and $\lim_{k\to\infty} \|y_{pm_k} - y_{pm_k+p}\| = 0$. Therefore, using the choice of m_k to be the smallest natural, so that to holds the inequality (14), we get

$$\begin{aligned}2d + \varepsilon_0 &\leq \lim_{k\to\infty}(\|x_{pm_k} - x_{pn_k+1}\| + \|y_{pm_k} - y_{pn_k+1}\|) \\ &\leq \lim_{k\to\infty}(\|x_{pm_k-p} - x_{pn_k+1}\| + \|y_{pm_k-p} - y_{pn_k+1}\|) \leq 2d + \varepsilon_0,\end{aligned}$$

i.e., $\lim_{k\to\infty} \|x_{pm_k} - x_{pn_k+1}\| + \lim_{k\to\infty} \|y_{pm_k} - y_{pn_k+1}\| = 2d + \varepsilon_0$.

From the inequality

$$\begin{aligned}2d + \varepsilon_0 &\leq \|x_{pm_k} - x_{pn_k+1}\| + \|y_{pm_k} - y_{pn_k+1}\| \\ &\leq \|x_{pm_k} - x_{pm_k+p}\| + \|x_{pm_k+p} - x_{pn_k+p+1}\| + \|x_{pn_k+p+1} - x_{pn_k+1}\| \\ &\quad + \|y_{pm_k} - y_{pm_k+p}\| + \|y_{pm_k+p} - y_{pn_k+p+1}\| + \|y_{pn_k+p+1} - y_{pn_k+1}\|.\end{aligned}$$

by using Lemma 4 we have $\lim_{k\to\infty} \|x_{pm_k} - x_{pm_k+p}\| = \lim_{k\to\infty} \|x_{pn_k+p+1} - x_{pn_k+1}\| = \lim_{k\to\infty} \|y_{pm_k} - y_{pm_k+p}\| = \lim_{k\to\infty} \|y_{pn_k+p+1} - y_{pn_k+1}\| = 0$ and thus

$$\begin{aligned}\varepsilon_0 &= \lim_{k\to\infty}(\|x_{pm_k} - x_{pn_k+1}\| + \|y_{pm_k} - y_{pn_k+1}\| - 2d) \\ &\leq \lim_{k\to\infty}(\|x_{pm_k+p} - x_{pn_k+p+1}\| + \|y_{pm_k+p} - y_{pn_k+p+1}\| - 2d) \\ &\leq \gamma^p \lim_{k\to\infty}(\|x_{pm_k} - x_{pn_k+1}\| + \|y_{pm_k} - y_{pn_k+1}\| - 2d) \\ &= \gamma^p \varepsilon_0.\end{aligned}$$

That is, $\varepsilon_0 \leq \gamma^p \varepsilon_0$, which is a contradiction, because $\gamma \in (0,1)$.

Hence $\{x_{np}\}_{n=1}^{\infty}$ and $\{y_{np}\}_{n=1}^{\infty}$ are Cauchy sequences, converging to some $(x, y) \in A_i \times A_i$ such that $\|x - T(x, y)\| = \|T(y, x) - y\| = d$.

From Lemma 7 it follows that (x, y), which is a limit of the iterated sequences is unique, for an arbitrary chosen initial guess, $(T^n(x, y), T^n(y, x))$ is a coupled best proximity point of T in $A_{i+n} \times A_{i+n}$, (x, y) is a p–periodic point of T.

It has remained to prove that $x = y$. It holds

$$\begin{aligned}S_8 &= \|x - T(y, x)\| + \|y - T(x, y)\| - 2d \\ &= \|T^p(x, y) - T(y, x)\| + \|T^p(y, x) - T(x, y)\| - 2d \\ &\leq \gamma(\|T^{p-1}(x, y) - y\| + \|T^{p-1}(y, x) - x\| - 2d) \\ &= \gamma(\|T^{p-1}(x, y) - T^p(y, x)\| + \|T^{p-1}(y, x) - T^p(x, y)\| - 2d) \\ &\leq \gamma^p(\|x - T(y, x)\| + \|y - T(x, y)\| - 2d).\end{aligned}$$

Consequently $\|x - T(y, x)\| = \|y - T(x, y)\| = d$. From $\|y - T(y, x)\| = \|x - T(x, y)\| = d$ and the uniform convexity of X it follows that $x = y$.

From (8) there holds the inequality

$$\|x_{n+1} - x_n\| + \|y_{n+1} - y_n\| - 2d \leq \gamma^k(\|x_{n+1-k} - x_{n-k}\| + \|y_{n+1-k} - y_{n-k}\| - 2d).$$

Thus we get

$$\max\{\|x_{n+1} - x_n\|, \|y_{n+1} - y_n\|\} \leq \gamma^k(\|x_{n+1-k} - x_{n-k}\| + \|y_{n+1-k} - y_{n-k}\| - 2d) + d.$$

There hold the inequalities

$$\|x_{pn+1} - x_{pn}\| \le d + \gamma^l(\|x_{pn+1-l} - x_{pn-l}\| + \|y_{pn+1-l} - y_{pn-l}\| - 2d),$$

$$\begin{aligned}\|x_{pn+p} - x_{pn+p+1}\| &\le d + \gamma^{l+1}(\|x_{pn+1-l} - x_{pn-l}\| + \|y_{pn+1-l} - y_{pn-l}\| - 2d)\\ &\le d + \gamma^l(\|x_{pn+1-l} - x_{pn-l}\| + \|y_{pn+1-l} - y_{pn-l}\| - 2d)\end{aligned}$$

and

$$\begin{aligned}\|x_{pn+p} - x_{pn}\| &\le \|x_{pn+p} - x_{pn+1}\| + \|x_{pn+1} - x_{pn}\|\\ &\le 2\big(d + \gamma^l(\|x_{pn+1-l} - x_{pn-l}\| + \|y_{pn+1-l} - y_{pn-l}\| - 2d)\big).\end{aligned}$$

After a substitution in (1) with $x = x_{pn}$, $y = x_{pn+p}$, $z = x_{pn+1}$, $R = d + \gamma^l(\|x_{pn+1-l} - x_{pn-l}\| + \|y_{pn+1-l} - y_{pn-l}\| - 2d)$ and $r = \|x_{pn+p} - x_{pn}\|$ and using the convexity of the set A we get the chain of inequalities

$$d \le \left\|\frac{x_{pn} + x_{pn+p}}{2} - x_{pn+1}\right\| \le \left(1 - \delta_{\|\cdot\|}\left(\frac{\|x_{pn+p+k} - x_{2n}\|}{W}\right)\right)W, \quad (15)$$

where we have denoted $W = d + \gamma^l(\|x_{pn+1-l} - x_{pn-l}\| + \|y_{pn+1-l} - y_{pn-l}\| - 2d)$. From (15) we obtain the inequality

$$\delta_{\|\cdot\|}\left(\frac{\|x_{pn+p+k} - x_{pn+k}\|}{W}\right) \le \frac{\gamma^l W_{pn+1-l+k, pn-l+k}(x,y)}{W}. \quad (16)$$

From the uniform convexity of X is follows that $\delta_{\|\cdot\|}$ is strictly increasing and therefore there exists its inverse function $\delta_{\|\cdot\|}^{-1}$, which is strictly increasing too. From (16) we get

$$\|x_{pn} - x_{pn+p}\| \le W \delta_{\|\cdot\|}^{-1}\left(\frac{\gamma^l W_{pn+1-l, pn-l}(x,y)}{W}\right). \quad (17)$$

By the inequality $\delta_{\|\cdot\|}(t) \ge Ct^q$ it follows that $\delta_{\|\cdot\|}^{-1}(t) \le \left(\frac{t}{C}\right)^{1/q}$.
From (17) and the inequalities

$$\begin{aligned}d &\le d + \gamma^l(\|x_{pn+1-l} - x_{pn-l}\| + \|y_{pn+1-l} - y_{pn-l}\| - 2d)\\ &\le \|x_{pn+1-l} - x_{pn-l}\| + \|y_{pn+1-l} - y_{pn-l}\|\end{aligned}$$

we obtain

$$\begin{aligned}\|x_{pn} - x_{pn+p}\| &\le \big(d + \gamma^l W_{pn+1-l, pn-l}\big)\sqrt[q]{\frac{\gamma^l W_{pn+1-l, pn-l}}{C(d + \gamma^l W_{pn+1-l, pn-l})}}\\ &\le P_{pn+1-l, pn-l}\sqrt[q]{\frac{W_{pn+1-l, pn-l}}{Cd}}(\sqrt[q]{\gamma})^l.\end{aligned} \quad (18)$$

There exists a unique pair $(z, v) \in A_i \times A_i$, such that $\|z - T(z,v)\| = d$ and z is a limit of the sequence $\{x_{pn}\}_{n=1}^{\infty}$ for any $(x,y) \in A_i \times A_i$.
After a substitution with $l = pn$ and $k = 0$ in (18) we get the inequality

$$\begin{aligned}\sum_{n=1}^{\infty}\big(\|x_{pn} - x_{pn+p}\| + \|y_{pn} - y_{pn+p}\|\big) &\le P_{0,1}(x,y)\sqrt[q]{\frac{W_{0,1}(x,y)}{Cd}}\sum_{n=1}^{\infty}(\sqrt[q]{\gamma})^{pn}\\ &= P_{0,1}(x,y)\sqrt[q]{\frac{W_{0,1}(x,y)}{Cd}} \cdot \frac{\sqrt[q]{\gamma^p}}{1 - \sqrt[q]{\gamma^p}}\end{aligned}$$

and consequently the series $\sum_{n=1}^{\infty}(x_{pn} - x_{pn+p})$ is absolutely convergent. Thus for any $m \in \mathbb{N}$ there holds $z = x_{pm} - \sum_{n=m}^{\infty}(x_{pn} - x_{pn+p})$ and therefore we get the inequality

$$\|z - x_{pm}\| \le \sum_{n=m}^{\infty} \|x_{pn} - x_{pn+p}\| \le P_{0,1}(x,y) \sqrt[q]{\frac{W_{0,1}(x,y)}{Cd}} \cdot \frac{(\sqrt[q]{\gamma})^{pm}}{1 - \sqrt[q]{\gamma^p}}.$$

The proof for $\|v - y_{pm}\|$ can be done in a similar fashion.
After a substitution with $l = 1 + 2i$ in (18) we obtain

$$\|x_{pn+pi} - x_{pn+p(i+1)}\| \le P_{pn-1,pn}(x,y) \sqrt[q]{\frac{W_{pn-1,pn}(x,y)}{Cd}} (\sqrt[q]{\gamma})^{1+2i}. \tag{19}$$

From (19) we get that there holds the inequality

$$\begin{aligned}
\|x_{pn} - x_{p(n+m)}\| &\le \sum_{i=0}^{m-1} \|x_{pn+pi} - x_{pn+p(i+1)}\| \\
&\le \sum_{i=0}^{m-1} P_{pn-1,pn}(x,y) \sqrt[q]{\frac{W_{pn-1,pn}(x,y)}{Cd}} (\sqrt[q]{\gamma})^{1+pi} \\
&= P_{pn-1,pn}(x,y) \sqrt[q]{\frac{W_{pn-1,pn}(x,y)}{Cd}} \sum_{i=0}^{m-1} (\sqrt[q]{\gamma})^{1+pi} \\
&= P_{pn-1,pn}(x,y) \sqrt[q]{\frac{W_{pn-1,pn}(x,y)}{Cd}} \cdot \frac{1 - (\sqrt[q]{\gamma})^{pm}}{1 - \sqrt[q]{(\gamma)^p}} \sqrt[q]{\gamma}
\end{aligned} \tag{20}$$

and after letting $m \to \infty$ in (20) we obtain the inequality

$$\|x_{pn} - z\| \le P_{pn-1,pn}(x,y) \sqrt[q]{\frac{W_{pn-1,pn}(x,y)}{Cd}} \frac{\sqrt[q]{\gamma}}{1 - \sqrt[q]{\gamma^p}}.$$

The proof for $\|y_{pn} - v\|$ can be done in a similar fashion. □

5. Applications

Let $\varphi, \psi : [1, +\infty) \to [1, +\infty)$ be such that $\max\{\varphi(x), \psi(x)\} \le x$ for any $x \in [1, +\infty)$. Let us define the function $f(x,y) = \lambda + (1-\lambda)(\mu\varphi(x) + (1-\mu)\psi(y))$. Let us consider the system of equations

$$\begin{cases}
|x|^p + |\lambda + (1-\lambda)(\mu\varphi(x) + (1-\mu)\psi(y))|^p &= 2 \\
|y|^p + |\lambda + (1-\lambda)(\mu\psi(y) + (1-\mu)\varphi(x))|^p &= 2 \\
x - f(f(f(x,y), f(y,x)), f(f(y,x), f(x,y))) &= 0 \\
y - f(f(y,x), f(x,y)), f(f(x,y), f(y,x))) &= 0
\end{cases} \tag{21}$$

for $x, y \ge 0$ and $\lambda, \mu \in (0,1)$.

Let $A_1 = \{(x,0,0) : x \ge 1\}$, $A_2 = \{(0,x,0) : x \ge 1\}$, $A_3 = \{(0,0,x) : x \ge 1\}$ be subsets of $(\mathbb{R}^3, \|\cdot\|_p)$, $p \in (1, \infty)$. Let us define the map T by $T((x,0,0), (y,0,0)) = (0, f(x,y), 0)$; $T((0,x,0), (0,y,0)) = (0, 0, f(x,y))$; $T((0,0,x), (0,0,y)) = (f(x,y), 0, 0)$ for some $\lambda, \mu \in (0,1)$. It is easy to see that for any $x, y \ge 1$ there holds $f(x,y) \ge 1$ and therefore $T : A_i \times A_i \to A_{i+1}$.

From the inequality, using that $(1 - (1-\lambda)\mu - (1-\lambda)(1-\mu)) = \lambda$

$$\begin{aligned}
S_9 &= \|T((x,0,0),(y,0,0)) - T((0,u,0),(0,v,0))\|_p \\
&= \|(0, f(x,y), f(u,v))\|_p = \sqrt[p]{|f(x,y)|^p + |f(u,v)|^p} \\
&\leq \sqrt[p]{\lambda^p + \lambda^p} + (1-\lambda)\sqrt[p]{|\mu\varphi(x) + (1-\mu)\psi(y)|^p + |\mu\varphi(u) + (1-\mu)\psi(v)|^p} \\
&\leq \lambda\sqrt[p]{2} + (1-\lambda)\mu\sqrt[p]{|\varphi(x)|^p + |\varphi(u)|^p} + (1-\lambda)(1-\mu)\sqrt[p]{|\psi(y)|^p + |\psi(v)|^p} \\
&\leq \lambda\sqrt[p]{2} + (1-\lambda)\mu\sqrt[p]{|x|^p + |u|^p} + (1-\lambda)(1-\mu)\sqrt[p]{|y|^p + |v|^p} \\
&= \lambda\,\mathrm{dist}(A_1, A_2) + (1-\lambda)\mu\|x-u\|_2 + (1-\lambda)(1-\mu)\|y-v\|_p
\end{aligned}$$

and

$$\begin{aligned}
S_{10} &= \|T((x,0,0),(y,0,0)) - T((0,u,0),(0,v,0))\|_p \\
&= \|T((0,x,0),(0,y,0)) - T((0,0,u),(v,0,0))\|_p \\
&= \|T((0,0,x),(0,0,y)) - T((u,0,0),(v,0,0))\|_p
\end{aligned}$$

it follows that T satisfies the conditions of Theorem 2. Therefore there exist (z,z), which is a coupled best proximity point of T in $A_1 \times A_1$ and it is easy to see that $z = (1,0,0)$. Consequently (z,z) is the unique solution of the system of equations

$$\begin{vmatrix} \|x - T(x,y)\|_p^p &=& 2 \\ \|y - T(y,x)\|_p^p &=& 2 \\ x - T^3(x,y) &=& 0 \\ y - T^3(y,x) &=& 0, \end{vmatrix}$$

which is the solution of (21).

If we try to solve (21) with the use of some Computer Algebraic System, for example Maple, the software could not find the exact solution even for not too complicated functions ($p = 2$, $\varphi(x) = x^{1/2}$, $\psi(x) = x$). If we try to solve it numerically, Maple finds that $x = y = 1$, but could not find that this is a solution for every $\lambda, \mu \in (0,1)$ and presents two approximations of λ and μ.

If we consider the particular case $p = 3$, $\varphi(x) = \sqrt{x}$ and $\psi(x) = \sqrt{\log(x) + 1}$, then Maple could not solve (21) even numerically.

6. Discussion

It is interesting whether same conclusions can be made for existence of coupled fixed (or best proximity) points p-cyclic Meir–Keeler maps [16], Reich Maps p-cyclic maps [25].

We were not able to prove a uniqueness of the coupled best proximity points, as like as [13,16]. We were able to prove just uniqueness of the best proximity points, if obtained by the sequence of successive iterations, which is not the case of 2–cyclic maps. It will be interesting if this gap can be filled.

Author Contributions: The listed authors have made equal contributions to the presented research. All authors have read and agreed to the published version of the manuscript.

Funding: M.H. is partially supported by Shumen University Grant Number Rd-08-42/ 2021, and D.N. is partially supported by the Bulgarian National Fund for Scientific Research Grant Number KP-06-H22/4.

Institutional Review Board Statement: Not applicable.

Informed Consent Statement: Not applicable.

Data Availability Statement: Not applicable.

Acknowledgments: The authors would like to thank the anonymous reviewers for their comments and recommendations that have improved the value of the article.

Conflicts of Interest: The authors declare no conflict of interest.

References

1. Guo, D.; Lakshmikantham, V. Coupled fixed points of nonlinear operators with application. *Nonlinear Anal.* **1987**, *11*, 623–632. [CrossRef]
2. Das, A.; Hazarika, B.; Parvaneh, V.; Mursaleen, M. Solvability of generalized fractional order integral equations via measures of noncompactness. *Math. Sci.* **2021**. [CrossRef]
3. George, R.; Mitrovic, Z.D.; Radenovic, S. On some coupled fixed points of generalized T–contraction mappings in a $b(v)(s)$–metric space and its application. *Axioms* **2020**, *9*, 129. [CrossRef]
4. Mani, G.; Mishra, L.N.; Mishra, V.N.; Baloch, I.A. Application to coupled fixed–point theorems on complex partial b-metric space. *J. Math.* **2020**, *2020*, 8881859. [CrossRef]
5. Shateri, T.L. Coupled fixed points theorems for non-linear contractions in partially ordered modular spaces. *Int. J. Nonlinear Anal. Appl.* **2020**, *11*, 133–147. [CrossRef]
6. Dzhabarova, Y.; Kabaivanov, S.; Ruseva, M.; Zlatanov, B. Existence, Uniqueness and Stability of Market Equilibrium in Oligopoly Markets. *Adm. Sci.* **2020**, *10*, 70. [CrossRef]
7. Kabaivanov, S.; Zlatanov, B. A variational principle, coupled fixed points and market equilibrium. *Nonlinear Anal. Model. Control* **2021**, *26*, 169–185. [CrossRef]
8. Kirk, W.; Srinivasan, P.; Veeramani, P. Fixed Points for Mapping Satisfying Cyclical Contractive Conditions. *Fixed Point Theory* **2003**, *4*, 79–89.
9. Eldred, A.; Veeramani, P. Existence and Convergence of Best Proximity Points. *J. Math. Anal. Appl.* **2006**, *323*, 1001–1006. [CrossRef]
10. Goswami, N. Some coupled best proximity point results for weak GKT cyclic Φ-contraction mappings on metric spaces. *Proc. Jangjeon Math. Soc.* **2021**, *23*, 485–502. [CrossRef]
11. Gabeleh, M.; Künzi, H.-P.A. Mappings of Generalized Condensing Type in Metric Spaces with Busemann Convex Structure. *Bull. Iran. Math. Soc.* **2020**, *46*, 1465–1483. [CrossRef]
12. Gopi, R.; Pragadeeswarar, V. Determining fuzzy distance via coupled pair of operators in fuzzy metric space. *Soft Comput.* **2020**, *24*, 9403–9412. [CrossRef]
13. Karpagam, S.; Agrawal, S. Existence of best proximity Points of p-cyclic contractions. *Fixed Point Theory* **2012**, *13*, 99–105.
14. Felicit, J.M.; Eldred, A.A. Best proximity points for cyclical contractive mappings. *Appl. Gen. Topol.* **2005**, *16*, 119–126. [CrossRef]
15. Hemalatha, P.K.; Gunasekar, T.; Karpagam, S. Existence of fixed point and best proximity point of p-cyclic orbital contraction of Boyd—Wong type, *Int. J. Appl. Math.* **2018**, *31*, 805–814. [CrossRef]
16. Karpagam, S.; Agrawal, S. Best proximity point theorems for p-cyclic Meir–Keeler contractions. *Fixed Point Theory Appl.* **2009**, *2009*, 197308. [CrossRef]
17. Vetro, C. Best proximity points: Convergence and existence theorems for p-cyclic mappings. *Nonlinear Anal.* **2010**, *73*, 2283–2291. [CrossRef]
18. Zlatanov, B. Error estimates for approximating best proximity points for cyclic contractive maps. *Carpathian J. Math.* **2016**, *32*, 265–270.
19. Ilchev, A.; Zlatanov, B. Error estimates for approximation of coupled best proximity points for cyclic contractive maps. *Appl. Math. Comput.* **2016**, *290*, 412–425. [CrossRef]
20. Clarkson, J. Uniformly convex space. *Trans. Amer. Math. Soc.* **1936**, *40*, 394–414. [CrossRef]
21. Beauzamy, B. *Introduction to Banach Spaces and Their Geometry*; North—Holland Publishing Company: Amsterdam, The Netherlands, 1979; ISBN 978-0-444-86416-1.
22. Deville, R.; Godefroy, G.; Zizler, V. *Smoothness and Renormings in Banach Spaces*; Longman Scientific & Technical: Harlow, UK, 1993; ISBN 978-0-582-07250-3.
23. Fabian, M.; Habala, P.; Hájek, P.; Montesinos, V.; Pelant, J.; Zizler, V. *Functional Analysis and Infinite-Dimensional Geometry*; Springer: New York, NY, USA, 2011; ISBN 978-1-4757-3480-5.
24. Sintunavarat, W.; Kumam, P. Coupled best proximity point theorem in metric spaces. *Fixed Point Theory Appl.* **2012**, *2012*, 1–16. [CrossRef]
25. Ilchev, A.; Zlatanov, B. Error Estimates of Best Proximity Points for Reich Maps in Uniformly Convex Banach Spaces. *Annu. Konstantin Preslavsky Univ. Shumen Fac. Math. Inform.* **2018**, *XIX*, 3–20.

Applications of Coupled Fixed Points for Multivalued Maps in the Equilibrium in Duopoly Markets and in Aquatic Ecosystems

Gana Gecheva [1], Miroslav Hristov [2,*], Diana Nedelcheva [3], Margarita Ruseva [4] and Boyan Zlatanov [5]

1. Department of Ecology and Environmental Conservation, Faculty of Biology, University of Plovdiv "Paisii Hilendarski", 24 Tzar Assen Str., 4000 Plovdiv, Bulgaria; ggecheva@uni-plovdiv.bg
2. Department of Algebra and Geometry, Faculty of Mathematics and Computer Science, University of Shumen "Konstantin Preslavsky", 115 Universitetska Str., 9700 Shumen, Bulgaria
3. Department of Mathematics and Physics, Faculty of Electrical Engineering, Technical University of Varna, 1 Studentska Str., 9000 Varna, Bulgaria; diana.nedelcheva@tu-varna.bg
4. Department of Management and Quantitative Economicsx, Faculty of Economics and Social Sciences, University of Plovdiv "Paisii Hilendarski", 24 Tzar Assen Str., 4000 Plovdiv, Bulgaria; m_ruseva@uni-plovdiv.bg
5. Department of Mathematical Analysis, Faculty of Mathematics and Informatics, University of Plovdiv "Paisii Hilendarski", 24 Tzar Assen Str., 4000 Plovdiv, Bulgaria; bobbyz@uni-plovdiv.bg
* Correspondence: miroslav.hristov@shu.bg; Tel.: +359-89-336-5666

Abstract: We have obtained a new class of ordered pairs of multivalued maps that have pairs of coupled fixed points. We illustrate the main result with two examples that cover a wide range of models. We apply the main result in models in duopoly markets to get a market equilibrium and in aquatic ecosystems, also to get an equilibrium.

Keywords: multivalued maps; coupled fixed points; equilibrium

MSC: Primary 47H04; 47H10; 91B26

1. Introduction

Fixed point theory has been extensively researched and widely applied in a multitude of directions for many years. The "Banach Contraction Principle" states that under certain conditions a self map T on a set X admits one or more fixed points $x = Tx$. The "Banach Contraction Principle" and its numerous generalizations are widely used in many branches of mathematics because it requires only the structure of a complete metric space with conditions on the map which are easily tested.

We will mention just a few directions of the generalizations (fixed points for set-valued maps, coupled fixed points, fixed points for cyclic maps) of "Banach Contraction Principle" that initiate the present investigation.

Following the "Banach Contraction Principle", Nadler introduced the concept of set-valued contractions in [1]. He also proved that a set-valued contraction possesses a fixed point in a complete metric space. In the late twentieth century Dontchev and Hager successfully presented an extension of Nadler's result in [2]. They determined the location of a fixed point with respect to an initial value of the set-valued mapping. Their conclusion was obtained under two modified conditions and it has since been playing an important role in the development of the metric fixed point theory. We would like just to mention a few recent results about fixed points for set-valued maps and their applications [3–6].

A different direction is the notion of coupled fixed points introduced in [7]. There are a lot of recent results about coupled fixed points [8–11].

Another kind of generalization of the Banach contraction principle is the notation of cyclic maps [12] and later its generalization to the best proximity point, introduced in [13]. The definition presented in [13] is more general than the one in [12], in the sense that if the sets intersect, then every best proximity point is a fixed point. Some very recent results in this field are presented in [14–17].

It seems that recently all mentioned directions of research in fixed point theory are of interest.

By combining the notions of coupled fixed (or best proximity) points for cyclic maps, a model of duopoly market was built in [18,19].

We will try to enrich the notions of set-valued maps, coupled fixed points, cyclic maps and to get results that we will apply in economics and ecology.

2. Preliminaries

We will recall basic notions and facts that we will need for investigation of coupled fixed points for multi-valued maps. Let (X, ρ) and (Y, σ) be two metric spaces. We will denote by $B_{X,r}(\bar{x})$ the open ball and by $B_{X,r}[\bar{x}]$ the closed ball with a radius r and a center x in the metric space X. If no confusion arises, we will denote them with $B_r(\bar{x})$ and $B_r[\bar{x}]$, respectively. Let $x \in X$ and $C \subset X$. We will denote the distance from x to C by $d(x, C) = \inf\{\rho(x, z) : z \in C\}$. If $C = \emptyset$ then we put $d(x, \emptyset) = \infty$.

Let $A, B \subset X$ be two subsets. An excess of A beyond B is called $e(A, B) = \sup\{d(x, B) : x \in A\}$, where the convention is used that

$$e(\emptyset, B) = \begin{cases} 0, & B \neq \emptyset \\ \infty, & B = \emptyset \end{cases}.$$

Let (X, ρ) and (Y, σ) be two metric spaces. Let us denote by $F : X \rightrightarrows Y$ a set-valued mapping defined on the metric space (X, ρ) with values in the metric space (Y, σ). Let F be a set-valued map: Its graph is the set $\text{gph } F = \{(x, y) \in X \times Y \mid y \in F(x)\}$, its effective domain is the set $\text{dom } F = \{x \in X \mid F(x) \neq \emptyset\}$ and its effective range is $\text{rge } F = \{y \in Y \mid \text{there exists } x \text{ such that } y \in F(x)\}$.

Definition 1. ([1]) *A point $x \in X$ is said to be a fixed point of the set-valued map $F : X \rightrightarrows X$ if $x \in F(x)$.*

Definition 2. ([7]) *A point $(x, y) \in X \times X$ is said to be a coupled fixed point of the map $F : X \times X \to X$ if $x = F(x, y)$ and $y = F(y, x)$.*

Definition 3. ([20]) *A point $(x, y) \in X \times X$ is said to be a coupled fixed point of the set-valued map $F : X \times X \rightrightarrows X$ if $x \in F(x, y)$ and $y \in F(y, x)$.*

The model that will be constructed in the application section will be of two set-valued maps $F_1 : X \times Y \rightrightarrows X$ and $F_2 : X \times Y \rightrightarrows Y$ and we will be interested in the existence of ordered pairs (x, y), such that $x \in F_1(x, y)$ and $y \in F_2(x, y)$, which are called generalized coupled fixed points for the ordered pair of set-valued maps (F_1, F_2).

3. Main Results

We will present a result, which extends the result from [2] and establishes a solution of the generalized coupled fixed point problem for an ordered pair of set-valued maps.

Theorem 1. *Let (X, ρ) and (Y, σ) be complete metric spaces, $F_1 : X \times Y \rightrightarrows X$ and $F_2 : X \times Y \rightrightarrows Y$ be multi-valued maps and $\bar{x} \in X, \bar{y} \in Y$. Let there exist a constant $r > 0$ and non-negative constants $\alpha, \beta, \gamma, \delta$, satisfying $\max\{\alpha + \gamma, \beta + \delta\} < 1$, such that the following assumptions hold:*

(a) *$F_1(x, y)$ and $F_2(x, y)$ are nonempty closed subsets of X and Y for all $(x, y) \in B_r(\bar{x}) \times B_r(\bar{y})$*

(b) *the inequality $d(\bar{x}, F_1(\bar{x}, \bar{y})) + d(\bar{y}, F_2(\bar{x}, \bar{y})) < r(1 - \lambda)$ holds, where $\lambda = \max\{\alpha + \gamma, \beta + \delta\}$*

(c) the inequality

$$\begin{aligned} S_1 &= e(F_1(x,y) \cap B_r(\bar{x}), F_1(u,v)) + e(F_2(z,w) \cap B_r(\bar{y}), F_2(t,s)) \\ &\leq \alpha\rho(x,u) + \beta\sigma(y,v) + \gamma\rho(z,t) + \delta\sigma(w,s) \end{aligned} \quad (1)$$

holds for all $(x,y), (u,v), (z,w), (t,s) \in B_r(\bar{x}) \times B_r(\bar{y})$.

Then, the generalized coupled fixed point problem has at least one solution $(x,y) \in B_r(\bar{x}) \times B_r(\bar{y})$.

Proof. Let $\bar{x} = x_0$ and $\bar{y} = y_0$.

We will construct by induction a sequence $\{(x_n, y_n)\}_{n=1}^{\infty}$, which will satisfy the inclusions $x_{n+1} \in F_1(x_n, y_n) \cap B_r(x_0)$ and $y_{n+1} \in F_2(x_n, y_n) \cap B_r(y_0)$ for $n \geq 0$.

Step one of the induction: We will choose (x_1, y_1). By assumption (b), there exist $x_1 \in F_1(x_0, y_0)$ and $y_1 \in F_2(x_0, y_0)$ such that $\rho(x_1, x_0) + \sigma(y_1, y_0) < r(1-\lambda) < r$. Therefore, $x_1 \in F_1(x_0, y_0) \cap B_r(x_0)$ and $y_1 \in F_2(x_0, y_0) \cap B_r(y_0)$.

We will proceed with the choice of (x_2, y_2). From (1) we have the chain of inequalities

$$\begin{aligned} S_2 &= d(x_1, F_1(x_1, y_1)) + d(y_1, F_2(x_1, y_1)) \\ &\leq e(F_1(x_0, y_0) \cap B_r(x_0), F_1(x_1, y_1)) + e(F_2(x_0, y_0) \cap B_r(y_0), F_2(x_1, y_1)) \\ &\leq \alpha\rho(x_1, x_0) + \beta\sigma(y_1, y_0) + \gamma\rho(x_1, x_0) + \delta\sigma(y_1, y_0) \\ &= (\alpha + \gamma)\rho(x_1, x_0) + (\beta + \delta)\sigma(y_1, y_0) \\ &\leq \max\{\alpha + \gamma, \beta + \delta\}(\rho(x_1, x_0) + \sigma(y_1, y_0)) \\ &< r(1-\lambda)\lambda. \end{aligned}$$

The above inequalities imply the existence of $x_2 \in F_1(x_1, y_1)$ and $y_2 \in F_2(x_1, y_1)$, such that $\rho(x_2, x_1) + \sigma(y_2, y_1) < r(1-\lambda)\lambda$. Using the triangular inequality we get

$$\begin{aligned} \rho(x_2, x_0) + \sigma(y_2, y_0) &\leq \rho(x_2, x_1) + \sigma(y_2, y_1) + \rho(x_1, x_0) + \sigma(y_1, y_0) \\ &\leq r(1-\lambda)\lambda + r(1-\lambda) < r. \end{aligned}$$

Consequently $x_2 \in F_1(x_1, y_1) \cap B_r(x_0)$ and $y_2 \in F_2(x_1, y_1) \cap B_r(y_0)$.

Step two of the induction: Let us suppose that we have already chosen $\{(x_k, y_k)\}_{k=1}^{n}$, satisfying for each $k = 1, 2, \ldots, n$

$$x_k \in F_1(x_{k-1}, y_{k-1}) \cap B_r(x_0) \ , \ y_k \in F_2(x_{k-1}, y_{k-1}) \cap B_r(y_0)$$

and

$$\rho(x_k, x_{k-1}) + \sigma(y_k, y_{k-1}) < r(1-\lambda)\lambda^{k-1}.$$

Step three of the induction: We will prove that we can choose (x_{n+1}, y_{n+1}), provided that we have already chosen $\{(x_k, y_k)\}_{k=1}^{n}$. By assumption (1) we have the chain of inequalities

$$\begin{aligned} S_3 &= d(x_n, F_1(x_n, y_n)) + d(y_n, F_2(x_n, y_n)) \\ &\leq e(F_1(x_{n-1}, y_{n-1}) \cap B_r(x_0), F_1(x_n, y_n)) + e(F_2(x_{n-1}, y_{n-1}) \cap B_r(y_0), F_2(x_n, y_n)) \\ &\leq \alpha\rho(x_n, x_{n-1}) + \beta\sigma(y_n, y_{n-1}) + \gamma\rho(x_n, x_{n-1}) + \delta\sigma(y_n, y_{n-1}) \\ &= (\alpha + \gamma)\rho(x_n, x_{n-1}) + (\beta + \delta)\sigma(y_n, y_{n-1}) \\ &\leq \max\{\alpha + \gamma, \beta + \delta\}(\rho(x_n, x_{n-1}) + \sigma(y_n, y_{n-1})) \\ &< \lambda r(1-\lambda)\lambda^{n-1} = r(1-\lambda)\lambda^n. \end{aligned}$$

The above inequalities imply that there exist

$$x_{n+1} \in F_1(x_n, y_n) \text{ and } y_{n+1} \in F_2(x_n, y_n),$$

such that
$$\rho(x_n, x_{n+1}) + \sigma(y_n, y_{n+1}) < r(1-\lambda)\lambda^n. \qquad (2)$$

Using the triangular inequality we get

$$\rho(x_{n+1}, x_0) + \sigma(y_{n+1}, y_0) \le \sum_{k=0}^{n} \rho(x_{k+1}, x_k) + \sigma(y_{k+1}, y_k) < r(1-\lambda) \sum_{k=0}^{n} \lambda^k < r.$$

Consequently, $x_{n+1} \in F_1(x_n, y_n) \cap B_r(x_0)$ and $y_{n+1} \in F_2(x_n, y_n) \cap B_r(y_0)$ and this completes the induction.

It follows from (2) that $\max\{\rho(x_n, x_{n+1}), \sigma(y_n, y_{n+1})\} < r(1-\lambda)\lambda^n$ and thus the inequality

$$\rho(x_n, x_m) \le \sum_{k=m}^{n-1} \rho(x_{k+1}, x_k) < r(1-\lambda)\lambda^m \sum_{k=0}^{n-1} \lambda^k < r\lambda^m$$

holds for any $n > m$.

Therefore, $\{x_n\}_{n=0}^{\infty}$ is a Cauchy sequence and from the assumption that X is a complete metric space it follows that $\{x_n\}_{n=0}^{\infty}$ converges to some $x^* \in B_r(x_0)$. By similar arguments we get that the sequence $\{y_n\}_{n=0}^{\infty}$ is a Cauchy sequence and converges to some $y^* \in B_r(y_0)$.

By using of assumption (1) we get the chain of inequalities

$$\begin{aligned}
S_4 &= d(x_n, F_1(x^*, y^*)) + d(y_n, F_2(x^*, y^*)) \\
&\le e(F_1(x_{n-1}, y_{n-1}) \cap B_r(x_0), F_1(x^*, y^*)) + e(F_2(x_{n-1}, y_{n-1}) \cap B_r(y_0), F_2(x^*, y^*)) \\
&\le \alpha\rho(x^*, x_{n-1}) + \beta\sigma(y^*, y_{n-1}) + \gamma\rho(x^*, x_{n-1}) + \delta\sigma(y^*, y_{n-1}) \\
&\le (\alpha+\gamma)\rho(x^*, x_{n-1}) + (\beta+\delta)\sigma(y^*, y_{n-1}) \\
&\le \lambda(\rho(x^*, x_{n-1}) + \sigma(y^*, y_{n-1})).
\end{aligned}$$

Applying the triangle inequality we obtain

$$\begin{aligned}
S_5 &= d(x^*, F_1(x^*, y^*)) + d(y^*, F_2(x^*, y^*)) \\
&\le \rho(x^*, x_n) + d(x_n, F_1(x^*, y^*)) + \sigma(y^*, y_n) + d(y_n, F_2(x^*, y^*)) \\
&\le \rho(x^*, x_n) + \lambda((\rho(x^*, x_{n-1}) + \sigma(y^*, y_{n-1})) + \sigma(y^*, y_n).
\end{aligned} \qquad (3)$$

After taking a limit as $n \to \infty$ in (3) we get that $d(x^*, F_1(x^*, y^*)) + d(y^*, F_2(x^*, y^*)) = 0$. From the assumption that $F_1(x^*, y^*)$ and $F_2(x^*, y^*)$ are closed it follows that $x^* \in F_1(x^*, y^*)$ and $y^* \in F_2(x^*, y^*)$, i.e (x^*, y^*) is a generalized coupled fixed point. □

Remark 1. *If F_1 and F_2 are single-valued, then assumption (1) implies that (x^*, y^*) is the unique coupled fixed point of (F_1, F_2) in $B_r(x_0) \times B_r(y_0)$.*

4. Examples and Applications

We will illustrate Theorem 1 with two examples. We will use these two examples to construct models in economics and ecology.

4.1. Examples

Example 1: Let us choose $0 \le \alpha < \beta < \gamma < \delta \le \eta < +\infty$, $n, m \in (0, 1]$, so that

$$\max\left\{\frac{n(\gamma-\beta) + m(\gamma-\beta)}{2((\eta+1)^n - (\alpha+1)^n)}, \frac{n(\gamma-\beta) + m(\gamma-\beta)}{2(\delta+1)^n}\right\} < 1.$$

Let us define the maps

$$f: [0, \delta] \to \left[\frac{\beta+\gamma}{2}, \gamma\right], \quad g: [\alpha, \eta] \to \left[\beta, \frac{\beta+\gamma}{2}\right],$$

$$\varphi: [0, \delta] \to \left[\frac{\beta+\gamma}{2}, \gamma\right], \quad \psi: [\alpha, \eta] \to \left[\beta, \frac{\beta+\gamma}{2}\right].$$

by

$$f(x) = \frac{\gamma - \beta}{2(\delta+1)^n}(x+1)^n + \frac{\beta+\gamma}{2},$$

$$g(x) = \frac{\gamma - \beta}{2((\eta+1)^n - (\alpha+1)^n)}(x+1)^n + \beta - (\alpha+1)^n \frac{\gamma - \beta}{2((\eta+1)^n - (\alpha+1)^n)},$$

$$\varphi(x) = \frac{\gamma - \beta}{2(\delta+1)^m}(x+1)^m + \frac{\beta+\gamma}{2},$$

$$\psi(x) = \frac{\gamma - \beta}{2((\eta+1)^m - (\alpha+1)^m)}(x+1)^m + \beta - (\alpha+1)^m \frac{\gamma - \beta}{2((\eta+1)^m - (\alpha+1)^m)}.$$

Let us denote $\bar{x} = \bar{y} = \frac{\beta+\gamma}{2}$ and $\theta = \min\{|\delta - \bar{x}|, |\alpha - \bar{x}|\}$. Let us endow $\mathbb{R}$ with the absolute value metrics $|\cdot - \cdot|$. Let us consider the sets $X = [0, \delta]$, $Y = [\alpha, \eta]$. Let us define the multivalued maps $F: X \times Y \rightrightarrows X$ and $G: X \times Y \rightrightarrows Y$ by

$$F(x, y) = \{\xi : g(y) \leq \xi \leq f(x)\}$$

and

$$G(x, y) = \{\xi : \psi(y) \leq \xi \leq \varphi(x)\}.$$

We will check that F and G satisfy Theorem 1.

It is easy to see that for any $(x, y) \in B_r(\bar{x}) \times B_r(\bar{y})$ the sets $F(x, y) = [g(y), f(x)]$ and $G(x, y) = [\psi(y), \varphi(x)]$ are non empty and closed subsets of X or Y, respectively.

From $g(\bar{x}) \leq \bar{x} \leq f(\bar{x})$ and $\psi(\bar{x}) \leq \bar{x} \leq \varphi(\bar{x})$ we get that $F(\bar{x}, \bar{x}) \rightrightarrows [g(\bar{x}), f(\bar{x})]$ and $G(\bar{x}, \bar{x}) \rightrightarrows [\psi(\bar{x}), \varphi(\bar{x})]$. Then

$$d(\bar{x}, F(\bar{x}, \bar{x})) + d(\bar{x}, G(\bar{x}, \bar{x})) = 0 < r(1 - \lambda)$$

for any $r > 0$ and any $\lambda \in [0, 1)$.

From $F(x, y) \rightrightarrows [g(y), f(x)] \subseteq B_r(\bar{x})$ and $G(x, y) \rightrightarrows [\psi(y), \varphi(x)] \subseteq B_r(\bar{y})$ it follows that $F(x, y) \cap B_r(\bar{x}) = F(x, y) = [g(y), f(x)]$ and $G(x, y) \cap B_r(\bar{y}) = G(x, y) = [\psi(y), \varphi(x)]$ for $r = \theta$.

Consequently,

$$\begin{aligned} e(F(x, y) \cap B_r(\bar{x}), F(u, v)) &= e([g(y), f(x)], [g(v), f(u)]) \\ &= \sup\nolimits_{x \in [g(y), f(x)]} d(x, [g(v), f(u)]). \end{aligned}$$

There are four cases:
(I) $x \leq u$ and $y \leq v$; (II) $x \leq u$ and $y \geq v$; (III) $x \geq u$ and $y \leq v$; (IV) $x \geq u$ and $y \geq v$.
We will need the inequalities:

$$|f(x) - f(y)| = f'(\zeta)|x - y| \leq n \frac{\gamma - \beta}{2((\eta+1)^n - (\alpha+1)^n)}|x - y|,$$

and

$$|g(x) - g(y)| = g'(\zeta)|x - y| \leq n \frac{\gamma - \beta}{2(\delta+1)^n}|x - y|.$$

Case (I). We will illustrate this case with a figure for easier reading (Figure 1). The other three cases are similar.

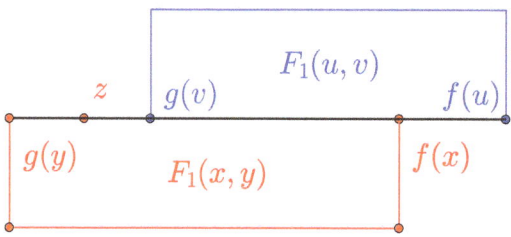

Figure 1. Case (I) $x \leq u$ and $y \leq v$.

$$\begin{aligned} S_5 &= \sup_{z \in [g(y), f(x)]} d(z, [g(v), f(u)]) \\ &= \sup_{g(y) \leq z \leq g(v)} |z - g(v)| = |g(y) - g(v)| = n \frac{\gamma - \beta}{2(\delta + 1)^n} |y - v|. \end{aligned}$$

Case (II). In this case $F_1(x, y) \subseteq F_1(u, v)$ and we get

$$\sup_{z \in [g(y), f(x)]} d(z, [g(v), f(u)]) = 0.$$

Case (III). In this case $F_1(u, v) \subseteq F_1(x, y)$ and we get

$$\begin{aligned} S_6 &= \sup_{z \in [g(y), f(x)]} d(z, [g(v), f(u)]) = \max\{|g(y) - g(v)|, |f(x) - f(u)|\} \\ &= \max\left\{ n \frac{\gamma - \beta}{2(\delta + 1)^n} |y - v|, n \frac{\gamma - \beta}{2((\eta + 1)^n - (\alpha + 1)^n)} |x - u| \right\}. \end{aligned}$$

Case (IV). This case is very similar to case I) and we get

$$\sup_{z \in [g(y), f(x)]} d(z, [g(v), f(u)]) = |f(x) - f(u)| = n \frac{\gamma - \beta}{2((\eta + 1)^n - (\alpha + 1)^n)} |x - u|.$$

Therefore by combining the four Cases (I) to (IV) we get

$$\begin{aligned} S_7 &= e(F(x, y) \cap B_r(\overline{x}), F(u, v)) \\ &\leq \max\left\{ n \frac{\gamma - \beta}{2(\delta + 1)^n} |y - v|, n \frac{\gamma - \beta}{2((\eta + 1)^n - (\alpha + 1)^n)} |x - u| \right\} \\ &\leq n \frac{\gamma - \beta}{2(\delta + 1)^n} |y - v| + n \frac{\gamma - \beta}{2((\eta + 1)^n - (\alpha + 1)^n)} |x - u|. \end{aligned}$$

By similar calculations we can get that

$$\begin{aligned} S_8 &= e(G(z, w) \cap B_r(\overline{y}), G(t, s)) \\ &\leq m \frac{\gamma - \beta}{2(\delta + 1)^m} |w - s| + m \frac{\gamma - \beta}{2((\eta + 1)^m - (\alpha + 1)^m)} |z - t|. \end{aligned}$$

Thus there holds the inequality

$$\begin{aligned} S_9 &= e(F(x, y) \cap B_r(\overline{x}), F(u, v)) + e(G(z, w) \cap B_r(\overline{y}), G(t, s)) \\ &\leq n \frac{(\gamma - \beta)|x - u|}{2((\eta + 1)^n - (\alpha + 1)^n)} + m \frac{(\gamma - \beta)|z - t|}{2((\eta + 1)^m - (\alpha + 1)^m)} \\ &\quad + n \frac{\gamma - \beta}{2(\delta + 1)^n} |y - v| + m \frac{\gamma - \beta}{2(\delta + 1)^m} |w - s|. \end{aligned}$$

A particular case can be obtained if $n = m = 1$, $\alpha = 0$, $\beta = 2$, $\gamma = 4$, $\delta = 6$ and $\eta = 8$. We get $f(x) = \varphi(x) = \frac{x}{7} + \frac{22}{7}$, $g(y) = \psi(y) = \frac{y}{8} + 2$, $r = 3$, $\bar{x} = \bar{y} = 3$ and

$$\begin{aligned} S_{10} &= e(F(x,y) \cap B_r(\bar{x}), F(u,v)) + e(G(z,w) \cap B_r(\bar{y}), G(t,s)) \\ &\leq \tfrac{1}{8}|x-u| + \tfrac{1}{7}|y-v| + \tfrac{1}{8}|z-t| + \tfrac{1}{7}|w-s|. \end{aligned}$$

Example 2. Let us consider the space $\mathbb{R}^2$. Let us choose $0 < \alpha_i < \beta_i < \gamma_i < \delta_i < \eta_i < +\infty$, $n_i, m_i \in (0,1]$ for $i = 1,2$, so that

$$\max_{i=1,2}\left\{\frac{n_i(\gamma_i - \beta_i)}{2(\delta_i + 1)^{n_i}}\right\} + \max_{i=1,2}\left\{\frac{m_i(\gamma_i - \beta_i)}{2(\delta_i + 1)^{m_i}}\right\} < 1$$

and

$$\max_{i=1,2}\left\{\frac{n_i(\gamma_i - \beta_i)}{2((\eta_i + 1)^{n_i} - (\alpha_i + 1)^{n_i})}\right\} + \max_{i=1,2}\left\{\frac{m_i(\gamma_i - \beta_i)}{2((\eta_i + 1)^{m_i} - (\alpha_i + 1)^{m_i})}\right\} < 1.$$

Let us define the maps

$$f_i : [0, \delta_i] \to \left[\frac{\beta_i + \gamma_i}{2}, \gamma_i\right], \quad g_i : [\alpha_i, \eta_i] \to \left[\beta_i, \frac{\beta_i + \gamma_i}{2}\right],$$

$$\varphi_i : [0, \delta_i] \to \left[\frac{\beta_i + \gamma_i}{2}, \gamma_i\right], \quad \psi_i : [\alpha_i, \eta_i] \to \left[\beta_i, \frac{\beta_i + \gamma_i}{2}\right]$$

for $i = 1, 2$ by

$$f_i(x) = \frac{\gamma_i - \beta_i}{2(\delta_i + 1)^{n_i}}(x+1)^{n_i} + \frac{\beta_i + \gamma_i}{2},$$

$$g_i(x) = C(x+1)^{n_i} + \beta_i - (\alpha_i + 1)^{n_i} C,$$

$$\varphi_i(x) = \frac{\gamma_i - \beta_i}{2(\delta_i + 1)^{m_i}}(x+1)^{m_i} + \frac{\beta_i + \gamma_i}{2},$$

$$\psi_i(x) = D(x+1)^{m_i} + \beta_i - (\alpha_i + 1)^{m_i} D,$$

where $C = \frac{\gamma_i - \beta_i}{2((\eta_i+1)^{n_i} - (\alpha_i+1)^{n_i})}$ and $D = \frac{\gamma_i - \beta_i}{2((\eta_i+1)^{m_i} - (\alpha_i+1)^{m_i})}$.

Let us denote $\bar{x}_i = \frac{\beta_i + \gamma_i}{2}$ and $\theta_i = \min\{|\delta_i - \bar{x}_i|, |\alpha_i - \bar{x}_i|\}$ for $i = 1, 2$. Let us endow $\mathbb{R}^2$ with the metrics $\rho((x,y),(u,v)) = \left(\left|\frac{x-u}{\theta_1}\right|^p + \left|\frac{y-v}{\theta_2}\right|^p\right)^{1/p}$, $p \in (1, +\infty)$. Let us consider the sets $X_i = [0, \delta_i]$, $Y_i = [\alpha_i, \eta_i]$ for $i = 1,2$ and let $X = X_1 \times X_2$, $Y = Y_1 \times Y_2$. Let us define the multivalued maps $F : X \times Y \rightrightarrows X$ and $G : X \times Y \rightrightarrows Y$ by

$$F((x_1, x_2), (y_1, y_2)) = \{(\xi_1, \xi_2) : g_i(y_i) \leq \xi_i \leq f_i(x_i)\}$$

and

$$G((x_1, x_2), (y_1, y_2)) = \{(\xi_1, \xi_2) : \psi_i(y_i) \leq \xi_i \leq \varphi_i(x_i)\}.$$

We will check that the pair (F, G) satisfies Theorem 1.
Let us choose $r = 2$ and $\bar{x} = \bar{y} = (\bar{x}_1, \bar{x}_2)$. By definition

$$B_r(\bar{x}) \equiv B_r(\bar{y}) = \{x = (x_1, x_2) : \rho(x, \bar{x}) \leq 2\}.$$

It is easy to see that $R_2 \subseteq B_2(\bar{x}) \subseteq R_1$, where R_1 be the rectangular with vertices (α_1, α_2), (δ_1, α_2), (δ_1, δ_2), (α_1, δ_2) and R_2 be the rectangular with vertices (β_1, β_2), (γ_1, β_2), (γ_1, γ_2), (β_1, γ_2) and $B_2(\bar{x})$ (Figure 2) is an ellipse for $p = 2$.

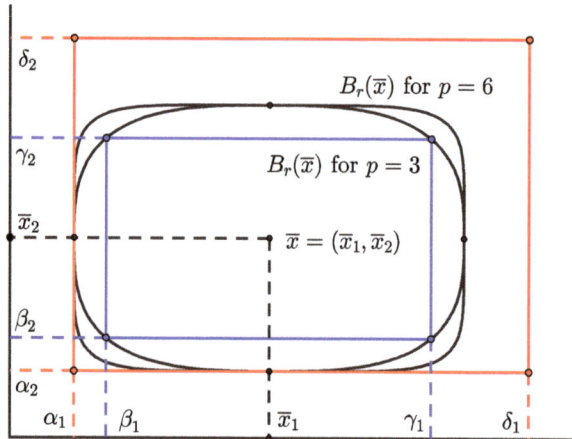

Figure 2. $R_2 \subseteq B_2(\overline{x}) \subseteq R_1$.

Indeed, let $(x_1, x_2) \in R_2$. Then $\beta_i \leq x_i \leq \gamma_i$. Thereafter it holds for $i = 1, 2$

$$\left| \frac{\beta_i + \gamma_i}{2} - x_i \right| = |\overline{x}_i - x_i| \leq \min\{|\delta_i - \overline{x}_i|, |\alpha_i - \overline{x}_i|\} = \theta_i$$

and consequently we can write the inequalities

$$\rho((x_1, x_2), (\overline{x}_1, \overline{x}_2)) = \left(\left| \frac{\overline{x}_1 - x_1}{\theta_1} \right|^p + \left| \frac{\overline{x}_2 - x_2}{\theta_2} \right|^p \right)^{1/p} \leq 2^{1/p} < 2.$$

From $g_i(\overline{x}_i) \leq \overline{x}_i \leq f_i(\overline{x}_i)$ it follows that $\overline{x} = (\overline{x}_1, \overline{x}_2) \in F(\overline{x}_1, \overline{x}_2)$ and from $\psi_i(\overline{x}_i) \leq \overline{x}_i \leq \varphi_i(\overline{x}_i)$ it follows that $\overline{x} = (\overline{x}_1, \overline{x}_2) \in G(\overline{x}_1, \overline{x}_2)$ and therefore $d(\overline{x}, F(\overline{x}, \overline{y})) = d(\overline{y}, G(\overline{x}, \overline{y})) = 0 \leq r(1 - \lambda)$ holds for any $r \geq 1$ and $\lambda \in [0, 1)$.

We observe that there hold $F(x, y) \cap B_r(\overline{x}) = F(x, y)$ and $G(x, y) \cap B_r(\overline{y}) = G(x, y)$. Therefore we will need to calculate $e(F(x, y), F(u, v))$ and $e(G(z, w), G(t, s))$.

The set $F(x, y) = F((x_1, x_2), (y_1, y_2))$ is a rectangular with vertexes $(g(y_1), g(y_2))$, $(f(x_1), g(y_2))$, $(f_1(x_1), f(x_2))$ and $(g(y_1), f(x_2))$. There are several possible cases: $g(y_i) \leq g(v_i)$, or $g(v_i) \leq g(y_i)$ and $f(x_i) \leq f(u_i)$ or $f(u_i) \leq f(x_i)$ with all the possible combinations of $i = 1, 2$.

Let us first consider the case: $g(y_1) \leq g(v_1) \leq f(u_1) \leq f(x_1)$ and $g(y_2) \leq g(v_2) \leq f(u_2) \leq f(x_2)$. It is easy to observe that $e(F(x, y), F(u, v)) = \max\{A_i : i = 1, 2, 3, 4\}$ (Figure 3), where

$$A_1 = \text{dist}((g(y_1), g(y_2)), F(u, v)),$$
$$A_2 = \text{dist}((f(x_1), g(y_2)), F(u, v)),$$
$$A_3 = \text{dist}((f(x_1), f(x_2)), F(u, v)),$$
$$A_4 = \text{dist}((g(y_1), f(x_2)), F(u, v)).$$

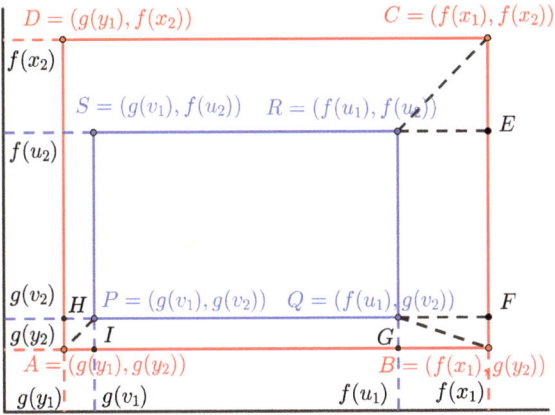

Figure 3. $e(F(x,y), F(u,v)) = \max\{A_i : i = 1,2,3,4\}$.

We will need the inequalities:

$$|f_i(x) - f_i(y)| = f'_i(\zeta)|x-y| \leq n_i \frac{\gamma_i - \beta_i}{2(\delta_i + 1)^{n_i}}|x-y|,$$

$$|g_i(x) - g_i(y)| = g'_i(\zeta)|x-y| \leq n_i \frac{\gamma_i - \beta_i}{2((\eta_i + 1)^{n_i} - (\alpha_i + 1)^{n_i})}|x-y|,$$

where $n_i \in (0,1)$ and

$$f'(\zeta) \leq n_i \frac{\gamma_i - \beta_i}{2(\delta_i + 1)^{n_i}} \max\{(x+1)^{n_i - 1} : x \in [0, \delta_i]\} = n_i \frac{\gamma_i - \beta_i}{2(\delta_i + 1)^{n_i}}$$

and

$$\begin{aligned} g'_i(\zeta) &\leq n_i \frac{\gamma_i - \beta_i}{2((\eta_i+1)^{n_i} - (\alpha_i+1)^{n_i})} \max\{(x+1)^{n_i-1} : x \in [\alpha_i, \eta_i]\} \\ &= n_i \frac{\gamma_i - \beta_i}{2((\eta_i+1)^{n_i} - (\alpha_i+1)^{n_i})} = n_i C_i. \end{aligned}$$

We calculate

$$\begin{aligned} A_1 &= \sqrt[p]{\left|\frac{g_1(v_1)-g_1(y_1)}{\theta_1}\right|^p + \left|\frac{g_2(v_2)-g_2(y_2)}{\theta_2}\right|^p} \\ &= \sqrt[p]{\frac{1}{\theta_1^p}|n_1 C_1|^p |v_1-y_1|^p + \frac{1}{\theta_2^p}|n_2 C_2|^p |v_2-y_2|^p} \\ &\leq \left(\max_{i=1,2}\left\{\frac{n_i(\gamma_i-\beta_i)}{2((\eta_i+1)^{n_i}-(\alpha_i+1)^{n_i})}\right\}\right)\rho((v_1,v_2),(y_1,y_2)) \end{aligned}$$

and

$$\begin{aligned} A_3 &= \sqrt[p]{\left|\frac{f_1(u_1)-f_1(x_1)}{\theta_1}\right|^p + \left|\frac{f_2(u_2)-f_2(x_2)}{\theta_2}\right|^p} \\ &= \sqrt[p]{\frac{1}{\theta_1^p}\left|\frac{n_1(\gamma_1-\beta_1)}{2(\delta_1+1)^{n_1}}\right|^p |u_1-x_1|^p + \frac{1}{\theta_2^p}\left|\frac{n_2(\gamma_2-\beta_2)}{2(\delta_2+1)^{n_2}}\right|^p |u_2-x_2|^p} \\ &\leq \left(\max_{i=1,2}\left\{\frac{n_i(\gamma_i-\beta_i)}{2(\delta_i+1)^{n_i}}\right\}\right)\rho((x_1,x_2),(u_1,u_2)). \end{aligned}$$

For the estimation of A_2 and A_4 let us denote (Figure 3)

$$A = (g_1(y_1), g_2(y_2)), B = (f_1(x_1), g_2(y_2)),$$
$$C = (f_1(x_1), f_2(x_2)), P = (g_1(v_1), g_2(v_2)),$$
$$Q = (f_1(u_1), g_2(v_2)), R = (f_1(u_1), f_2(u_2)),$$
$$I = (g_1(v_1), g_2(y_2)), G = (f_1(u_1), g_2(y_2)),$$
$$F = (f_1(x_1), g_2(v_2)), E = (f_1(x_1), f_2(u_2)).$$

There holds

$$\begin{aligned} A_2 &= \rho(B,Q) \leq \rho(G,Q) + \rho(Q,F) = \rho(P,I) + \rho(R,E) \\ &\leq \rho(A,P) + \rho(R,C) \\ &\leq \left(\max_{i=1,2}\{n_i C_i\}\right)\rho(v,y) + \left(\max_{i=1,2}\left\{\frac{n_i(\gamma_i - \beta_i)}{2(\delta_i + 1)^{n_i}}\right\}\right)\rho(x,u). \end{aligned}$$

By similar observation we can prove that

$$A_4 \leq \left(\max_{i=1,2}\{n_i C_i\}\right)\rho(v,y) + \left(\max_{i=1,2}\left\{\frac{n_i(\gamma_i - \beta_i)}{2(\delta_i + 1)^{n_i}}\right\}\right)\rho(x,u).$$

Consequently $e(F(x,y), F(u,v)) \leq \alpha\rho(x,u) + \beta\rho(y,v)$, where

$$\alpha = \max_{i=1,2}\left\{\frac{n_i(\gamma_i - \beta_i)}{2(\delta_i + 1)^{n_i}}\right\} \text{ and } \beta = \max_{i=1,2}\left\{\frac{n_i(\gamma_i - \beta_i)}{2((\eta_i + 1)^{n_i} - (\alpha_i + 1)^{n_i})}\right\}.$$

Let us denote the two rectangles $F(x,y)$ and $F(u,v)$ by $ABCD$ and $PQRS$, respectively. We have just investigated the case $PQRS \subseteq ABCD$. All the other cases are variants of Figure 4 and we can get that

$$\begin{aligned} e(F(x,y), F(u,v)) &\leq \max\{\rho(A,P), \rho(B,Q), \rho(C,R), \rho(D,S)\} \\ &\leq \alpha\rho(x,u) + \beta\rho(y,v). \end{aligned}$$

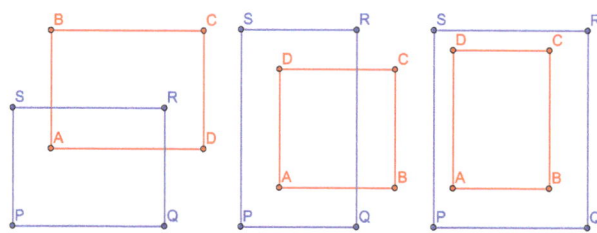

Figure 4. $ABCD$ and $PQRS$.

By similar calculations for the multivalued map G we get

$$e(G(z,w), G(t,s)) \leq \gamma\rho(z,t) + \delta\rho_p(w,s),$$

where

$$\gamma = \max_{i=1,2}\left\{\frac{m_i(\gamma_i - \beta_i)}{2(\delta_i + 1)^{m_i}}\right\} \text{ and } \delta = \max_{i=1,2}\left\{\frac{m_i(\gamma_i - \beta_i)}{2((\eta_i + 1)^{m_i} - (\alpha_i + 1)^{m_i})}\right\}$$

and thus

$$e(F(x,y), F(u,v)) + e(G(z,w), G(t,s)) \leq \alpha\rho(x,u) + \beta\rho(y,v) + \gamma\rho(z,t) + \delta\rho(w,s).$$

4.2. Examples for the Existence of an Equilibrium in Oligopoly (Duopoly) Markets

The theory of oligopoly (duopoly) markets was initiated in [21]. Following [22,23] we present the main features of an oligopoly model in economics. The oligopoly is a market structure in the presence of imperfect competition in which a limited number of large companies control the production and sale of a predominant part of the product in a particular sector of the economy. It is believed that oligopolies are the result of the trend in the economy towards concentration of capital and labor. The oligopoly is characterized by product differentiation; high barriers preventing the emergence of new "players"; limited access to information; non-price competition through advertising and other marketing activities, as well as price control. In the oligopolistic structure, large companies determine the behavior of competitors and take it into account when developing their strategy, which can be rivalry, even "trade wars" in terms of production volume, sales, and prices; the strategic interaction results in agreements (through secret or open collusion or without collusion) in order to guarantee stability and ensure high profits. They contribute to raising economic and organizational barriers, making it difficult for new "players" to emerge. This is the nature of the large initial capital costs for entering the business and achieving minimum effective production and sales capacity in view of economies of scale and resilience against competitors, the development of own research and development for product innovation, industrial and trade secrets. Oligopolistic market structures arise and are imposed by three key points (1) the concentration of assets; (2) inter-firm agreements; and (3) fencing off activities in order to gain market power, restrict competition, and generate large profits.

The distinctive feature of the oligopoly is that in determining individual supply and market price, companies are interdependent. The change in the market behavior of each of them can lead to a change in market conditions and possibly cause a change in the behavior of other companies.

The equilibrium quantity and price in the oligopoly will depend on the number of firms on the market, the information available, the strategy chosen by competitors and whether the firms in the market act independently of each other or in concert. The latter factor is the basis for two types of oligopolistic equilibrium-coherent and inconsistent, which differ significantly in end results and economic efficiency.

The classic model of uncoordinated oligopolistic behavior is the Cournot duopoly model, which considers the problem of the interdependence of firms in the market. The duopoly is a market structure in which two companies, protected from the emergence of other sellers, act as the only producers of standardized products that have no close substitutes, in which there are only two sellers of a particular product that are not interconnected by monopolistic agreement for prices market for selling products and quotas. The participants in the model try to maximize their payoff functions $\Pi_1(x,y) = xP(x+y) - C_1(x)$ and $\Pi_2(x,y) = yP(x+y) - C_2(x)$, respectively, where $P(Z) = P(x+y)$ be the inverse of the demand function and $C_1(x)$ and $C_2(y)$ be the cost functions of the two players. By maximizing its payoff functions, the players get their response functions $F: X \times Y \to X$ and $f: X \times Y \to X$, respectively.

The company equilibrium would become market if the supply of one company is equal to the supply of the other company, so that none of the companies is motivated to change their positions condition for market equilibrium. This condition is present if each of the companies produces one third of the total market supply under conditions of perfect competition and both companies sell at a specific market price, which is one third of the market price. Taking into account the strategic considerations of the companies, their behavior will depend on the decisions of their competitors.

Cournot's theory is based on competition and the fact that buyers announce prices and sellers adjust their products to those prices. Each company evaluates the product search function and then sets the quantity that will be sold, assuming that the competitor's output remains constant.

Deeper research on the oligopoly market can be found in [22–25].

Cournot's classical model deals with a maximization of the payoff functions of each of the players. A different approach is presented in [18], where attention is paid to the response functions of the players. The benefits of this approach are commented on in [18].

We will just say that as far as the players do not have a perfect knowledge of the market they react in some sense by not maximizing their payoff function, but rather by choosing a strategy based on their production and their rival's production levels. The solution of the maximization of the payoff function is actually the coupled fixed points (x, y), such that $x = F(x, y)$ and $y = f(x, y)$.

Focusing on response functions allows us to put Cournot and Bertand's models together. Indeed let the first company reaction be $F(X, Y)$ and the second one $f(X, Y)$, where $X = (x, p)$ and $Y = (y, q)$. Here x and y denote the output quantity and (p, q) are the prices set by players. In this, companies can compete in terms of both price and quantity.

A disadvantage of the presented model is that players do not choose a fixed production of a fixed price. Actually, the response of each player is any quality from a set of possible productions or a price from possible prices. Therefore we will consider the response functions $F : X \times Y \rightrightarrows U \subset X$ and $f : X \times Y \rightrightarrows V \subset Y$ be multivalued maps and a market equilibrium will be the pair (x, y), such that $x \in F(x, y)$ and $y \in f(x, y)$.

Now we can restate Theorem 1 in terms of oligopoly.

Theorem 2. *Let us assume that two companies are offering products that are perfect substitutes. The first one can produce qualities from the set X and the second firm can produce qualities from the set Y, where X and Y be nonempty subsets of a partially ordered complete metric space (Z, ρ) and $\bar{x} \in X, \bar{y} \in Y$. Consider $F : X \times Y \rightrightarrows X$ and $G : X \times Y \rightrightarrows Y$ to be the response function of players one and two, respectively. Let F and G satisfy all the conditions in Theorem 1.*

Then there exists at least one market equilibrium point $(x, y) \in B_r(\bar{x}) \times B_r(\bar{y})$, which is a coupled fixed point for the ordered pair of response functions (F, G).

Example 3: Let us consider in Example 1 two firms, producing one commodity, which is a perfect substitute. Let us put $\alpha = 10$, $\beta = 30$, $\gamma = 50$, $\delta = 80$ and $\eta = 100$ in Example 1. We may consider the interval $[0, \eta]$ as the set of the total production. Let the first firm be a smaller one and its production set is $[0, \delta]$ and the second one be a larger firm with a production set $[\alpha, \eta]$. Let $n = 1$ and $m = 1/2$. Then for any initial start $[x, y]$ in the market the first firm chooses a production from the set

$$\left[\frac{y}{9} + \frac{260}{9}, \frac{10}{81}x + \frac{3250}{81} \right],$$

and the second firm from the set

$$\left[\frac{10\sqrt[2]{y+1} + 30\sqrt[2]{101} - 40\sqrt[2]{11}}{\sqrt[2]{101} - \sqrt[2]{11}}, \frac{10}{9}\sqrt[2]{x+1} + 40 \right],$$

and

$$S_{11} = e(F(x,y) \cap B_r(\bar{x}), F(u,v)) + e(G(z,w) \cap B_r(\bar{y}), G(t,s))$$
$$\leq \tfrac{10}{81}|x-u| + \tfrac{1}{9}|y-v| + \tfrac{5}{9}|z-t| + \gamma|w-s|,$$

where $\gamma = \frac{5}{\sqrt{101}-\sqrt{11}} < \frac{5}{6}$. From $\max\left\{\frac{10}{81} + \frac{5}{9}, \frac{1}{9} + \frac{5}{6}\right\} = \max\left\{\frac{55}{81}, \frac{17}{18}\right\} = \frac{17}{18} < 1$ it follows that the pair of response functions satisfies Theorem 2 and consequently there exists an equilibrium pair of productions (x, y), such that $x \in F(x, y)$ and $y \in G(x, y)$.

Example 4. Let us consider a model of a duopoly with two players, producing one good, which is a complete substitute, and let they compete on qualities and prices simultaneously. Let us choose in Example 2, $\alpha_1 = 10$, $\beta_1 = 30$, $\gamma_1 = 40$, $\delta_1 = 60$, $\eta_1 = 100$, $\alpha_2 = 1$, $\beta_2 = 3$, $\gamma_2 = 4$, $\delta_2 = 5$, $\eta_2 = 8$ $n_1 = 1$, $n_2 = 1/2$, $m_1 = 1/2$, $m_2 = 1/4$. Let us consider the sets $X_i = [0, \delta_i]$, $Y_i = [\alpha_i, \eta_i]$ for $i = 1, 2$ and let $X = X_1 \times X_2$, $Y = Y_1 \times Y_2$ and the multivalued maps $F : X \times Y \rightrightarrows X$ and $G : X \times Y \rightrightarrows Y$ from Example 2, which are the response functions of the two players, respectively, where the first coordinates are the response on the qualities and the second coordinate is the response on the price. Let us endow $\mathbb{R}^2$ with the metrics $\rho((x,y),(u,v)) = \left(\left|\frac{x-u}{\theta_1}\right|^p + \left|\frac{y-v}{\theta_2}\right|^p\right)^{1/p}$, $p \in (1,+\infty)$ from Example 2.

We get that

$$\begin{aligned} S_{12} &= e(F(x,y), F(u,v)) + e(G(z,w), G(t,s)) \\ &\leq 0.2\rho(x,u) + 0.4\rho(y,v) + 0.2\rho(z,t) + 0.4\rho(w,s). \end{aligned}$$

From the inequality $\max\{0.2 + 0.2, 0.4 + 0.4\} < 1$ it follows that we can apply Theorem 2. Consequently there exists an equilibrium pair of productions and prices $((x,p),(y,q))$, such that $(x,p) \in F((x,p),(y,q))$ and $(y,q) \in G((x,p),(y,q))$. The actual values of α, β, γ and δ are smaller.

4.3. Example for the Existence of an Equilibrium in Ecology

Despite the long history of aquatic ecosystems contamination and numerous extensive research undertaken, there are still open questions that remain to be explored. Revealing the relationship between the pollutant, pathway (water) and biota will help water bodies assessment and management. Heavy metals and other contaminants can bioaccumulate in aquatic organisms depending on their bioavailability and concentration in the water media. Among the most applied biomonitors for evaluating sources and releases of contaminants are aquatic bryophytes. Many studies reported a positive correlation between contaminants in aqueous environment and in mosses, for example for Cu [26]. Nevertheless, numerous research have reported that aquatic mosses often accumulate toxic elements in concentrations much higher than those reached in their ambient water media [27] or even when the contaminant in water samples is below the LOD [28].

Now we can restate Theorem 1 in terms of ecology.

Theorem 3. *Let us assume that in an aquatic ecosystem there is one pollutant and a kind of aquatic organisms that accumulate the pollutant. The pollutant can have qualities from the set Y and the aquatic organisms can accumulate qualities from the pollutant from the set X, where X and Y be nonempty subsets of a partially ordered complete metric space (Z, ρ) and $\bar{x} \in X, \bar{y} \in Y$. Consider $F: X \times Y \rightrightarrows X$ and $G: X \times Y \rightrightarrows Y$ to be the response function of the aquatic organisms and the pollutant, respectively. Let F and G satisfy all the conditions in Theorem 1.*

Then there exists at least one point $(x,y) \in B_r(\bar{x}) \times B_r(\bar{y})$, which is a generalized coupled fixed point for the ordered pair of response functions (F, G).

Example 5. Let us consider in Example 1 two media (water and biota), the first one of which is an aquatic ecosystem (e.g. river water), which is polluted continuously and the second one is a bryophyte species that accumulates the contaminant. Let the pollution be from the set $Y = [\alpha, \eta]$ and the accumulated substance in the bryophyte be from the set $X = [0, \delta]$. Let us consider Example 1 with $\alpha = 1, \beta = 3, \gamma = 7, \delta = 8, \eta = 10, n = 3/4$ and $m = 4/5$. If the pollution is y and the accumulated substance in the bryophyte is estimated as x, then, due to the new inflow of pollution and the accumulation of the substance in the bryophyte, which are also reproduced, the pollution and the accumulation change in time to be the multivalued maps

$$F \rightrightarrows [g(y), f(x)] \text{ for all } (x,y) \in X \times Y$$

and

$$G \rightrightarrows [\psi(y), \varphi(x)] \text{ for all } (x,y) \in X \times Y,$$

respectively, where

$$f(x) = \frac{2}{\sqrt[4]{9^3}}\sqrt[4]{(x+1)^3} + 5, \quad g(y) = \frac{2}{\sqrt[4]{11^3} - \sqrt[4]{2^3}}\sqrt[4]{(y+1)^3} + 3 - \frac{2\sqrt[4]{2^3}}{\sqrt[4]{11^3} - \sqrt[4]{2^3}}.$$

and

$$\varphi(x) = \frac{2}{\sqrt[5]{9^4}}\sqrt[4]{(x+1)^5} + 5, \quad \psi(y) = \frac{2}{\sqrt[5]{11^4} - \sqrt[5]{2^4}}\sqrt[5]{(y+1)^4} + 3 - \frac{2\sqrt[5]{2^4}}{\sqrt[5]{11^4} - \sqrt[5]{2^4}}.$$

From Example 1 we get the inequality

$$\begin{aligned} S_{13} &= e(F(x,y), F(u,v)) + e(G(z,w), G(t,s)) \\ &\leq 0.3\rho(x,u) + 0.4\rho(y,v) + 0.3\rho(z,t) + 0.4\rho(w,s). \end{aligned}$$

and consequently there exists an equilibrium pair of (x, y), such that $x \in F(x, y)$ and $y \in G(x, y)$. The actual values of α, β, γ and δ are smaller.

Author Contributions: The authors are listed in an alphabetical order. All authors have read and agreed to the published version of the manuscript.

Funding: M.H. is partially supported by Shumen University Grant Number Rd-08-42/ 2021, and D.N. is partially supported by the Bulgarian National Fund for Scientific Research Grant Number KP-06-H22/4.

Institutional Review Board Statement: Not applicable.

Informed Consent Statement: Not applicable.

Data Availability Statement: Not applicable.

Acknowledgments: The authors would like to thank the anonymous reviewers for their comments and recommendations that have improved the value of the article.

Conflicts of Interest: The authors declare no conflict of interest.

References

1. Nadler, S.B. Multi–Valued Contraction Mappings. *Pac. J. Math.* **1969**, *30*, 475–488. [CrossRef]
2. Dontchev, A.; Hager, W. An inverse mapping theorem for set-valued maps. *Proc. Am. Math. Soc.* **1994**, *121*, 481–489. [CrossRef]
3. Ahmad, W.; Sarwar, M.; Abdeljawad, T.; Rahmat, G. Multi–Valued Versions of Nadler, Banach, Branciari and Reich Fixed Point Theorems in Double Controlled Metric Type Spaces with Applications. *AIMS Math.* **2021**, *6*, 477–499. [CrossRef]
4. Alsaedi, A.; Broom, A.; Ntouyas, S.K.; Ahmad, B. Nonlocal Fractional Boundary Value Problems Involving Mixed Right and Left Fractional Derivatives and Integrals. *Axioms* **2020**, *9*, 50. [CrossRef]
5. Chen, L.; Yang, N.; Zhou, J. Common Attractive Points of Generalized Hybrid Multi–Valued Mappings and Applications. *Mathematics* **2020**, *8*, 1307. [CrossRef]
6. Shoaib, M.; Sarwar, M.; Kumam, P. Multi-Valued Fixed Point Theorem via F–Contraction of Nadler Type and Application to Functional and Integral Equations. *Bol. Soc. Paran. Mat.* **2021**, *39*, 83–95. [CrossRef]
7. Bhaskar, T.G.; Lakshmikantham, N. Fixed Point Theorems in Partially Ordered Metric Spaces and Applications. *Nonlinear Anal.* **2006**, *65*, 1379–1393. [CrossRef]
8. George, R.; Mitrović, Z.D.; Radenovixcx, S. On Some Coupled Fixed Points of Generalized T–Contraction Mappings in a $b_v(s)$–Metric Space and Its Application. *Axioms* **2020**, *9*, 129. [CrossRef]
9. Kim, K.S. A Constructive Scheme for a Common Coupled Fixed Point Problems in Hilbert Space. *Mathematics* **2020**, *8*, 1717. [CrossRef]
10. Kishore, G.; Rao, B.S.; Radenović, S.; Huang, H.P. Caristi Type Cyclic Contraction and Coupled Fixed Point Results in Bipolar Metric Spaces. *Sahand Commun. Math. Anal.* **2020**, *17*, 1–22. [CrossRef]
11. Shateri, T. Coupled fixed points theorems for non-linear contractions in partially ordered modular spaces. *Int. J. Nonlinear Anal. Appl.* **2020**, *11*, 133–147. [CrossRef]
12. Kirk, W.; Srinivasan, P.; Veeramani, P. Fixed points for mappings satisfying cyclical contractive conditions. *Fixed Point Theory* **2003**, *4*, 79–189.
13. Eldred, A.; Veeramani, P. Existence and convergence of best proximity points. *J. Math. Anal. Appl.* **2006**, *323*, 1001–1006. [CrossRef]
14. Abdeljawad, T.; Ullah, K.; Ahmad, J.; Sen, M.D.L.; Ulhaq, A. Approximation of Fixed Points and Best Proximity Points of Relatively Nonexpansive Mappings. *J. Math.* **2020**, *2020*, 8821553. [CrossRef]
15. Debnath, P.; Srivastava, H.M. Global Optimization and Common Best Proximity Points for Some Multivalued Contractive Pairs of Mappings. *Axioms* **2020**, *9*, 102. [CrossRef]
16. Mishra, L.N.; Dewangan, V.; Mishra, V.N.; Karateke, S. Best proximity points of admissible almost generalized weakly contractive mappings with rational expressions on b–metric spaces. *J. Math. Comput. Sci.* **2021**, *22*, 97–109. [CrossRef]

17. Pant, R.; Shukla, R.; Rakocevic, V. Approximating best proximity points for Reich type non-self nonexpansive mappings. *Rev. R. Acad. Cienc. Exactas Fis. Nat. Ser. A Mat. RACSAM* **2020**, *114*, 197. [CrossRef]
18. Dzhabarova, Y.; Kabaivanov, S.; Ruseva, M.; Zlatanov, B. Existence, Uniqueness and Stability of Market Equilibrium in Oligopoly Markets. *Adm. Sci.* **2020**, *10*, 70. [CrossRef]
19. Kabaivanov, S.; Zlatanov, B. A variational principle, coupled fixed points and market equilibrium. *Nonlinear Anal. Model. Control.* **2021**, *26*, 169–185. [CrossRef]
20. Zhang, X. Fixed Point Theorems of Multivalued Monotone Mappings in Ordered Metric Spaces. *Appl. Math. Lett.* **2010**, *23*, 235–240. [CrossRef]
21. Cournot, A.A. *Researches into the Mathematical Principles of the Theory of Wealth*, Translation ed.; Macmillan: New York, NY, USA, 1897.
22. Friedman, J.W. *Oligopoly Theory*; Cambradge University Press: Cambradge, UK, 2007.
23. Matsumoto, A.; Szidarovszky, F. *Dynamic Oligopolies with Time Delays*; Springer Nature Singapore Pte Ltd.: Singapore, 2018.
24. Okuguchi, K.; Szidarovszky, F. *The Theory of Oligopoly with Multi–Product Firms*; Springer: Berlin/Heidelberg, Germany, 1990.
25. Smith, A. *A Mathematical Introduction to Economics*; Basil Blackwell Limited: Oxford, UK, 1982.
26. Empain, A.M. A posteriori detection of heavy metal pollution of aquatic habitats. In *Methods in Bryology. Proc. Bryol. Method. Workshop, Mainz*; Glime, J.M., Ed.; Hattori Bot Lab.: Nichinan, Japan, 1988; pp. 213–220.
27. Gecheva, G.; Yurukova, L. Water pollutant monitoring with aquatic bryophytes: A review. *Environ. Chem. Lett.* **2014**, *12*, 49–61. [CrossRef]
28. Gecheva, G.; Yancheva, V.; Velcheva, I.; Georgieva, E.; Stoyanova, S.; Arnaudova, D.; Stefanova, V.; Georgieva, D.; Genina, V.; Todorova, B.; et al. Integrated Monitoring with Moss-Bag and Mussel Transplants in Reservoirs. *Water* **2020**, *12*, 1800. [CrossRef]

Article

A Relation-Theoretic Matkowski-Type Theorem in Symmetric Spaces

Based Ali [1], Mohammad Imdad [1] and Salvatore Sessa [2,*]

[1] Department of Mathematics, Aligarh Muslim University, Aligarh 202002, India; basedaliamu@gmail.com (B.A.); mhimdad@gmail.com (M.I.)
[2] Dipartimento di Architettura, Universitá Degli Studi di Napoli Federico II, 80134 Napoli, Italy
* Correspondence: sessa@unina.it

Abstract: In this paper, we present a fixed-point theorem in $\mathcal{R}$-complete regular symmetric spaces endowed with a locally T-transitive binary relation $\mathcal{R}$ using comparison functions that generalizes several relevant existing results. In addition, we adopt an example to substantiate the genuineness of our newly proved result. Finally, as an application of our main result, we establish the existence and uniqueness of a solution of a periodic boundary value problem.

Keywords: fixed points; symmetric spaces; binary relations; T-transitivity; regular spaces

MSC: 47H10; 54H25

1. Introduction

In 1922, one of the most pivotal results in analysis was proved by Banach [1] in his doctoral thesis, which asserts that every contraction mapping on a complete metric space admits a unique fixed point. This principle continues to inspire generations of researchers in metric fixed-point theory. Thus far, this classical result has been generalized and improved in various ways, and by now there exists an extensive literature on and around this premier result. Over the last several decades, there have been many interesting generalizations of this classical result in various directions.

There exist several extensions of the Banach contraction principle to various spaces obtained by lightening the underlying involved metric conditions. In doing so, we are in receipt of several spaces, namely: rectangular metric spaces, generalized metric spaces, partial metric spaces, b-metric spaces, partial b-metric spaces, symmetric spaces, quasi-metric spaces, quasi-partial metric spaces, and many more. In 1976, Cicchese [2] established the first ever fixed-point theorem in the framework of symmetric spaces. The idea of such spaces was coined by Wilson [3] by relaxing the triangle inequality from metric conditions. By now, there exists a considerable literature on fixed-point theory in symmetric spaces. For work of this kind, one can be referred to [4–11].

On the other hand, there have been various generalizations that were obtained by varying the class of contractions (e.g., see [12–14]). In 2004, Ran and Reurings [15] obtained a very useful generalization of the Banach fixed-point theorem in a partially ordered metric space by taking a relatively weaker contraction condition that is required to hold only on those elements that were comparable in the underlying ordering. In doing so, they were essentially motivated by Turinici [16]. This result was further generalized by Nieto and Rodríguez-López in [17,18] in 2005 and 2007, respectively. Subsequently, in 2015, Alam and Imdad [19] furnished a natural extension of the Banach contraction principle in a complete metric space endowed with a binary relation that generalizes all of the above-mentioned results [15,17,18].

The existing literature contains several results on nonlinear contractions, which were initiated by Browder [13] and were followed by similar works by Boyd and Wong [14]

and Matkowski [20]. In 2014, Bessenyei and Páles [21] extended Matkowski's result in symmetric spaces, which required an additional regularity condition.

The intent of this paper is to prove a relation-theoretic version of a theorem due to Bessenyei and Páles [21]. In doing so, we are essentially motivated by [15,17–19].

2. Preliminaries

In this section, we recall some definitions, propositions, and lemmas that will be utilized in our subsequent discussions. The following are taken from Wilson's paper [3] on symmetric space. Throughout the paper, $\mathbb{R}, \mathbb{R}^+, \mathbb{N}, \mathbb{N}_0$, and $\mathbb{Q}$ denote the sets of reals, nonnegative reals, natural numbers, whole numbers, and rational numbers respectively.

Definition 1. *Let X be a nonempty set and let $d : X \times X \to \mathbb{R}^+$ be a mapping satisfying the following axioms: for each $a, b \in X$,*
(i) $d(a,b) = 0$ *if and only if* $a = b$;
(ii) $d(a,b) = d(b,a)$.

Then, d is called symmetric on X and the pair (X, d) is a symmetric space.

In such spaces, the notions of convergent and Cauchy sequences are as usual.

- A sequence $(x_n) \subset X$ is said to converge to $x \in X$ if $\lim_{n \to \infty} d(x_n, x) = 0$.
- A sequence $(x_n) \subset X$ is said to be Cauchy if for each $\epsilon > 0$, there exists $N \in \mathbb{N}$ such that $d(x_n, x_m) < \epsilon \ \forall n, m \geq N$.

The space is said to be complete if every Cauchy sequence converges. For an open ball centered at p with radius r, the notation $B(p,r)$ is used. The diameter of $B(p,r)$ is the supremum of distances taken over the pairs of points of the ball. The topology of such spaces is the topology induced by the open balls.

Because of the unavailability of the triangle inequality, the following problems are obvious:

- There is nothing to assure that limits are unique (thus, the space need not be Hausdorff);
- A convergent sequence need not be a Cauchy sequence;
- The mapping $d(a,.) : X \to \mathbb{R}$ need not be continuous.

Definition 2. *Consider a symmetric space (X, d). A function $\psi : \mathbb{R}_+^2 \to \mathbb{R}_+$ is a triangle function [21] for d if the following hold:*
(i) $\psi(u,v) = \psi(v,u) \ \forall u,v \in \mathbb{R}_+$;
(ii) ψ *is monotone increasing in both of its arguments;*
(iii) $\psi(0,0) = 0$;
(iv) $d(x,y) \leq \psi\big(d(x,z), d(y,z)\big) \ \forall x, y, z \in X$ *for all* $x, y, z \in X$.

It has been shown in [21] that every symmetric space (X, d) admits a unique triangle function Φ_d, which has the property that if ψ is any other triangle function for d, then $\Phi_d \leq \psi$. Such a triangle function Φ_d is called the basic triangle function.

Definition 3. *A symmetric space (X, d) is called a regular space if the basic triangle function with respect to the symmetric d is continuous at $(0,0)$.*

Throughout this paper, we shall restrict our attention to regular spaces only. The utility of such spaces is enlightened by the next important result.

Lemma 1 ([21]). *The topology of a regular symmetric space is Hausdorff. A convergent sequence in a regular symmetric space has a unique limit and it has the Cauchy property. Moreover, a symmetric space (X, d) is regular if and only if*

$$\limsup_{r \to 0} \operatorname{diam} B(p, r) = 0.$$
$$p \in X$$

Definition 4. *A monotone increasing function $\varphi : \mathbb{R}_+ \to \mathbb{R}_+$ is called a comparison function if $\lim_{n \to \infty} \varphi^n(t) = 0$ for each $t \in \mathbb{R}^+$. A mapping $T : X \to X$ is called a φ-contraction if*

$$d(T(x), T(y)) \leq \varphi(d(x,y)) \text{ for each } x, y \in X.$$

These concepts are due to Matkowski [20]. The following is the main result of [21].

Theorem 1. *If (X, d) is a complete regular symmetric space and φ is a comparison function, then every φ-contraction on X has a unique fixed point.*

3. Relation-Theoretic Notions and Related Results

Definition 5 ([22]). *Let X be a nonempty set. A subset $\mathcal{R}$ of $X \times X$ is called a binary relation on X. For $x, y \in X$ when $(x, y) \in \mathcal{R}$, we say that x is related to y, or in other words, x relates to y under $\mathcal{R}$. Sometimes, we write $x\mathcal{R}y$ instead of $(x, y) \in \mathcal{R}$. If $(x, y) \notin \mathcal{R}$, we say x is not related to y.*

Definition 6. *Let $\mathcal{R}$ be a binary relation on a nonempty set X and $x, y \in X$. We say that x and y are $\mathcal{R}$-comparative if either $(x, y) \in \mathcal{R}$ or $(y, x) \in \mathcal{R}$. When x and y are $\mathcal{R}$-comparative, we write it as $[x, y] \in \mathcal{R}$.*

Proposition 1 ([19]). *If (X, d) is a symmetric space, $\mathcal{R}$ is a binary relation on X, T is a self-mapping on X, and $\varphi : \mathbb{R}_+ \to \mathbb{R}_+$ is a comparison function, then the following conditions are equivalent:*

(i) $d(Tx, Ty) \leq \varphi(d(x, y)) \ \forall (x, y) \in \mathcal{R}$;
(ii) $d(Tx, Ty) \leq \varphi(d(x, y)) \ \forall [x, y] \in \mathcal{R}$.

The proof is simple and follows from symmetry of d.

Definition 7. *A binary relation $\mathcal{R}$ defined on a nonempty set X is called*

- *reflexive if $(x, x) \in R \ \forall x \in X$;*
- *transitive if $(x, y) \in \mathcal{R}$ and $(y, z) \in \mathcal{R}$ implies $(x, z) \in \mathcal{R}$;*
- *complete, connected, or dichotomous if $[x, y] \in \mathcal{R} \ \forall x, y \in X$.*

Definition 8 ([19]). *Let X be a nonempty set endowed with a binary relation $\mathcal{R}$. A sequence $(x_n) \subset X$ is called $\mathcal{R}$-preserving if $(x_n, x_{n+1}) \in \mathcal{R} \ \forall n \in \mathbb{N}$.*

Definition 9 ([19]). *Let X be a nonempty set and let T be a self-mapping on X. A binary relation $\mathcal{R}$ on X is called T-closed if, for any $x, y \in X$,*

$$(x, y) \in \mathcal{R} \Rightarrow (Tx, Ty) \in \mathcal{R}.$$

Definition 10 ([23]). *Let X be a nonempty set and let T be a self-mapping on X. A binary relation $\mathcal{R}$ on X is said to be T-transitive if, for any $x, y, z \in X$,*

$$(Tx, Ty), (Ty, Tz) \in \mathcal{R} \Rightarrow (Tx, Tz) \in \mathcal{R}.$$

Definition 11 ([22]). *Let X be a nonempty set endowed with a binary relation $\mathcal{R}$ and $E \subset X$. The restriction of $\mathcal{R}$ to E, denoted as $\mathcal{R}|_E$, is the set $\mathcal{R} \cap E^2$. Indeed, $\mathcal{R}|_E$ is a relation on E induced by $\mathcal{R}$.*

Definition 12 ([23]). *A binary relation $\mathcal{R}$ on a nonempty set X is called locally transitive if, for each $\mathcal{R}$-preserving sequence $(x_n) \subset X$ with range $E = \{x_n\}_{n \in \mathbb{N}_0}$, the binary relation $\mathcal{R}|_E$ is transitive.*

Definition 13 ([23]). *Let X be a nonempty set and let T be a self-mapping on X. A binary relation $\mathcal{R}$ on X is called locally T-transitive if, for each (effectively) $\mathcal{R}$-preserving sequence $(x_n) \subset T(X)$ with range $E = \{x_n\}_{n \in \mathbb{N}_0}$, the binary relation $\mathcal{R}|_E$ is transitive.*

Definition 14 ([24]). *Let X be a nonempty set and let $\mathcal{R}$ be a binary relation on X. For $x, y \in X$, a path of length k (where k is a natural number) in $\mathcal{R}$ from x to y is a finite sequence $\{x_0, x_1, x_2, \ldots, x_k\} \subset X$ satisfying the following conditions:*
(i) $x_0 = x$ and $x_k = y$;
(ii) $(x_i, x_{i+1}) \in \mathcal{R}$ for each i $(0 \le i \le k-1)$.

Definition 15 ([23]). *Let X be a nonempty set and let $\mathcal{R}$ be a binary relation on X. A subset E of X is called $\mathcal{R}$-connected if, for each pair $x, y \in X$, there exists a path (in $\mathcal{R}$) from x to y.*

Definition 16 ([23]). *Let X be a nonempty set and let $\mathcal{R}$ be a binary relation on X. A subset E of X is called $\mathcal{R}^S$-connected if, for each pair $x, y \in X$, there is a finite sequence $\{x_0, x_1, x_2, \ldots, x_k\} \subset X$ satisfying the following conditions:*
(i) $x_0 = x$ and $x_k = y$;
(ii) $[x_i, x_{i+1}] \in \mathcal{R}$ for each i $(0 \le i \le k-1)$.

Now, we define the analogue of the notion of *d*-self-closedness in metric space due to [23] in the framework of symmetric spaces.

Definition 17. *Let (X, d) be a symmetric space. A binary relation $\mathcal{R}$ defined on X is called d-self-closed if, for any $\mathcal{R}$-preserving sequence (x_n) converging to x, there exists a subsequence (x_{n_k}) of (x_n) with $(x_{n_k}, x) \in \mathcal{R}$.*

We will use the following notations in this paper:
- $F(T) := \{x \in X \mid T(x) = x\}$;
- $X(T, \mathcal{R}) := \{x \in X \mid (x, Tx) \in \mathcal{R}\}$.

4. Main Result

In an attempt to prove a relation-theoretic version of Matkowski's theorem [20] in symmetric spaces, we prove the following.

Theorem 2. *Let (X, d) be a regular symmetric space, $\mathcal{R}$ a binary relation on X, and T a self-mapping on X. Suppose that the following conditions hold:*
(a) (X, d) is $\mathcal{R}$-complete;
(b) $\mathcal{R}$ is T-closed and locally T-transitive;
(c) T is either $\mathcal{R}$-continuous or $\mathcal{R}$ is d-self-closed;
(d) $X(T, \mathcal{R})$ is nonempty;
(e) There is a comparison function φ such that
$$d(Tx, Ty) \le \varphi(d(x, y)) \ \forall (x, y) \in \mathcal{R}.$$

Then, T has a fixed point.

Moreover, if
(f) $F(T)$ is $\mathcal{R}^S$-connected, then T has a unique fixed point.

Proof. As $X(T, \mathcal{R})$ is nonempty, let x_0 be such that $(x_0, Tx_0) \in \mathcal{R}$. If $Tx_0 = x_0$, then we are done. Suppose that $Tx_0 \ne x_0$. Since $(x_0, Tx_0) \in \mathcal{R}$ and $\mathcal{R}$ is T-closed, we obtain by induction that
$$(T^n x_0, T^{n+1} x_0) \in \mathcal{R} \ \forall n \in \mathbb{N}.$$

Construct the sequence (x_n) of Picard iterates with initial point x_0, i.e., $x_n = T^n(x_0)$. So, $(x_n, x_{n+1}) \in \mathcal{R} \ \forall n \in \mathbb{N}_0$, i.e., the sequence is $\mathcal{R}$-preserving. As $\mathcal{R}$ is locally T-transitive, we have $(x_n, x_m) \in \mathcal{R} \ \forall m > n$. Observe that the sequence $d(x_n, x_{n+k})$ tends to zero for all fixed $k \in \mathbb{N}$;

$$\begin{aligned} d(x_n, x_{n+k}) &= d(Tx_{n-1}, Tx_{n+k-1}) \\ &\leq \varphi\big(d(x_{n-1}, x_{n+k-1})\big) \\ &\leq \varphi^2\big(d(x_{n-2}, x_{n+k-2})\big) \\ &\vdots \\ &\leq \varphi^n d(x_0, x_k) \to 0 \text{ as } n \to \infty. \end{aligned}$$

Now, we are going to prove that $\{x_n\}$ is a Cauchy sequence. Let $\epsilon > 0$ be any positive number. As (X, d) is regular, the basic triangle function Φ_d is continuous at $(0,0)$. So, there exists a neighborhood U of the origin such that $\Phi_d(u, v) < \epsilon \ \forall (u, v) \in U$. In other words, $\exists \delta > 0$ such that $\Phi_d(u, v) < \epsilon \ \forall u, v : 0 \leq u, v \leq \delta$. We take $\delta < \epsilon$. As φ is a comparison function, $\varphi^n(t) \to 0 \ \forall t > 0$; so there exists $N \in \mathbb{N}$ such that $\varphi^N(\epsilon) < \delta$. Set $S = T^N$. We can see that

$$d(Sx, Sy) = d(T^N x, T^N y) \leq \varphi^N d(x, y) \text{ when } (x, y) \in \mathcal{R}.$$

Define $n_k : d(x_n, T^k S x_n) < \delta \ \forall n \geq n_k$ and set $M = \max\{n_0, n_1, \ldots, n_N\}$.
If $V = \{x_M, x_{M+1}, x_{M+2}, \ldots x_{M+k}, \ldots\}$ then for any $y \in B(x_M, \epsilon) \cap V, y \neq x_M$, we have

$$\begin{aligned} d(T^k S x_M, T^k S y) = d(S T^k x_M, S T^k y) &\leq \varphi^N d(T^k x_M, T^k y) \text{ as } (T^k x_M, T^k y) \in \mathcal{R} \\ &\leq \varphi^N \varphi^k d(x_M, y) < \varphi^N d(x_M, y) < \varphi^N(\epsilon) < \delta. \end{aligned}$$

So,

$$\begin{aligned} d(T^k S y, x_M) &\leq \Phi_d\big(d(T^k S y, T^k S x_M), d(T^k S x_M, x_M)\big) \\ &\leq \Phi_d(\delta, \delta) \ \forall k = 0, 1, 2, \ldots, N; \\ \implies d(T^k S y, x_M) &< \epsilon, \forall k = 0, 1, 2, \ldots, N. \end{aligned}$$

and for $y = x_M$, $d(T^k S x_M, x_M) < \delta < \epsilon, \forall k = 0, 1, 2, \ldots, N$.

Thus we see that $T^k S$ maps $V \cap B(x_M, \epsilon)$ into itself. In particular, each iteration of S maps $V \cap B(x_M, \epsilon)$ into itself. Now, if $n > M$ is any arbitrarily given natural number, i.e., $n = Nk + p$ where $k \in \mathbb{N}_0$ and $0 \leq p < N$, then

$$T^n S = T^{Nk+p} S = T^p S^{k+1},$$

and hence,

$$\begin{aligned} T^n S(V \cap B(x_M, \epsilon)) &= T^p S^{k+1}(V \cap B(x_M, \epsilon)) \\ &= T^p S(S^k(V \cap B(x_M, \epsilon)) \\ &\subset T^p S(V \cap B(x_M, \epsilon)) \\ &\subset V \cap B(x_M, \epsilon); \text{ as } 0 \leq p < N. \end{aligned}$$

Therefore, $T^n S(x_M) \in B(x_M, \epsilon) \ \forall n > M$, i.e., $x_{M+N+k} \in B(x_M, \epsilon) \ \forall k \in \mathbb{N}$. As the space is regular, $\text{diam}(x_M, \epsilon) \to 0$ when $\epsilon \to 0$, and from this, we conclude that the sequence $\{x_n\}$ is a Cauchy sequence. The completeness of the space (X, d) gives some element $x \in X$ such that $x_n \to x$.

Now, if T is $\mathcal{R}$-continuous, then $T(x_n) \to T(x)$, i.e., $x_{n+1} \to T(x)$. As the space is regular, we conclude that $T(x) = x$, as the limit is unique in regular spaces. So, the limit of the sequence constructed above is a fixed point.

If $\mathcal{R}$ is d-self-closed, then there is a subsequence (x_{n_k}) of (x_n) such that $[x_{n_k}, x] \in \mathcal{R}$ $\forall k \in \mathbb{N}_0$. So,

$$d(x, Tx) \leq \Phi_d\big(d(x, x_{n_k+1}), d(x_{n_k+1}, Tx)\big) \leq \Phi_d\big[d(x_{n_k+1}, x), \varphi\big(d(x_{n_k}, x)\big)\big].$$

Now, for $\epsilon > 0$, there exists $\delta > 0$ such that $\Phi_d(u, v) < \epsilon$ $\forall u, v : 0 \leq u, v \leq \delta$, and for $\delta > 0$, there exists $K \in \mathbb{N}$ such that $d(x_n, x) \leq \delta$ $\forall n \geq K$. Therefore, if we take $n_k \geq K$, we have

$$d(x, Tx) \leq \Phi_d(\delta, \varphi(\delta)) \leq \Phi_d(\delta, \delta) < \epsilon.$$

So, $T(x) = x$ i.e., x is a fixed point.

To show that T has a unique fixed point, let y be any other fixed point of T. Now, $F(T)$ is $\mathcal{R}^S$-connected and $x, y \in F(T)$; so, there is a finite sequence of elements $\{z_0, z_1, z_2, \ldots, z_k\} \subset X$ satisfying the following conditions:

(i) $z_0 = x, z_k = y$;
(ii) $[z_i, z_{i+1}] \in \mathcal{R}$ for each i ($0 \leq i \leq k-1$).

Now, as T is a φ-contraction on $\mathcal{R}$, $d(Tz_i, Tz_{i+1}) \leq \varphi(d(z_i, z_{i+1}))$. Using induction, we get $d(T^n z_i, T^n z_{i+1}) \leq \varphi^n d(z_i, z_{i+1})$. We already have, for $\epsilon > 0, \exists \delta > 0$ such that $\Phi_d(u, v) < \epsilon$ $\forall u, v : 0 \leq u, v < \delta$.

Let $\delta_1 = \delta$, define δ_i ($2 \leq i \leq k-1$) : $\Phi_d(u, v) < \delta_{i-1}$ $\forall u, v : 0 \leq u, v < \delta_i$, and set $\alpha = \min\{\delta_1, \delta_2, \ldots, \delta_{k-1}\}$.

In addition, set $M = \max\{N_1, N_2, \ldots, N_{k-1}\}$, where $N_i : d(T^n z_i, T^n z_{i+1}) \leq \varphi^n d(z_i, z_{i+1}) < \alpha$ $\forall n \geq N_i$. Hence, for $n \geq M$, we have

$$\begin{aligned}
d(T^n z_{k-1}, T^n y) &< \alpha \leq \delta_{k-1} \\
d(T^n z_{k-2}, T^n y) &\leq \Phi_d[d(T^n z_{k-2}, T^n z_{k-1}), d(T^n z_{k-1}, T^n y)] \\
&\leq \Phi_d(\alpha, \delta_{k-1}) \leq \Phi_d(\delta_{k-1}, \delta_{k-1}) < \delta_{k-2} \\
d(T^n z_{k-3}, T^n y) &\leq \Phi_d[d(T^n z_{k-3}, T^n z_{k-2}), d(T^n z_{k-2}, T^n y)] \\
&\leq \Phi_d(\alpha, \delta_{k-2}) \leq \Phi_d(\delta_{k-2}, \delta_{k-2}) < \delta_{k-3} \\
&\vdots \\
d(T^n z_1, T^n y) &\leq \Phi_d(d(T^n z_1, T^n z_2), d(T^n z_2, T^n y)) \\
&\leq \Phi_d(\alpha, \delta_2) \leq \Phi_d(\delta_2, \delta_2) < \delta_1 \\
d(T^n x, T^n y) &\leq \Phi_d(d(T^n x, T^n z_1), d(T^n z_1, T^n y)) \\
&\leq \Phi_d(\alpha, \delta_1) \leq \Phi_d(\delta_1, \delta_1) < \epsilon.
\end{aligned}$$

Therefore, $d(T^n x, T^n y) = d(x, y) = 0$, i.e., $x = y$. Hence, the fixed point of T is unique. □

Now, we consider some special cases, where our result deduces some well-known results from the existing literature.

(1) Under the universal relation $\mathcal{R} = X^2$, our theorem deduces the result by M. Bessenyei and Z. Pàles [21]. Clearly, under the universal relation, the hypotheses of our result hold trivially.

(2) As every metric space is a symmetric space, the result of Alam and Imdad [19], which is a generalization of the classical Banach contraction principle, is yielded immediately. In this case, we take $\varphi(t) = ct$ as the comparison function, where $c \in [0, 1)$ is such that $d(Tx, Ty) \leq c(d(x, y))$ $\forall x \mathcal{R} y$.

(3) The fixed-point result of Ran and Reurings [15] can be obtained from our result, as every partially ordered complete metric space is automatically a symmetric space, and the associated relation to the partial order satisfies all the hypotheses of our result if we take the comparison function φ as the same as the earlier case (2), i.e., $\varphi(t) = ct$.

(4) The result of Neito and Rodríguez-López becomes a corollary of our result because of the same reasons as the earlier one. Notice that the d-self-closedness property is a generalization of the ICU (increasing-convergence upper bound) property.

Finally, we produce an illustrative example to substantiate the utility of our result, which does not satisfy the hypotheses of the existing results [1,15,17–19,21,23], but satisfies the hypotheses of our result, and hence has a fixed point.

Example 1. *Let $X = \mathbb{R}$ and $d(x,y) = (x-y)^2$; then, (X,d) is a complete regular symmetric space. Consider the binary relation*

$$\mathcal{R} = \{(x,y) \in R^2 : x \geq y \geq 0, x \in \mathbb{Q}\}.$$

We define a mapping $T : X \to X$ as follows:

$$T(x) = \begin{cases} 2x, & \text{if } x \leq 0, \\ \frac{x}{3}, & \text{if } x > 0. \end{cases}$$

We see that the self-mapping T on X is not a φ-contraction on the whole space X for any comparison function φ. So, the result of Bessenyei and Páles [21] does not apply here. However, when we consider the elements x,y such that $(x,y) \in \mathcal{R}$, then T is a φ-contraction on $\mathcal{R}$ for $\varphi(t) = \frac{t}{2}$, and all the other hypotheses of our result hold.

In addition, we see that the fixed-point results of [1,15,17–19,23] do not apply here, as the space is not a metric space.

5. Application to Ordinary Differential Equations

In this section, we study the existence and uniqueness of a first-order periodic boundary value problem as an application of our main fixed-point theorem.

Consider the first-order periodic boundary value problem

$$\begin{aligned} x'(t) &= g(t, x(t)), t \in [0, \lambda] \\ x(0) &= x(\lambda), \end{aligned} \quad (1)$$

where $\lambda > 0$ and $g : [0, \lambda] \times \mathbb{R} \to \mathbb{R}$ is a continuous function.

We consider the space $X = C[0, \lambda]$ of all continuous functions on $[0, \lambda]$ under the symmetric given by

$$d(x,y) = \sup_{t \in [0,\lambda]} (x(t) - y(t))^2.$$

We define a relation $\mathcal{R}$ on X as

$$x\mathcal{R}y \iff x(t) \leq y(t) \; \forall t \in [0, \lambda].$$

Now, we give the following definition, which will be useful in the subsequent theorem.

Definition 18. *A function z is said to be a lower solution of (1) if*

$$\begin{aligned} z'(t) &\leq g(t, z(t)) \text{ for } t \in [0, \lambda] \\ z(0) &\leq z(\lambda). \end{aligned}$$

Theorem 3. *Consider problem (1) with $g : I \times \mathbb{R} \to \mathbb{R}$, a continuous function, and suppose that there exists some $k > 0$ such that for s_1, s_2 in $\mathbb{R}$ with $s_1 \geq s_2$,*

$$0 \leq g(t, s_1) + ks_1 - [g(t, s_2) + ks_2] \leq k\sqrt{\varphi(s_1 - s_2)^2},$$

where φ is a comparison function. Then, the existence of a lower solution for (1) guarantees the existence of a unique solution of (1).

Proof. Problem (1) can be rewritten as

$$x'(t) + kx(t) = g(t, x(t)) + kx(t), t \in [0, \lambda]$$
$$x(0) = x(\lambda).$$

This problem is equivalent to the integral equation

$$x(t) = \int_0^\lambda G(t,s)[g(s,x(s)) + kx(s)]ds,$$

where

$$G(t,s) = \begin{cases} \frac{e^{k(\lambda+s-t)}}{e^{k\lambda}-1}, 0 \leq s < t \leq \lambda \\ \frac{e^{k(s-t)}}{e^{k\lambda}-1}, 0 \leq t < s \leq \lambda. \end{cases}$$

Consider the self-mapping T on X defined as

$$(Tx)(t) = \int_0^\lambda G(t,s)[g(s,x(s)) + kx(s)]ds.$$

Here, it is apparent that a fixed point of T is, in fact, a solution of the above problem (1). Now, we will show that the hypotheses in Theorem 2 are satisfied.
To prove that the relation $\mathcal{R}$ is T-closed, take $x, y \in X$ such that $x\mathcal{R}y$, i.e.,

$$x(t) \leq y(t) \,\forall t \in [0, \lambda].$$

As $y(t) \geq x(t)$, from the hypothesis, we obtain

$$g(t, x(t)) + kx(t) \leq g(t, y(t)) + ky(t) \,\forall t \in [0, \lambda].$$

As $G(t,s) > 0 \,\forall t, s \in [0, \lambda]$, we have

$$(Tx)(t) = \int_0^\lambda G(t,s)[g(s,x(s)) + kx(s)]ds$$
$$\leq \int_0^\lambda G(t,s)[g(s,y(s)) + ky(s)]ds$$
$$= (Ty)(t).$$

Hence, $\mathcal{R}$ is T-closed. In addition, for $x\mathcal{R}y$, we have

$$\begin{aligned}
\sqrt{d(Tx, Ty)} &= \sup_{t \in [0,\lambda]} |(Tx)(t) - (Ty)(t)| \\
&\leq \sup_{t \in [0,\lambda]} \int_0^\lambda G(t,s)|g(s,x(s)) + kx(s) - g(s,y(s)) - ky(s)|ds \\
&\leq \sup_{t \in [0,\lambda]} \int_0^\lambda G(t,s)\sqrt{\varphi(y(s) - x(s))^2}ds \\
&\leq \sqrt{\varphi d(x,y)} \sup_{t \in [0,\lambda]} \int_0^\lambda G(t,s)ds \\
&= \sqrt{\varphi d(x,y)} \sup_{t \in [0,\lambda]} \frac{1}{e^{k\lambda}-1}\left(\left[\frac{1}{k}e^{k(\lambda+s-t)}\right]_0^t + \left[\frac{1}{k}e^{k(s-t)}\right]_t^\lambda\right) \\
&= \sqrt{\varphi d(x,y)} \sup_{t \in [0,\lambda]} \frac{1}{e^{k\lambda}-1}(e^{k\lambda}-1) \\
&= \sqrt{\varphi d(x,y)}.
\end{aligned}$$

Thus, we have
$$d(Tx, Ty) \leq \varphi d(x,y).$$

Hence, the required contraction condition (2) holds.

Now, as there is some lower solution, say $x_0 \in X$, we have
$$x_0'(t) \leq g(t, x_0(t)),$$

which can be rewritten as
$$x_0'(t) + k x_0(t) \leq g(t, x_0(t)) + k x_0(t) \text{ for } t \in [0, \lambda].$$

Multiplying both the sides by e^{kt}, we obtain
$$(x_0(t) e^{kt})' \leq [g(t, x_0(t)) + k x_0(t)] e^{kt} \text{ for } t \in [0, \lambda],$$

and thus, we get
$$x_0(t) e^{kt} \leq x_0(0) + \int_0^t [g(s, x_0(s)) + k x_0(s)] e^{ks} ds \text{ for } t \in [0, \lambda], \qquad (2)$$

which implies that
$$x_0(0) e^{k\lambda} \leq x_0(\lambda) e^{k\lambda} \leq x_0(0) + \int_0^\lambda [g(s, x_0(s)) + k x_0(s)] e^{ks} ds,$$

thereby yielding
$$x_0(0) \leq \int_0^\lambda \frac{e^{ks}}{e^{k\lambda} - 1} [g(s, x_0(s)) + x_0(s)] ds.$$

Using the above inequality (2), we get
$$x_0(t) e^{kt} \leq \int_0^t [g(s, x_0(s)) + x_0(s)] e^{ks} ds + \int_0^\lambda \frac{e^{ks}}{e^{k\lambda} - 1} [g(s, x_0(s)) + x_0(s)] ds$$
$$= \int_0^t [g(s, x_0(s)) + x_0(s)] \frac{e^{k(s+\lambda)}}{e^{k\lambda} - 1} ds + \int_t^0 [g(s, x_0(s)) + x_0(s)] \frac{e^{ks}}{e^{k\lambda} - 1} ds$$
$$+ \int_0^\lambda \frac{e^{ks}}{e^{k\lambda} - 1} [g(s, x_0(s)) + x_0(s)] ds$$
$$= \int_0^t [g(s, x_0(s)) + x_0(s)] \frac{e^{k(s+\lambda)}}{e^{k\lambda} - 1} ds + \int_t^\lambda [g(s, x_0(s)) + x_0(s)] \frac{e^{ks}}{e^{k\lambda} - 1} ds.$$

Hence,
$$x_0(t) \leq \int_0^t \frac{e^{k(s+\lambda-t)}}{e^{k\lambda} - 1} [g(s, x_0(s)) + x_0(s)] ds + \int_t^\lambda \frac{e^{k(s-t)}}{e^{k\lambda} - 1} [g(s, x_0(s)) + x_0(s)] ds,$$

i.e.,
$$x_0(t) \leq \int_0^\lambda G(t,s) [g(s, x_0(s)) + x_0(s)] ds = (T x_0)(t).$$

Thus, the existence of some element $x_0 \in X$ such that $x_0 \mathcal{R} T x_0$ is ensured.

To show that $\mathcal{R}$ is d-self-closed, let (x_n) be an $\mathcal{R}$-preserving Cauchy sequence converging to $x \in X$. As (x_n) is $\mathcal{R}$-preserving, we have
$$x_0(t) \leq x_1(t) \leq x_2(t) \leq \cdots \leq x_n(t) \leq x_{n+1}(t) \leq \cdots \leq x(t) \ \forall t \in [0, \lambda],$$

thereby yielding $x_n \mathcal{R} x \ \forall n \in \mathbb{N}$. Therefore, $\mathcal{R}$ is d-self-closed.

The remaining hypotheses of Theorem 2 also hold and are easy to check. Hence, T possesses a fixed point in X. □

Author Contributions: The authors B.A., M.I. and S.S. contributed equally and significantly in writing this article. All authors read and approved the final manuscript.

Funding: This research received no external funding.

Institutional Review Board Statement: Not applicable.

Informed Consent Statement: Not applicable.

Data Availability Statement: Not applicable.

Acknowledgments: All the authors are grateful to the anonymous referees for their excellent suggestions, which greatly improved the presentation of the paper.

Conflicts of Interest: The authors declare no conflict of interest.

References

1. Banach, S. Sur les opérations dans les ensembles abstraits et leur application aux équations intégrales. *Fund. Math.* **1922**, *3*, 133–181. [CrossRef]
2. Cicchese, M. Questioni di completezza e contrazioni in spazi metrici generalizzati. *Boll. Unione Mat. Ital.* **1976**, *5*, 175–179.
3. Wilson, W.A. On semi-metric spaces. *Am. J. Math.* **1931**, *53*, 361–373. [CrossRef]
4. Kirk, W.A.; Shahzad, N. Fixed points and Cauchy sequences in semimetric spaces. *J. Fixed Point Theory Appl.* **2015**, *17*, 541–555. [CrossRef]
5. Jachymski, J.; Matkowski, J.; Swiatkowski, T. Nonlinear contractions on semimetric spaces. *J. Appl. Anal.* **1995**, *1*, 125–134. [CrossRef]
6. Hicks, T.L.; Rhoades, B.E. Fixed point theory in symmetric spaces with applications to probabilistic spaces. *Nonlinear Anal.* **1999**, *36*, 331–344. [CrossRef]
7. Aamri, M.; Moutawakil, D.E. Common fixed points under contractive conditions in symmetric spaces. *Appl. Math. E-Notes* **2003**, *3*, 156–162.
8. Aamri, M.; Bassou, A.; El Moutawakil, D. Common fixed points for weakly compatible maps in symmetric spaces with application to probabilistic spaces. *Appl. Math. E-Notes* **2005**, *5*, 171–175.
9. Imdad, M.; Ali, J.; Khan, L. Coincidence and fixed points in symmetric spaces under strict contractions. *J. Math. Anal. Appl.* **2006**, *320*, 352–360. [CrossRef]
10. Cho, S.; Lee, G.; Bae, J. On coincidence and fixed-point theorems in symmetric spaces. *Fixed Point Theory Appl.* **2008**, *2008*, 562130. [CrossRef]
11. Imdad, M.; Chauhan, S.; Soliman, A.H.; Ahmed, M.A. Hybrid fixed point theorems in semimetric spaces via common limit range property. *Demonstr. Math.* **2014**, *47*, 949–962.
12. Jleli, M.; Samet, B. A new generalization of the Banach contraction principle. *J. Inequalities Appl.* **2014**, *2014*, 38. [CrossRef]
13. Browder, F.E. On the convergence of successive approximations for nonlinear functional equations. *Indag. Math.* **1968**, *30*, 27–35. [CrossRef]
14. Boyd, D.W.; Wong, J.S.W. On nonlinear contractions. *Proc. Am. Math. Soc.* **1969**, *20*, 458–464. [CrossRef]
15. Ran, A.C.; Reurings, M.C. A fixed point theorem in partially ordered sets and some applications to matrix equations. *Proc. Am. Math. Soc.* **2004**, *132*, 1435–1443. [CrossRef]
16. Turinici, M. Fixed points for monotone iteratively local contractions. *Demonstr. Math.* **1986**, *19*, 171–180.
17. Nieto, J.J.; Rodríguez-López, R. Contractive mapping theorems in partially ordered sets and applications to ordinary differential equations. *Order* **2005**, *22*, 223–239. [CrossRef]
18. Nieto, J.J.; Rodríguez-López, R. Existence and uniqueness of fixed point in partially ordered sets and applications to ordinary differential equations. *Acta Math. Sin. Engl. Ser.* **2007**, *23*, 2205–2212. [CrossRef]
19. Alam, A.; Imdad, M. Relation-theoretic contraction principle. *J. Fixed Point Theory Appl.* **2015**, *11*, 693–702. [CrossRef]
20. Matkowski, J. Integrable solutions of functional equations. *Diss. Math. Rozpr. Mat.* **1975**, *127*, 5–63.
21. Bessenyei, M.; Páles, Z. A contraction principle in symmetric spaces. *arXiv* **2014**, arXiv:1401.1709.
22. Lipschutz, S. *Schaum's Outlines of Theory and Problems of Set Theory and Related Topics*; McGraw-Hill: New York, NY, USA, 1964.
23. Alam, A.; Imdad, M. Nonlinear contractions in metric spaces under locally T-transitive binary relations. *Fixed Point Theory* **2018**, *19*, 13–24. [CrossRef]
24. Kolman, B.; Busby, R.C.; Ross, S. *Discrete Mathematical Structures*, 3rd ed.; PHI Pvt. Ltd.: New Delhi, India, 2000.

Article

Some Fixed Point Results in b-Metric Spaces and b-Metric-Like Spaces with New Contractive Mappings

Kapil Jain * and Jatinderdeep Kaur

School of Mathematics, Thapar Institute of Engineering and Technology, Patiala 147001, India; jkaur@thapar.edu
* Correspondence: kjain_phdp16@thapar.edu

Abstract: The aim of our paper is to present a new class of functions and to define some new contractive mappings in b-metric spaces. We establish some fixed point results for these new contractive mappings in b-metric spaces. Furthermore, we extend our main result in the framework of b-metric-like spaces. Some consequences of main results are also deduced. We present some examples to illustrate and support our results. We provide an application to solve simultaneous linear equations. In addition, we present some open problems.

Keywords: b-metric space; b-metric-like spaces; Cauchy sequence; fixed point

MSC: 47H10; 54H25

1. Introduction

The well-known concept of metric space was introduced by M. Frechet [1] as an extension of usual distance. In the theory of metric space, Banach's contraction principle [2] is one of the most important theorems and a powerful tool. A mapping $T : X \to X$, where (X, d) is a metric space, is called a contraction mapping if there exists $\alpha < 1$ such that for all $x, y \in X$, $d(Tx, Ty) \leq \alpha d(x, y)$. If the metric space (X, d) is complete, then T has a unique fixed point. Contraction mappings are continuous. In [3], Kannan proved the following result which gives the fixed point for discontinuous mapping: let $T : X \to X$, be a mapping on a complete metric space (X, d) with

$$d(Tx, Ty) \leq \alpha(d(x, Tx) + d(y, Ty)),$$

where $\alpha \in [0, \frac{1}{2})$ and $x, y \in X$. Then, T has a unique fixed point. Contraction mappings have been extended or generalized in several directions by various authors (see, for example, [4–10]). Not only contraction mappings but the concept of metric space is also extended in many ways in the literature (see, for example, [11–19]).

The concept of b-metric spaces was initiated by Bakhtin [11] and Czerwik [13,14] as an extension of metric spaces by weakening the triangular inequality.

Definition 1 ([11,13,14]). *Let X be a non-empty set. Then, a mapping $d : X \times X \to [0, +\infty)$ is called a b-metric if there exists a number $s \geq 1$ such that for all $x, y, z \in X$,*

(d1) $d(x, y) = 0$ if and only if $x = y$;
(d2) $d(x, y) = d(y, x)$;
(d3) $d(x, z) \leq s(d(x, y) + d(y, z))$.

Then triplet (X, d, s) is called a b-metric space. Clearly, every metric space is a b-metric space with $s = 1$, but the converse is not true in general. In fact, the class of b-metric spaces is larger than the class of metric spaces.

In [14], Banach's contraction principle is proved in the framework of b-metric spaces. In 2013, Kir and Kiziltunc established the results in b-metric spaces, which generalized the Kannan and Chatterjea type mappings. In [20], the authors introduced the following result that improves Theorem 1 in [21].

Theorem 1 ([20]). *Let (X, d) be a complete b-metric space with a constant $s \geq 1$. If $T : X \to X$ satisfies the inequality:*

$$d(Tx, Ty) \leq \lambda_1 d(x, y) + \lambda_2 d(x, Tx) + \lambda_3 d(y, Ty) + \lambda_4 (d(x, Ty) + d(Tx, y)),$$

where $\lambda_i \geq 0$ for all $i = 1, 2, 3, 4$ and $\lambda_1 + \lambda_2 + \lambda_3 + 2\lambda_4 < 1$ for $s \in [1, 2]$ and $\frac{2}{s} < \lambda_1 + \lambda_2 + \lambda_3 + 2\lambda_4 < 1$ for $s \in (2, +\infty)$; then, T has a unique fixed point.

In [6], the author introduced quasi-contraction mappings in metric spaces (X, d): A mapping $T : X \to X$ is said to be a quasi-contraction if there exists $0 \leq q < 1$ such that for any $x, y, \in X$,

$$d(Tx, Ty) \leq q \max\{d(x, y), d(x, Tx), d(y, Ty), d(x, Ty), d(Tx, y)\}.$$

Many authors proved fixed point theorems for quasi-contraction mappings in b-metric spaces with some more restriction on values of q (see, for example, [20,22–25]). More on b-metric spaces can be found in [26–37].

In the present work, we define a new class of functions. After that, we define some new contractive mappings which combine the terms $d(x, y)$, $d(x, Tx)$, $d(y, Ty)$, $d(x, Ty)$ and $d(Tx, y)$ by means of the member of a newly defined class. We also prove some fixed point results. To prove our results, we need the following concepts and results from the literature.

Definition 2 ([27]). *Let $(X, d, s \geq 1)$ be a b-metric space. Then, a sequence $\{x_n\}$ in X is called:*

(i) *Cauchy sequence if for each $\epsilon > 0$ there exist $n_0 \in \mathbb{N}$ such that $d(x_n, x_m) < \epsilon$ for all $n, m \geq n_0$.*
(ii) *convergent if there exists $l \in X$ such that for each $\epsilon > 0$ there exist $n_0 \in \mathbb{N}$ such that $d(x_n, l) < \epsilon$ for all $n \geq n_0$. In this case, the sequence $\{x_n\}$ is said to converge to l.*

Definition 3 ([27]). *A b-metric space $(X, d, s \geq 1)$ is said to be complete if every Cauchy sequence is convergent in it.*

Lemma 1 ([29]). *Let $(X, d, s \geq 1)$ be a b-metric space and suppose that sequences $\{x_n\}$ and $\{y_n\}$ converge to x and $y \in X$, respectively. Then,*

$$\frac{1}{s^2} d(x, y) \leq \liminf_{n \to +\infty} d(x_n, y_n) \leq \limsup_{n \to +\infty} d(x_n, y_n) \leq s^2 d(x, y).$$

In particular, if $x = y$, then $\lim_{n \to +\infty} d(x_n, y_n) = 0$.
Moreover, for any $z \in X$, we have

$$\frac{1}{s} d(x, z) \leq \liminf_{n \to +\infty} d(x_n, z) \leq \limsup_{n \to +\infty} d(x_n, z) \leq s d(x, z).$$

Lemma 2 ([31]). *Every sequence $\{x_n\}$ of elements from a b-metric space $(X, d, s \geq 1)$, having the property that there exists $\lambda \in [0, 1)$ such that $d(x_n, x_{n+1}) \leq \lambda d(x_{n-1}, x_n)$ for every $n \in \mathbb{N}$, is Cauchy.*

2. Fixed Point Results in b-Metric Spaces

In this section, we first define a new class of functions, and then we define a new contractive mapping in b-metric spaces as follows.

Definition 4. *For any $m \in \mathbb{N}$, we define Ξ_m to be the set of all functions $\xi : [0, +\infty)^m \to [0, +\infty)$ such that*

(ξ_1) $\xi(t_1, t_2, ..., t_m) < \max\{t_1, t_2, ..., t_m\}$ *if* $(t_1, t_2, ..., t_m) \neq (0, 0, ..., 0)$;

(ξ_2) *if* $\{t_i^{(n)}\}_{n \in \mathbb{N}}$, $1 \leq i \leq m$, *are m sequences in $[0, +\infty)$ such that* $\limsup\limits_{n \to +\infty} t_i^{(n)} = t_i$

$< +\infty$ *for all*

$i = 1$ *to m, then* $\liminf\limits_{n \to +\infty} \xi\left(t_1^{(n)}, t_2^{(n)}, ..., t_m^{(n)}\right) \leq \xi(t_1, t_2, ..., t_m).$

2.1. First Main Result

Definition 5. *Let $(X, d, s \geq 1)$ be a b-metric space. The mapping $T : X \to X$ is said to be an ξ-contractive mapping of type-I if there exists $\xi \in \Xi_4$ and*

$$d(Tx, Ty) \leq \frac{1}{s}\xi\left(d(x, y), d(x, Tx), d(y, Ty), \frac{d(x, Ty) + d(Tx, y)}{2s}\right), \quad (1)$$

for all $x, y \in X$.

Now, the first result of this paper is as follows:

Theorem 2. *Let $(X, d, s \geq 1)$ be a complete b-metric space and $T : X \to X$ be an ξ-contractive mapping of type-I. Then, T has a unique fixed point.*

Proof. Let $x_0 \in X$. Define a sequence $\{x_n\}$ in X as $x_n = Tx_{n-1}$ for all $n \geq 1$. Assume that any two consecutive terms of the sequence $\{x_n\}$ are distinct; otherwise, T has a fixed point. First, we prove that $\{x_n\}$ is a Cauchy sequence. For this, let $n \in \mathbb{N}$.
Consider

$$d(x_n, x_{n+1}) \leq \frac{1}{s}\xi\left(d(x_{n-1}, x_n), d(x_{n-1}, x_n), d(x_n, x_{n+1}), \frac{d(x_{n-1}, x_{n+1})}{2s}\right) \quad (2)$$

$$< \frac{1}{s}\max\left\{d(x_{n-1}, x_n), d(x_{n-1}, x_n), d(x_n, x_{n+1}), \frac{d(x_{n-1}, x_{n+1})}{2s}\right\}$$

$$= \frac{1}{s}\max\left\{d(x_{n-1}, x_n), \frac{d(x_{n-1}, x_{n+1})}{2s}\right\}$$

$$\leq \frac{1}{s}\max\left\{d(x_{n-1}, x_n), \frac{d(x_{n-1}, x_n) + d(x_n, x_{n+1})}{2}\right\},$$

which implies that

$$d(x_n, x_{n+1}) < \frac{1}{s}d(x_{n-1}, x_n) \quad \text{for all } n \geq 1. \quad (3)$$

Case 1: If $s > 1$, then by Lemma 2 in view of (3), $\{x_n\}$ is a Cauchy sequence.
Case 2: If $s = 1$, then by (3), the sequence $\{d(x_n, x_{n+1})\}$ is monotonically decreasing and bounded below. Therefore, $d(x_n, x_{n+1}) \to k$ for some $k \geq 0$. Suppose that $k > 0$; now, taking liminf $n \to +\infty$ in (2), we have $k \leq \xi(k, k, k, k')$, where

$$k' = \limsup_{n \to +\infty} \frac{d(x_{n-1}, x_{n+1})}{2} \leq \limsup_{n \to +\infty} \frac{d(x_{n-1}, x_n) + d(x_n, x_{n+1})}{2} = k.$$

Now,
$$k \leq \xi(k,k,k,k') < \max\{k,k,k,k'\} = k,$$ which is a contradiction; therefore,
$$\lim_{n \to +\infty} d(x_n, x_{n+1}) = 0. \tag{4}$$

Suppose that $\{x_n\}$ is not a Cauchy sequence; then, there exists $\varepsilon > 0$ such that for any $r \in \mathbb{N}$, there exists $m_r > n_r \geq r$ such that
$$d(x_{m_r}, x_{n_r}) \geq \varepsilon. \tag{5}$$

Furthermore, assume that m_r is the smallest natural number greater than n_r such that (5) holds. Then,
$$\begin{aligned}
\varepsilon &\leq d(x_{m_r}, x_{n_r}) \\
&\leq d(x_{m_r}, x_{m_r-1}) + d(x_{m_r-1}, x_{n_r}) \\
&< d(x_{m_r}, x_{m_r-1}) + \varepsilon \\
&< d(x_r, x_{r-1}) + \varepsilon,
\end{aligned}$$

thus, using (4) and taking $\lim r \to +\infty$, we get
$$\lim_{r \to +\infty} d(x_{m_r}, x_{n_r}) = \varepsilon. \tag{6}$$

Now, consider
$$d(x_{m_r+1}, x_{n_r+1}) \leq \xi\left(d(x_{m_r}, x_{n_r}), d(x_{m_r}, x_{m_r+1}), d(x_{n_r}, x_{n_r+1}), \frac{d(x_{m_r}, x_{n_r+1}) + d(x_{m_r+1}, x_{n_r})}{2}\right).$$

Therefore, we have
$$\begin{aligned}
d(x_{m_r}, x_{n_r}) &\leq d(x_{m_r}, x_{m_r+1}) + d(x_{m_r+1}, x_{n_r+1}) + d(x_{n_r+1}, x_{n_r}) \\
&\leq d(x_{m_r}, x_{m_r+1}) + d(x_{n_r+1}, x_{n_r}) + \\
&\quad \xi\left(d(x_{m_r}, x_{n_r}), d(x_{m_r}, x_{m_r+1}), d(x_{n_r}, x_{n_r+1}), \frac{d(x_{m_r}, x_{n_r+1}) + d(x_{m_r+1}, x_{n_r})}{2}\right).
\end{aligned}$$

Thus, by taking $\liminf r \to +\infty$ on both sides and also using (4) and (6), we get $\varepsilon \leq 0 + 0 + \xi(\varepsilon, 0, 0, \varepsilon')$, where
$$\begin{aligned}
\varepsilon' &= \limsup_{r \to +\infty} \frac{d(x_{m_r}, x_{n_r+1}) + d(x_{m_r+1}, x_{n_r})}{2} \\
&\leq \limsup_{r \to +\infty} \frac{d(x_{m_r}, x_{n_r}) + d(x_{n_r}, x_{n_r+1}) + d(x_{m_r+1}, x_{m_r}) + d(x_{m_r}, x_{n_r})}{2} \\
&= \frac{\varepsilon + 0 + 0 + \varepsilon}{2} = \varepsilon.
\end{aligned}$$

Thus, $\varepsilon \leq \xi(\varepsilon, 0, 0, \varepsilon') < \max\{\varepsilon, 0, 0, \varepsilon'\} = \varepsilon$, which is a contradiction. Thus, $\{x_n\}$ is a Cauchy sequence in $(X, d, s \geq 1)$.

Now, $(X, d, s \geq 1)$ is a complete b-metric space. Therefore, there exists $x \in X$ such that $x_n \to x$.

Now, consider
$$d(Tx_n, Tx) \leq \frac{1}{s}\xi\left(d(x_n, x), d(x_n, Tx_n), d(x, Tx), \frac{d(x_n, Tx) + d(x, Tx_n)}{2s}\right),$$

which implies that

$$d(x_{n+1}, Tx) \leq \frac{1}{s}\xi\left(d(x_n, x), d(x_n, x_{n+1}), d(x, Tx), \frac{d(x_n, Tx) + d(x, x_{n+1})}{2s}\right).$$

Taking lim inf $n \to +\infty$ on both sides and using Lemma 1, we get

$$\frac{1}{s}d(x, Tx) \leq \frac{1}{s}\xi(0, 0, d(x, Tx), l),$$

i.e.,

$$d(x, Tx) \leq \xi(0, 0, d(x, Tx), l),$$

where

$$l = \limsup_{n \to +\infty} \frac{d(x_n, Tx) + d(x, x_{n+1})}{2s} \leq \limsup_{n \to +\infty} \frac{sd(x, Tx) + 0}{2s} = \frac{d(x, Tx)}{2}.$$

Thus,

$$d(x, Tx) \leq \xi(0, 0, d(x, Tx), l) < \max\{0, 0, d(x, Tx), l\} = d(x, Tx),$$

which is a contradiction. Therefore, $Tx = x$.

Let $Ty = y$ for some $y \in X$ and suppose that $x \neq y$; then, consider

$$\begin{aligned}
d(x, y) = d(Tx, Ty) &\leq \frac{1}{s}\xi\left(d(x, y), d(x, Tx), d(y, Ty), \frac{d(x, Ty) + d(y, Tx)}{2s}\right) \\
&\leq \frac{1}{s}\xi\left(d(x, y), 0, 0, \frac{d(x, y)}{s}\right) \\
&< \frac{1}{s}\max\left\{d(x, y), 0, 0, \frac{d(x, y)}{s}\right\} \\
&= \frac{d(x, y)}{s},
\end{aligned}$$

which is a contradiction. Therefore, $x = y$. □

Now, the following remark improves our main result for Theorem 2.

Remark 1. *Theorem 2 is also valid if the term $\frac{d(x, Ty) + d(Tx, y)}{2s}$ in (1) is replaced by $\frac{d(x, Ty) + d(Tx, y)}{\delta s}$, where δ is a real number defined by*

$$\delta = \begin{cases} 2, & \text{if } s = 1, \\ \delta', & \text{if } 1 < s \leq 2, \\ 1, & \text{if } s > 2, \end{cases}$$

where δ' is any number in $\left(\frac{2}{s}, 1 + \frac{1}{s}\right)$.

Now, the following result is a consequence of Theorem 2.

Corollary 1. *Let $(X, d, s \geq 1)$ be a complete b-metric space and $T : X \to X$ be a mapping such that there exists $q \in [0, \frac{1}{s})$ and*

$$d(Tx, Ty) \leq q \max\left\{d(x, y), d(x, Tx), d(y, Ty), \frac{d(x, Ty) + d(Tx, y)}{2s}\right\}, \quad (7)$$

for all $x, y \in X$. Then T has a unique fixed point.

Proof. Let $\xi \in \Xi_4$ be defined by $\xi(t_1, t_2, t_3, t_4) = qs \max\{t_1, t_2, t_3, t_4\}$. Then, following Theorem 2, T has a unique fixed point. □

In the following example, we see that conditions of Theorem 2 are satisfied, but Corollary 1 is not applicable.

Example 1. Let $X = \left\{\frac{1}{\sqrt{n}} : n \in \mathbb{N}\right\} \cup \{0\}$. Define $d : X \times X \to [0, +\infty)$ by $d(x,y) = |x-y|^2$ for all $x,y \in X$. Then d is a b-metric on X with $s = 2$.

Define $T : X \to X$ by $T\left(\frac{1}{\sqrt{n}}\right) = \frac{1}{\sqrt{2(n+1)}}$ for all $n \in \mathbb{N}$ and $T(0)=0$. Define

$$\xi(t_1,t_2,t_3,t_4) = \begin{cases} \frac{\max\{t_1,t_2,t_3,t_4\}}{1+t_1}, & \text{if } t_1 > 0, \\ \frac{1}{2}\max\{t_2,t_3,t_4\}, & \text{otherwise.} \end{cases}$$

Now, for all $x,y \in X$, (1) is satisfied, and thus the conditions of Theorem 2 are satisfied. However, we see that if (7) is satisfied for all $x,y \in X$, we have

$$d(Tx,Ty) \leq qN(x,y),$$

for all $x,y \in X$, where $N(x,y) = \max\left\{d(x,y), d(x,Tx), d(y,Ty), \frac{d(x,Ty)+d(Tx,y)}{2s}\right\}$. So, in particular, we have

$$d\left(\frac{1}{\sqrt{2(n+1)}}, \frac{1}{\sqrt{2(m+1)}}\right) \leq qN\left(\frac{1}{\sqrt{n}}, \frac{1}{\sqrt{m}}\right) \quad \text{for all } m,n \in \mathbb{N}, m \neq n.$$

i.e.,

$$\frac{\left|\frac{1}{\sqrt{n+1}} - \frac{1}{\sqrt{m+1}}\right|^2}{N\left(\frac{1}{\sqrt{n}}, \frac{1}{\sqrt{m}}\right)} \leq 2q \quad \text{for all } m,n \in \mathbb{N}\ m \neq n.$$

Now, taking $\lim n,m \to +\infty$, we get $2q \geq 1$, which is a contradiction. Thus, Corollary 1 is not applicable for this example.

Remark 2. In view of Remark 1, Corollary 1 is also valid, if the term $\frac{d(x,Ty)+d(Tx,y)}{2s}$ is replaced by $\frac{d(x,Ty)+d(Tx,y)}{\delta s}$, where δ is the same as defined in Remark 1.

The following result is another consequence of Theorem 2.

Corollary 2. Let $(X,d,s \geq 1)$ be a complete b-metric space and $T : X \to X$ be a mapping such that

$$d(Tx,Ty) \leq \lambda_1 d(x,y) + \lambda_2 d(x,Tx) + \lambda_3 d(y,Ty) + \lambda_4(d(x,Ty) + d(Tx,y)), \tag{8}$$

for all $x,y \in X$, where $\lambda_1 + \lambda_2 + \lambda_3 + \delta s \lambda_4 < \frac{1}{s}$ and $\lambda_i \geq 0$ for all $i = 1$ to 4. Then, T has a unique fixed point.

Proof. Let $\xi \in \Xi_4$ be defined by $\xi(t_1,t_2,t_3,t_4) = s(\lambda_1 t_1 + \lambda_2 t_2 + \lambda_3 t_3 + \delta s \lambda_4 t_4)$. Then, by Theorem 2 and Remark 2, T has a unique fixed point. □

2.2. Second Main Result

Now, we define another contractive mapping in b-metric space.

Definition 6. Let $(X,d,s \geq 1)$ be a b-metric space. The mapping $T : X \to X$ is said to be an ξ-contractive mapping of type-II if there exists $\xi \in \Xi_5$ and

$$d(Tx,Ty) \leq \frac{1}{s}\xi\left(d(x,y), d(x,Tx), d(y,Ty), \frac{d(x,Ty)}{2s}, d(Tx,y)\right), \tag{9}$$

for all $x, y \in X$.

The proof of our next result proceeds in a similar manner as the proof of Theorem 2.

Theorem 3. *Let $(X, d, s \geq 1)$ be a complete b-metric space and $T : X \to X$ be an ζ-contractive mapping of type-II. Then T has a unique fixed point.*

The following remark improves Theorem 3.

Remark 3. *Theorem 3 is also valid, if the term $\frac{d(x, Ty)}{2s}$ in (9) is replaced by $\frac{d(x, Ty)}{\delta s}$, where δ is the same as in Remark 1.*

Corollary 3. *Let $(X, d, s \geq 1)$ be a complete b-metric space and $T : X \to X$ be a mapping such that there exists $q \in [0, \frac{1}{s})$ and*

$$d(Tx, Ty) \leq q \max \left\{ d(x, y), d(x, Tx), d(y, Ty), \frac{d(x, Ty)}{\delta s}, d(Tx, y) \right\}, \qquad (10)$$

for all $x, y \in X$. Then, T has a unique fixed point.

Proof. Let $\zeta \in \Xi_5$ be defined by $\zeta(t_1, t_2, t_3, t_4, t_5) = qs \max\{t_1, t_2, t_3, t_4, t_5\}$. Then, by Theorem 3, T has a unique fixed point. □

Corollary 4. *Let $(X, d, s \geq 1)$ be a complete b-metric space and $T : X \to X$ be a mapping such that*

$$d(Tx, Ty) \leq \lambda_1 d(x, y) + \lambda_2 d(x, Tx) + \lambda_3 d(y, Ty) + \lambda_4 d(x, Ty) + \lambda_5 d(Tx, y), \qquad (11)$$

for all $x, y \in X$, where $\lambda_1 + \lambda_2 + \lambda_3 + \delta s \lambda_4 + \lambda_5 < \frac{1}{s}$ and $\lambda_i \geq 0$ for all $i = 1$ to 5. Then, T has a unique fixed point.

Proof. Let $\zeta \in \Xi_5$ be defined by $\zeta(t_1, t_2, t_3, t_4, t_5) = s(\lambda_1 t_1 + \lambda_2 t_2 + \lambda_3 t_3 + \delta s \lambda_4 t_4 + \lambda_5 t_5)$. Then by Theorem 3, T has a unique fixed point. □

3. Fixed Point Results in b-Metric-Like Spaces

Partial metric spaces were introduced by Matthews (1992) as a generalization of metric spaces. The self-distance may be non-zero in partial metric space. In 2012, A. A. Harandi generalized the concept of the partial metric by establishing a new space named the metric-like-space. We notice that in metric-like space, the self-distance of a point may be greater than the distance of that point to any other point (see Example 2.2 in [15]). Later on, S. Shukla (2014) presented the idea of the partial b-metric as a generalization of the partial metric and b-metric. Meanwhile, in 2013, M.A. Alghamdi et al. introduced the concept of b-metric-like spaces that generalized the notions of partial b-metric space and metric-like space. Obviously, b-metric-like space generalizes all abstract spaces that we have mentioned in our paper. For the sake of clarity, we recall the definitions of these abstract spaces as follows.

Definition 7 ([12]). *Let X be a non-empty set. Then, a mapping $d : X \times X \to [0, +\infty)$ is called a partial metric if for all $x, y, z \in X$,*

(p1) $d(x, y) = 0 \Leftrightarrow d(x, x) = d(x, y) = d(y, y)$;
(p2) $d(x, x) \leq d(x, y)$;
(p3) $d(x, y) = d(y, x)$;
(p4) $d(x, z) \leq d(x, y) + d(y, z) - d(y, y)$.

Then, the pair (X, d) is called a partial metric space.

Definition 8 ([15,38]). *Let X be a non-empty set. Then, a mapping $d : X \times X \to [0, +\infty)$ is called a metric-like space if for all $x, y, z \in X$,*

(ml1) $d(x, y) = 0 \Rightarrow x = y$;
(ml2) $d(x, y) = d(y, x)$;
(ml3) $d(x, z) \leq d(x, y) + d(y, z)$.

Then, the pair (X, d) is called a metric-like space.

Definition 9 ([17]). *Let X be a non-empty set. Then, a mapping $d : X \times X \to [0, +\infty)$ is called a partial b-metric if there exists a number $s \geq 1$ such that for all $x, y, z \in X$,*

(pb1) $d(x, y) = 0 \Leftrightarrow d(x, x) = d(x, y) = d(y, y)$;
(pb2) $d(x, x) \leq d(x, y)$;
(pb3) $d(x, y) = d(y, x)$;
(pb4) $d(x, z) \leq s(d(x, y) + d(y, z)) - d(y, y)$.

Then, the triplet (X, d, s) is called a partial b-metric space.

Definition 10 ([16]). *Let X be a non-empty set. Then, a mapping $d : X \times X \to [0, +\infty)$ is called a b-metric-like if there exists a number $s \geq 1$ such that for all $x, y, z \in X$,*

(bml1) $d(x, y) = 0 \Rightarrow x = y$;
(bml2) $d(x, y) = d(y, x)$;
(bml3) $d(x, z) \leq s(d(x, y) + d(y, z))$.

Then, the triplet (X, d, s) is called a b-metric-like space.

The following definitions and results related to b-metric-like spaces are required in the main results of this section.

Definition 11 ([16,39]). *Let $(X, d, s \geq 1)$ be a b-metric-like space and let $\{x_n\}$ be a sequence of points of X. A point $x \in X$ is said to be the limit of sequence $\{x_n\}$ if $\lim_{n \to +\infty} d(x, x_n) = d(x, x)$, and we say that the sequence $\{x_n\}$ is convergent to x and denote it by $x_n \to x$ as $n \to +\infty$.*

Definition 12 ([16,39]). *Let $(X, d, s \geq 1)$ be a b-metric-like space.*

(i)   *A sequence $\{x_n\}$ in X is called Cauchy sequence if $\lim_{n,m \to +\infty} d(x_n, x_m)$ exists and is finite.*

(ii)   *$(X, d, s \geq 1)$ is said to be complete if every Cauchy sequence $\{x_n\}$ in X converges to $x \in X$ so that*

$$\lim_{n,m \to +\infty} d(x_n, x_m) = d(x, x) = \lim_{n \to +\infty} d(x_n, x).$$

Proposition 1 ([16]). *Let $(X, d, s \geq 1)$ be a b-metric-like space and $\{x_n\}$ be a sequence in X such that for some $x \in X$, $\lim_{n \to +\infty} d(x_n, x) = 0$. Then,*

(i)   *x is unique.*

(ii) $\frac{1}{s} d(x, y) \leq \lim_{n \to +\infty} d(x_n, y) \leq s d(x, y)$ *for all $y \in X$.*

Lemma 3 ([40])**.** *Let* $(X, d, s \geq 1)$ *be a b-metric-like space and* $\{x_n\}$ *be a sequence in X such that*

$$d(x_n, x_{n+1}) \leq \lambda d(x_{n-1}, x_n)$$

for some $\lambda \in [0,1)$ *and for each* $n \in \mathbb{N}$. *Then,* $\{x_n\}$ *is a Cauchy sequence with* $\lim_{n,m \to +\infty} d(x_n, x_m) = 0$.

Now, we extend Theorem 2 in the framework of a b-metric-like space. At the end of the proof, we provide an example in support.

Theorem 4. *Let* $(X, d, s \geq 1)$ *be a complete b-metric-like space. Let* $T : X \to X$ *be a mapping such that there exists* $\zeta \in \Xi_4$ *and*

$$d(Tx, Ty) \leq \frac{1}{s}\zeta\left(d(x,y), d(x,Tx), d(y,Ty), \frac{d(x,Ty) + d(Tx,y) - d(y,y)}{2s}\right) \quad (12)$$

for all $x, y \in X$ *with* $d(x, Ty) + d(Tx, y) \geq d(y, y)$. *Then, T has a unique fixed point.*

Proof. Let $x_0 \in X$. Define a sequence $\{x_n\}$ in X as $x_n = Tx_{n-1}$ for all $n \geq 1$. Assume that any two consecutive terms of the sequence $\{x_n\}$ are distinct; otherwise, T has a fixed point. First, we prove that $\{x_n\}$ is a Cauchy sequence. For this, let $n \in \mathbb{N}$.
Now,

$$d(x_{n-1}, Tx_n) + d(Tx_{n-1}, x_n) = d(x_{n-1}, x_{n+1}) + d(x_n, x_n) \geq d(x_n, x_n);$$

therefore, using (12), we have

$$d(x_n, x_{n+1}) \leq \frac{1}{s}\zeta\left(d(x_{n-1}, x_n), d(x_{n-1}, x_n), d(x_n, x_{n+1}), \frac{d(x_{n-1}, x_{n+1}) + d(x_n, x_n) - d(x_n, x_n)}{2s}\right) \quad (13)$$

$$< \frac{1}{s}\max\left\{d(x_{n-1}, x_n), d(x_{n-1}, x_n), d(x_n, x_{n+1}), \frac{d(x_{n-1}, x_{n+1})}{2s}\right\}$$

$$= \frac{1}{s}\max\left\{d(x_{n-1}, x_n), \frac{d(x_{n-1}, x_{n+1})}{2s}\right\}$$

$$\leq \frac{1}{s}\max\left\{d(x_{n-1}, x_n), \frac{d(x_{n-1}, x_n) + d(x_n, x_{n+1})}{2}\right\},$$

which implies that

$$d(x_n, x_{n+1}) < \frac{1}{s}d(x_{n-1}, x_n) \quad \text{for all } n \geq 1. \quad (14)$$

Case 1: If $s > 1$, then by Lemma 3 and in view of (14), $\{x_n\}$ is a Cauchy sequence in $(X, d, s \geq 1)$ and $\lim_{n,m \to +\infty} d(x_n, x_m) = 0$.

Case 2: If $s = 1$, then by (14), the sequence $\{d(x_n, x_{n+1})\}$ is monotonically decreasing and bounded below. Therefore, $d(x_n, x_{n+1}) \to k$ for some $k \geq 0$. Suppose that $k > 0$; now, taking lim inf $n \to +\infty$ in (13), we have $k \leq \zeta(k, k, k, k')$, where

$$k' = \limsup_{n \to +\infty} \frac{d(x_{n-1}, x_{n+1})}{2} \leq \limsup_{n \to +\infty} \frac{d(x_{n-1}, x_n) + d(x_n, x_{n+1})}{2} = k.$$

Now, $k \leq \zeta(k, k, k, k') < \max\{k, k, k, k'\} = k$, a contradiction; therefore,

$$\lim_{n \to +\infty} d(x_n, x_{n+1}) = 0. \quad (15)$$

Furthermore,
$$d(x_n, x_n) \leq d(x_n, x_{n+1}) + d(x_{n+1}, x_n),$$
taking lim sup $n \to +\infty$, and using (15) we get
$$\lim_{n \to +\infty} d(x_n, x_n) = 0, \quad (16)$$

Suppose that $\lim_{n,m \to +\infty} d(x_n, x_m) \neq 0$; then, there exists $\varepsilon > 0$ such that for any $r \in \mathbb{N}$, there exists $m_r > n_r \geq r$ such that
$$d(x_{m_r}, x_{n_r}) \geq \varepsilon. \quad (17)$$

Furthermore, assume that m_r is the smallest natural number greater than n_r such that (17) holds. Then,
$$\begin{aligned}
\varepsilon &\leq d(x_{m_r}, x_{n_r}) \\
&\leq d(x_{m_r}, x_{m_r-1}) + d(x_{m_r-1}, x_{n_r}) \\
&< d(x_{m_r}, x_{m_r-1}) + \varepsilon \\
&< d(x_r, x_{r-1}) + \varepsilon.
\end{aligned}$$

Thus, using (15) and taking $\lim r \to +\infty$, we get
$$\lim_{r \to +\infty} d(x_{m_r}, x_{n_r}) = \varepsilon. \quad (18)$$

Now, suppose that there exist infinitely many r such that
$$d(x_{m_r}, T x_{n_r}) + d(T x_{m_r}, x_{n_r}) < d(x_{n_r}, x_{n_r}).$$

Taking lim sup $r \to +\infty$, and using (16), we get
$$\lim_{r \to +\infty} (d(x_{m_r}, T x_{n_r}) + d(T x_{m_r}, x_{n_r})) = 0,$$
which means that
$$\lim_{r \to +\infty} (d(x_{m_r}, x_{n_r+1}) = \lim_{r \to +\infty} d(x_{m_r+1}, x_{n_r})) = 0.$$

Now,
$$\varepsilon = \lim_{r \to +\infty} d(x_{m_r}, x_{n_r}) \leq \limsup_{r \to +\infty} ((d(x_{m_r}, x_{n_r+1}) + d(x_{n_r+1}, x_{n_r})) = 0,$$
which is a contradiction. Therefore, there exists $r_0 \in \mathbb{N}$ such that for all $r \geq r_0$, $d(x_{m_r}, T x_{n_r}) + d(T x_{m_r}, x_{n_r}) \geq d(x_{n_r}, x_{n_r})$. Thus, for all $r \geq r_0$, using (12),

$$d(x_{m_r+1}, x_{n_r+1}) \leq \xi\left(d(x_{m_r}, x_{n_r}), d(x_{m_r}, x_{m_r+1}), d(x_{n_r}, x_{n_r+1}), \frac{d(x_{m_r}, x_{n_r+1}) + d(x_{m_r+1}, x_{n_r}) - d(x_{n_r}, x_{n_r})}{2}\right).$$

Now,
$$\begin{aligned}
d(x_{m_r}, x_{n_r}) &\leq d(x_{m_r}, x_{m_r+1}) + d(x_{m_r+1}, x_{n_r+1}) + d(x_{n_r+1}, x_{n_r}) \\
&\leq d(x_{m_r}, x_{m_r+1}) + d(x_{n_r+1}, x_{n_r}) + \\
&\quad \xi\left(d(x_{m_r}, x_{n_r}), d(x_{m_r}, x_{m_r+1}), d(x_{n_r}, x_{n_r+1}), \frac{d(x_{m_r}, x_{n_r+1}) + d(x_{m_r+1}, x_{n_r}) - d(x_{n_r}, x_{n_r})}{2}\right).
\end{aligned}$$

Thus, by taking lim inf $r \to +\infty$ on both sides and also using (15) and (18), we get $\varepsilon \leq 0 + 0 + \xi(\varepsilon, 0, 0, \varepsilon')$, where

$$\varepsilon' = \limsup_{r \to +\infty} \frac{d(x_{m_r}, x_{n_r+1}) + d(x_{m_r+1}, x_{n_r}) - d(x_{n_r}, x_{n_r})}{2}$$

$$\leq \limsup_{r \to +\infty} \frac{d(x_{m_r}, x_{n_r}) + d(x_{n_r}, x_{n_r+1}) + d(x_{m_r+1}, x_{m_r}) + d(x_{m_r}, x_{n_r}) - 0}{2}$$

$$= \frac{\varepsilon + 0 + 0 + \varepsilon}{2} = \varepsilon.$$

Thus, $\varepsilon \leq \xi(\varepsilon, 0, 0, \varepsilon') < \max\{\varepsilon, 0, 0, \varepsilon'\} = \varepsilon$, which is a contradiction. Thus, $\{x_n\}$ is a Cauchy sequence in $(X, d, s \geq 1)$ with $\lim_{n,m \to +\infty} d(x_n, x_m) = 0$.

Now, $(X, d, s \geq 1)$ is a complete b-metric-like space; therefore, there exists $x \in X$ such that $x_n \to x$,

$$d(x, x) = \lim_{n \to +\infty} d(x_n, x) = \lim_{n,m \to +\infty} d(x_n, x_m) = 0.$$

Furthermore, according to Proposition 1, x is unique.

Suppose that $Tx \neq x$. Now, consider

$$d(Tx_n, Tx) \leq \frac{1}{s} \xi\left(d(x_n, x), d(x_n, Tx_n), d(x, Tx), \frac{d(x_n, Tx) + d(x, Tx_n) - d(x, x)}{2s}\right),$$

i.e.,

$$d(x_{n+1}, Tx) \leq \frac{1}{s} \xi\left(d(x_n, x), d(x_n, x_{n+1}), d(x, Tx), \frac{d(x_n, Tx) + d(x, x_{n+1})}{2s}\right).$$

Taking lim inf $n \to +\infty$ on both sides and using Proposition 1, we get

$$\frac{1}{s} d(x, Tx) \leq \frac{1}{s} \xi(0, 0, d(x, Tx), l);$$

i.e.,

$$d(x, Tx) \leq \xi(0, 0, d(x, Tx), l),$$

where

$$l = \limsup_{n \to +\infty} \frac{d(x_n, Tx) + d(x, x_{n+1})}{2s} \leq \limsup_{n \to +\infty} \frac{sd(x, Tx) + 0}{2s} = \frac{d(x, Tx)}{2}.$$

Thus,

$$d(x, Tx) \leq \xi(0, 0, d(x, Tx), l) < \max\{0, 0, d(x, Tx), l\} = d(x, Tx),$$

which is a contradiction. Therefore, $Tx = x$.

Let $Ty = y$ for some $y \in X$; then, by (12), $d(y, y) = 0$. Now, suppose that $x \neq y$, and consider

$$\begin{aligned}
d(x,y) &= d(Tx,Ty) \\
&\leq \frac{1}{s}\xi\left(d(x,y),d(x,Tx),d(y,Ty),\frac{d(x,Ty)+d(y,Tx)-d(x,x)}{2s}\right), \\
&= \frac{1}{s}\xi\left(d(x,y),d(x,Tx),d(y,Ty),\frac{d(x,Ty)+d(y,Tx)}{2s}\right), \\
&\leq \frac{1}{s}\xi\left(d(x,y),0,0,\frac{d(x,y)}{s}\right) \\
&< \frac{1}{s}\max\left\{d(x,y),0,0,\frac{d(x,y)}{s}\right\} \\
&= \frac{d(x,y)}{s},
\end{aligned}$$

which is a contradiction. Therefore, $x = y$. □

Example 2. *Let $X = [0,+\infty)$. Define $d : X \times X \to [0,+\infty)$ by $d(x,y) = (x+y)^2$ for all $x, y \in X$. Then, d is b-metric-like on X with $s = 2$, but d is not b-metric on X.*

Define $T : X \to X$ by $T(x) = \frac{x}{2}$. In addition, define $\xi(t_1,t_2,t_3,t_4) = \frac{1}{2}\max\{t_1,t_2,t_3,t_4\}$. Now, for all $x, y \in X$, with $d(x,Ty) + d(Tx,y) \geq d(y,y)$, (12) in Theorem 4 is satisfied and T has a unique fixed point 0.

4. Application

In this section, as an application of Theorem 2, we present the following result which provides a unique solution to simultaneous linear equations.

Theorem 5. *Consider a system of linear equations*

$$Ax = b \qquad (19)$$

where $A = [a_{ij}]_{n\times n}$ is an $n \times n$ matrix, $b = [b_i]_{1\times n}$ is a column vector of constants and $x = [x_i]_{1\times n}$ is a column matrix of n unknowns. If for each $x = [x_i]_{1\times n}$, $y = [y_i]_{1\times n}$ and $i = 1$ to n,

$$|(a_{ii}+1)(x_i-y_i) + \sum_{j=1, j\neq i}^{n} a_{ij}(x_j - y_j)|(1 + \max_{k=1}^{n}|x_k - y_k|) \leq |x_i - y_i|; \qquad (20)$$

then, the system has a unique solution.

Proof. Let $X = \{ [x_i]_{1\times n} \mid x_i$ is real for all $i = 1$ to n, n being fixed $\}$ and $d : X \times X \to [0,+\infty)$ be defined as

$$d(x,y) = \max_{i=1}^{n}|x_i - y_i|$$

for all $x = [x_i]_{1\times n}$, $y = [y_i]_{1\times n} \in X$. Then, clearly (X,d) is a complete b-metric space with constant $s = 1$ (i.e. (X,d) is a complete metric space).

Now, define a $n \times n$ matrix $C = [c_{ij}]$ by

$$c_{ij} = \begin{cases} a_{ij}+1, & \text{if } i = j \\ a_{ij}, & \text{if } i \neq j. \end{cases}$$

Then, the given system (19) reduces to

$$x = Cx - b. \qquad (21)$$

Condition (20) becomes

$$|\sum_{j=1}^{n} c_{ij}(x_j - y_j)|(1 + \max_{k=1}^{n}|x_k - y_k|) \leq |x_i - y_i| \qquad \text{for all } i = 1,2,...,n. \qquad (22)$$

Now, define a mapping $T : X \to X$ by

$$Tx = Cx - b, \quad \text{where } x \in X.$$

For $x = [x_i]_{1 \times n}$ and $y = [y_i]_{1 \times n}$, suppose that $Tx = u = [u_i]_{1 \times n}$ and $Ty = v = [v_i]_{1 \times n}$; then,

$$u_i = \sum_{j=1}^n c_{ij} x_j - b_i \quad (i = 1, 2, \ldots, n)$$

and

$$v_i = \sum_{j=1}^n c_{ij} y_j - b_i \quad (i = 1, 2, \ldots, n)$$

Define

$$\zeta(t_1, t_2, t_3, t_4) = \begin{cases} \frac{\max\{t_1, t_2, t_3, t_4\}}{1+t_1}, & \text{if } t_1 > 0, \\ \frac{1}{2} \max\{t_2, t_3, t_4\}, & \text{otherwise.} \end{cases}$$

Now, using condition (22),

$$\begin{aligned} d(Tx, Ty) &= \max_{i=1}^n |u_i - v_i| \\ &= \max_{i=1}^n \left| \sum_{j=1}^n c_{ij}(x_j - y_j) \right| \\ &\leq \max_{i=1}^n \left(\frac{|x_i - y_i|}{1 + \max_{k=1}^n |x_k - y_k|} \right) \\ &\leq \zeta\left(d(x,y), d(x, Tx), d(y, Ty), \frac{d(x, Ty) + d(Tx, y)}{2s} \right). \end{aligned}$$

Thus, it is straightforward to see that the hypothesis of Theorem 2 is satisfied. Therefore, T has a unique fixed point and system (19) has a unique solution. □

5. Conclusions

In this paper, we have defined a new class of functions, and with the help of this class of functions, we defined some new contractive mappings in b-metric spaces. Furthermore, we proved some fixed point results for these contractive mappings. One can easily extend these results to common fixed points for weakly compatible mappings (see [22,41,42]). We improve our main results in Theorems 2 and 3 with the help of Remarks 1 and 3, respectively. Can these results be further improved in terms of s? More precisely, we present here some open questions as follows.

Open Question 1: Does Theorem 2 hold also if the term $\frac{1}{s}$ (before ζ) in (1) is replaced by α, for some $\alpha \in [\frac{1}{s}, 1]$?

Open Question 2: Does Theorem 3 hold also if the term $\frac{1}{s}$ (before ζ) in (9) is replaced by α, for some $\alpha \in [\frac{1}{s}, 1]$?

Author Contributions: Both authors contributed equally in the planning, execution and analysis of the study. Both authors have read and agreed to the published version of the manuscript.

Funding: This research received no external funding.

Institutional Review Board Statement: Not applicable.

Informed Consent Statement: Not applicable.

Data Availability Statement: Not applicable.

Conflicts of Interest: The authors declare no conflict of interest.

References

1. Frechet, M. Sur quelques points duo calcul fonctionel. *Rendicouti Mahematicodi Palermo* **1906**, *22*, 1–74. [CrossRef]
2. Banach, S. Sur les operations dans les ensembles abstracts ET leur applications aux equations integrals. *Fund. Math.* **1922**, *3*, 133–181. [CrossRef]
3. Kannan, R. Some results on fixed points. *Bull. Calcutta Math. Soc.* **1972**, *25*, 727–730. [CrossRef]
4. Meir, A.; Keeler, E. A thorem on contractive mappings. *J. Math. Anal. Appl.* **1969**, *28*, 326–329. [CrossRef]
5. Nadler, J. Multivalued contraction mappings. *Pac. J. Math.* **1969**, *30*, 475–488. [CrossRef]
6. Ciric, L.B. A generalization of Banachs contraction principle. *Proceeding Am. Math. Soc.* **1974**, *45*, 267–273. [CrossRef]
7. Rhoades, B.E. A comparison of various definitions of contractive mappings. *Trans. Am. Math. Soc.* **1977**, *226*, 257–290. [CrossRef]
8. Rhoades, B.E. Some theorems on weakly contractive maps. *Non-Linear Anal. TMA* **2001**, *47*, 2683–2694. [CrossRef]
9. Samet, B.; Vetro, C.; Vetro, P. Fixed point theorems for α-ψ-contractive type mappings. *Nonlinear Anal. Theory Methods Appl.* **2012**, *75*, 2154–2165. [CrossRef]
10. Wardowski, D. Fixed points of a new type of contractive mappings in complete metric spaces. *Fixed Point Theory Appl.* **2012**, *2012*, 94. [CrossRef]
11. Bakhtin, I.A. The contraction mapping principle in almost metric spaces. *Funct. Anal.* **1989**, *30*, 26–37.
12. Matthews, G.S. *Partial Metric Topology*; Research Report 212; Dept. of Computer Science, University of Warwick: Coventry, UK, 1992.
13. Czerwik, S. Contraction mappings in b-metric spaces. *Acta Math. Inform. Univ. Ostrav.* **1993**, *30*, 5–11.
14. Czerwik, S. Nonlinear set-valued contraction mapping in b-metric spaces. *Atti Semin. Mat. Fis. Dell'Universita Modena Reeggio Emilia* **1998**, *46*, 263–276.
15. Haranadi, A.A. Metric-like spaces, partial metric spaces and fixed points. *Fixed Point Theory Appl.* **2012**, *2012*, 204. [CrossRef]
16. Alghamdi, M.A.; Hussain, N.; Salimi, P. Fixed point and coupled fixed point theorems on b-metric-like spaces. *J. Inequalities Appl.* **2013**, *402*, 1–25. [CrossRef]
17. Shukla, S. Partial b-metric spaces and fixed point theorems. *Mediter. J. Math.* **2014**, *11*, 703–711. [CrossRef]
18. Jleli, M.; Samet, B. A Generalized Metric Space and Related Fixed Point Theorems. *Fixed Point Theory Appl.* **2015**, *2015*, 61. [CrossRef]
19. Jleli, M.; Samet, B. On a new generalization of metric spaces. *J. Fixed Point Theory Appl.* **2018**, *20*, 128. [CrossRef]
20. Aleksic, S.; Mitrovic, Z.D.; Radenovic, S. On Some Recent Fixed Point Results For Single and Multivalued Mappings in b-metric spaces. *Fasc. Math.* **2018**, *61*, 5–16. [CrossRef]
21. Dubey, A.K.; Shukla, R.; Dubey, R. Some fixed point results in b-metric spaces. *Asian J. Math. Appl.* **2014**, *2014*, ama0147.
22. Jovanovic, M.; Kadelburg, Z.; Radenovic, S. Common fixed point results in metric type spaces. *Fixed Point Theory Appl.* **2010**, *210*, 978121. [CrossRef]
23. Aydi, H.; Bota, M.F.; Karapinar, E.; Mitrovic, S. A fixed point Theorem for set-valued quasi-contractions in b-metric spaces. *Fixed Point Theory Appl.* **2012**, *2012*, 88. [CrossRef]
24. Pant, R.; Panicker, R. Geraghty and Ciric type fixed point theorems in b-metric spaces. *J. Nonlinear Sci. Appl.* **2016**, *9*, 5741–5755. [CrossRef]
25. Mlaiki, N.; Dedovic, N.; Aydi, H.; Filipovic, M.G.; Mohsin, B.B.; Radenovic, S. Some new observations on Geraghty and Ciric type results in b-metric spaces. *Mathematics* **2019**, *7*, 643. [CrossRef]
26. Khamsi, M.A.; Hussain, N. KKM mapping in metric type space. *Nonlinear Anal.* **2010**, *73*, 3123–3129. [CrossRef]
27. Kir, M.; Kiziltunc, H. On some well known fixed point theorems in b-metric spaces. *Turk. J. Anal. Number Theory* **2013**, *1*, 13–16. [CrossRef]
28. Sintunavaat, W.; Plubtieng, S.; Katchang, P. Fixed point results and applications on b-metric space endowed with an arbitrary binary relation. *Fixed Point Theory Appl.* **2013**, *2013*, 296. [CrossRef]
29. Latif, A.; Parvaneh, V.; Salimi, P.; Al-Mazrooei, A.E. Various Suzuki type theorems in b-metric spaces. *J. Nonlinear Sci. Appl.* **2015**, *8*, 363–377. [CrossRef]
30. Ozturk, V.; Radenovic, S. Some remarks on b-(E.A)-property in b-metric spaces. *Springer Plus* **2016**, *5*, 544. [CrossRef]
31. Miculescu, R.; Mihail, A. New fixed point theorems for set-valued contractions in b-metric spaces. *J. Fixed Point Theory Appl.* **2017**, *19*, 2153–2163. [CrossRef]
32. Suzuki, T. Basic inequality on a b-metric space and its applications. *J. Inequalities Appl.* **2017**, *2017*, 256. [CrossRef]
33. Huang, H.; Deng, G.; Radenovic, S. Fixed point theorems in b-metric spaces with applications to differential equation. *J. Fixed Point Theory Appl.* **2018**, *20*, 52. [CrossRef]
34. Karapinar, E.; Czerrwik, S.; Aydi, H. (α, ψ)-Meir-Keeler Contraction Mappings in Generalized b-metric spaces. *J. Funct. Spaces* **2018**, *2*, 1–4. [CrossRef]
35. Qawaqneh, H.; Noorani, M.S.M.; Shatanawi, W.; Aydi, H.; Alsamir, H. Fixed point results for multivalued contractions in b-metric spaces and an application. *Mathematics* **2019**, *7*, 132. [CrossRef]
36. Karapinar, E.; Mitrovic, Z.; Ozturk, A.; Radenovic, S.N. On a theorem of Ciric in b-metric spaces. *Rendiconti Circolo Matematico Palermo Aeries 2* **2021**, *70*, 217–225. [CrossRef]
37. Mitrovic, Z.; Parvaneh, V.; Mlaiki, N.; Hussain, N.; Radenovic, S.N. On some new generalizations of Nadler contraction in b-metric spaces. *Cogent Math. Stat.* **2020**, *7*, 1760189.

38. Vujakovic, J.; Mitrovic, S.; Mitrovic, Z.D.; Radenovic, S. On **F**-Contractions for Weak α-Admissible Mappings in Metric-Like Spaces. *Mathematics* **2020**, *8*, 1629. [CrossRef]
39. Chen, C.; Dong, J.; Zhu, C. Some fixed point theorems in *b*-metric-like spaces. *Fixed Point Theory Appl.* **2015**, *2015*, 122. [CrossRef]
40. Sen, M.D.; Nicolic, N.; Dosenovic, T.; Pavlovic, M.; Radenovic, S. Some Results on (s-q)-Graphic Contraction Mapping in *b*-Metric-like Spaces. *Mathematics* **2019**, *7*, 1190. [CrossRef]
41. Abbas, M.; Jungck, G. Common fixed point results for noncommuting mappings without continiuty in cone metric spaces. *J. Math. Anal. Appl.* **2008**, *341*, 416-420. [CrossRef]
42. Jungck, G.; Radenovic, S.; Radojovic, S.; Rakocevic, V. Common fixed point theorems for weakly compatible pairs on cone metric spaces. *Fixed Point Theory Appl.* **2009**, *2009*, 643840. [CrossRef]

Article

Banach Contraction Principle and Meir–Keeler Type of Fixed Point Theorems for Pre-Metric Spaces

Hsien-Chung Wu

Department of Mathematics, National Kaohsiung Normal University, Kaohsiung 802, Taiwan; hcwu@nknucc.nknu.edu.tw

Abstract: The fixed point theorems in so-called pre-metric spaces is investigated in this paper. The main issue in the pre-metric space is that the symmetric condition is not assumed to be satisfied, which can result in four different forms of triangle inequalities. In this case, the fixed point theorems in pre-metric space will have many different styles based on the different forms of triangle inequalities.

Keywords: Cauchy sequence; fixed point; pre-metric space; triangle inequality; weakly uniformly strict contraction

MSC: 47H10; 54H25

Citation: Wu, H.-C. Banach Contraction Principle and Meir–Keeler Type of Fixed Point Theorems for Pre-Metric Spaces. *Axioms* **2021**, *10*, 57. https://doi.org/10.3390/axioms10020057

Academic Editor: Erdal Karapinar

Received: 6 March 2021
Accepted: 5 April 2021
Published: 9 April 2021

Publisher's Note: MDPI stays neutral with regard to jurisdictional claims in published maps and institutional affiliations.

Copyright: © 2021 by the author. Licensee MDPI, Basel, Switzerland. This article is an open access article distributed under the terms and conditions of the Creative Commons Attribution (CC BY) license (https://creativecommons.org/licenses/by/4.0/).

1. Introduction

In this paper, we consider a so-called pre-metric space that does not assume the symmetric condition. We first recall the basic concept of (conventional) metric space as follows.

Given a nonempty universal set X, let $d : X \times X \to \mathbb{R}_+$ be a nonnegative real-valued function defined on the product set $X \times X$. Recall that (X, d) is a metric space when the following conditions are satisfied:

- $d(x, y) = 0$ implies $x = y$ for any $x, y \in X$;
- $d(x, x) = 0$ for any $x \in X$;
- $d(x, y) = d(y, x)$ for any $x, y \in X$;
- $d(x, z) \leq d(x, y) + d(y, z)$ for any $x, y, z \in X$.

Different kinds of spaces that weaken the above conditions have been proposed. Wilson [1] says that (X, d) is a quasi-metric space when the symmetric condition is not satisfied. More precisely, (X, d) is a quasi-metric space when the following conditions are satisfied:

- $d(x, y) = 0$ if and only if $x = y$ for any $x, y \in X$;
- $d(x, z) \leq d(x, y) + d(y, z)$ for any $x, y, z \in X$.

Many authors (by referring to [2–16] and the references therein) also defined the different type of quasi-metric space as follows:

- $d(x, y) = 0 = d(y, x)$ if and only if $x = y$ for any $x, y \in X$;
- $d(x, z) \leq d(x, y) + d(y, z)$ for any $x, y, z \in X$.

In the Wilson's sense, it is obvious that we also have $d(y, x) = 0$ if and only if $y = x$. Wilson [17] also says that (X, d) is a semi-metric space when the triangle inequality is not satisfied. More precisely, the following conditions are satisfied:

- $d(x, y) = 0$ if and only if $x = y$ for any $x, y \in X$;
- $d(x, y) = d(y, x)$ for any $x, y \in X$.

Matthews [11] proposed the concept of partial metric space that satisfies the following conditions:

- $x = y$ if and only if $d(x, x) = d(x, y) = d(y, y)$ for any $x, y \in X$;

- $d(x,x) \leq d(x,y)$ for any $x, y \in X$;
- $d(x,y) = d(y,x)$ for any $x, y \in X$.
- $d(x,z) \leq d(x,y) + d(y,z) - d(y,y)$ for any $x,y,z \in X$.

The partial metric space does not assume the self-distance condition $d(x,x) = 0$. In this paper, we consider a so-called pre-metric space by assuming that

$$d(x,y) = 0 \text{ implies } x = y \text{ for any } x, y \in X.$$

The triangle inequality always plays a very important role in the study of metric space. Without considering the symmetric condition, the triangle inequalities can be considered in four different forms by referring to Wu [18]. The purpose of this paper is to establish the fixed point theorems in pre-metric space based on the different forms of triangle inequalities. We separately study the Banach contraction principle and Meir–Keeler type of fixed point theorems for pre-metric spaces. On the other hand, three types of contraction functions are considered in this paper. We also mention that the Meir–Keeler type of fixed point theorems in the context of b-metric spaces have been studied by Pavlović and Radenović [19].

This paper is organized as follows. In Section 2, four different forms of triangle inequalities in pre-metric space are presented. Many basic properties are also provided for further study. In Section 3, based on the different forms of triangle inequalities, many concepts of Cauchy sequences in pre-metric space are proposed in order to establish the fixed point theorems in pre-metric space. In Section 4, three different types of contraction functions are considered to establish the fixed point theorems using the different forms of triangle inequalities.

2. Pre-Metric Spaces

We formally introduce the basic concept of pre-metric space by considering four different forms of triangle inequalities as follows.

Definition 1. *Given a nonempty universal set X, let d be a mapping from $X \times X$ into $\mathbb{R}_+$.*

- *The metric d is said to satisfy the $\bowtie$-triangle inequality when the following inequality is satisfied:*

$$d(x,y) + d(y,z) \geq d(x,z) \text{ for all } x,y,z \in X.$$

- *The metric d is said to satisfy the $\triangleright$-triangle inequality when the following inequality is satisfied:*

$$d(x,y) + d(z,y) \geq d(x,z) \text{ for all } x,y,z \in X.$$

- *The metric d is said to satisfy the $\triangleleft$-triangle inequality when the following inequality is satisfied:*

$$d(y,x) + d(y,z) \geq d(x,z) \text{ for all } x,y,z \in X.$$

- *The metric d is said to satisfy the $\diamond$-triangle inequality when the following inequality is satisfied:*

$$d(y,x) + d(z,y) \geq d(x,z) \text{ for all } x,y,z \in X.$$

Suppose that d satisfies the symmetric condition. It is clear to see that all the concepts of $\bowtie$-triangle inequality, $\triangleright$-triangle inequality, $\triangleleft$-triangle inequality and $\diamond$-triangle inequality described above are all equivalent. This means that the pre-metric space extends the concept of (conventional) metric space.

Remark 1. *Now, we represent some interesting observations that are used in the study.*

- *Suppose that the metric d satisfies the $\bowtie$-triangle inequality. Then, we have*

$$d(a,b) + d(b,c) + d(c,d) \geq d(a,c) + d(c,d) \geq d(a,d).$$

We also see that
$$d(b,a) + d(c,b) = d(c,b) + d(b,a) \geq d(c,a),$$
which implies
$$d(b,a) + d(c,b) + d(d,c) \geq d(d,a).$$
In general, we can have the following inequalities
$$d(x_1,x_2) + d(x_2,x_3) + \cdots + d(x_p,x_{p+1}) \geq d(x_1,x_{p+1})$$
and
$$d(x_2,x_1) + d(x_3,x_2) + \cdots + d(x_{p+1},x_p) \geq d(x_{p+1},x_1).$$

- Suppose that the metric d satisfies the ▷-triangle inequality. Since
$$d(a,b) + d(c,b) \geq d(a,c) \text{ and } d(c,b) + d(a,b) \geq d(c,a),$$
we see that
$$d(a,b) + d(c,b) = d(c,b) + d(a,b) \geq \max\{d(a,c), d(c,a)\}.$$
Therefore, we obtain
$$d(a,b) + d(c,b) + d(d,c) \geq \max\{d(a,d), d(d,a)\}. \tag{1}$$
In general, we can have the following inequalities
$$d(x_1,x_2) + d(x_3,x_2) + d(x_4,x_3) + \cdots + d(x_{p+1},x_p) \geq \max\{d(x_1,x_{p+1}), d(x_{p+1},x_1)\}.$$

- Suppose that the metric d satisfies the ◁-triangle inequality. Since
$$d(b,a) + d(b,c) = d(b,c) + d(b,a) \geq \max\{d(a,c), d(c,a)\},$$
we see that
$$d(b,a) + d(b,c) + d(c,d) \geq \max\{d(a,d), d(d,a)\}. \tag{2}$$
In general, we can have the following inequalities
$$d(x_2,x_1) + d(x_2,x_3) + d(x_3,x_4) + \cdots + d(x_p,x_{p+1}) \geq \max\{d(x_1,x_{p+1}), d(x_{p+1},x_1)\}.$$

- Suppose that the metric d satisfies the ⋄-triangle inequality. Then, we have
$$d(a,b) + d(b,c) + d(d,c) = d(b,c) + d(a,b) + d(d,c) \geq d(c,a) + d(d,c) \geq d(a,d) \tag{3}$$
and
$$d(b,a) + d(c,b) + d(c,d) \geq d(a,c) + d(c,d) = d(c,d) + d(a,c) \geq d(d,a). \tag{4}$$
In general, we consider the following cases.

(a) Suppose that p is an even number. Then, we have the following inequalities
$$d(x_1,x_2) + d(x_2,x_3) + d(x_4,x_3) + d(x_4,x_5) + d(x_6,x_5)$$
$$+ d(x_6,x_7) + \cdots + d(x_p,x_{p+1}) \geq d(x_{p+1},x_1)$$
and
$$d(x_2,x_1) + d(x_3,x_2) + d(x_3,x_4) + d(x_5,x_4) + d(x_5,x_6)$$
$$+ d(x_7,x_6) + \cdots + d(x_{p+1},x_p) \geq d(x_1,x_{p+1}).$$

(b) Suppose that p is an odd number. Then, we have the following inequalities

$$d(x_1, x_2) + d(x_2, x_3) + d(x_4, x_3) + d(x_4, x_5) + d(x_6, x_5)$$
$$+ d(x_6, x_7) + \cdots + d(x_p, x_{p+1}) \geq d(x_1, x_{p+1})$$

and

$$d(x_2, x_1) + d(x_3, x_2) + d(x_3, x_4) + d(x_5, x_4) + d(x_5, x_6)$$
$$+ d(x_7, x_6) + \cdots + d(x_{p+1}, x_p) \geq d(x_{p+1}, x_1).$$

Definition 2 (Wu [18]). *Given a nonempty universal set X, let d be a mapping from $X \times X$ into $\mathbb{R}_+$. We say that (X, d) is a pre-metric space when $d(x, y) = 0$ implies $x = y$ for any $x, y \in X$.*

Proposition 1 (Wu [18]). *Given a nonempty universal set X, let d be a mapping from $X \times X$ into $\mathbb{R}_+$. Suppose that the following conditions are satisfied:*
- $d(x, x) = 0$ for all $x \in X$;
- *d satisfies the $\triangleright$-triangle inequality or the $\triangleleft$-triangle inequality or the $\diamond$-triangle inequality.*

Then d satisfies the symmetric condition.

We also remark that Proposition 4.4 in Wu [18] is redundant and it can be omitted.

3. Cauchy Sequences in Pre-Metric Space

Let (X, d) be a pre-metric space. Many different concepts of limit are proposed below because of lacking the symmetric condition.

Definition 3. *Let (X, d) be a pre-metric space, and let $\{x_n\}_{n=1}^\infty$ be a sequence in X.*
- *We write $x_n \xrightarrow{d^\triangleright} x$ as $n \to \infty$ when $d(x_n, x) \to 0$ as $n \to \infty$.*
- *We write $x_n \xrightarrow{d^\triangleleft} x$ as $n \to \infty$ when $d(x, x_n) \to 0$ as $n \to \infty$.*
- *We write $x_n \xrightarrow{d} x$ as $n \to \infty$ when*

$$\lim_{n \to \infty} d(x_n, x) = \lim_{n \to \infty} d(x, x_n) = 0.$$

The uniqueness of limits are given below.

Proposition 2 (Wu [18]). *Let (X, d) be a pre-metric space, and let $\{x_n\}_{n=1}^\infty$ be a sequence in X.*
(i) *Suppose that the metric d satisfies the $\bowtie$-triangle inequality or $\diamond$-triangle inequality. If $x_n \xrightarrow{d^\triangleleft} x$ and $x_n \xrightarrow{d^\triangleright} y$, then $x = y$.*
(ii) *Suppose that the metric d satisfies the $\triangleleft$-triangle inequality. If $x_n \xrightarrow{d^\triangleright} x$ and $x_n \xrightarrow{d^\triangleright} y$, then $x = y$. In other words, the $d^\triangleright$-limit is unique.*
(iii) *Suppose that the metric d satisfies the $\triangleright$-triangle inequality. If $x_n \xrightarrow{d^\triangleleft} x$ and $x_n \xrightarrow{d^\triangleleft} y$, then $x = y$. In other words, the $d^\triangleleft$-limit is unique.*

Without the symmetric condition, the different concepts of Cauchy sequences are also presented below.

Definition 4. *Let (X, d) be a pre-metric space, and let $\{x_n\}_{n=1}^\infty$ be a sequence in X.*
- *We say that $\{x_n\}_{n=1}^\infty$ is a $>$-Cauchy sequence when, given any $\epsilon > 0$, there exists an integer N such that $d(x_m, x_n) < \epsilon$ for all pairs (m, n) of integers m and n with $m > n \geq N$.*
- *We say that $\{x_n\}_{n=1}^\infty$ is a $<$-Cauchy sequence when, given any $\epsilon > 0$, there exists an integer N such that $d(x_n, x_m) < \epsilon$ for all pairs (m, n) of integers m and n with $m > n \geq N$.*

- We say that $\{x_n\}_{n=1}^{\infty}$ is a Cauchy sequence *when, given any $\epsilon > 0$, there exists an integer N such that $d(x_m, x_n) < \epsilon$ and $d(x_n, x_m) < \epsilon$ for all pairs (m, n) of integers m and n with $m, n \geq N$ and $m \neq n$.*

We can also consider the different concepts of completeness for pre-metric space.

Definition 5. *Let (X, d) be a pre-metric space.*
- *We say that (X, d) is $(>, \triangleright)$-complete when each $>$-Cauchy sequence is convergent in the sense of $x_n \xrightarrow{d^{\triangleright}} x$.*
- *We say that (X, d) is $(>, \triangleleft)$-complete when each $>$-Cauchy sequence is convergent in the sense of $x_n \xrightarrow{d^{\triangleleft}} x$.*
- *We say that (X, d) is $(<, \triangleright)$-complete when each $<$-Cauchy sequence is convergent in the sense of $x_n \xrightarrow{d^{\triangleright}} x$.*
- *We say that (X, d) is $(<, \triangleleft)$-complete when each $<$-Cauchy sequence is convergent in the sense of $x_n \xrightarrow{d^{\triangleleft}} x$.*
- *We say that (X, d) is $\triangleleft$-complete when each Cauchy sequence is convergent in the sense of $x_n \xrightarrow{d^{\triangleleft}} x$.*
- *We say that (X, d) is $\triangleright$-complete when each Cauchy sequence is convergent in the sense of $x_n \xrightarrow{d^{\triangleright}} x$.*

Based on the above different concepts of completeness, we establish many fixed point theorems in pre-metric space by using the different types of triangle inequalities. Next, we present some examples to demonstrate the completeness.

Let S be a bounded subset S of $\mathbb{R}^k$ containing infinitely many points. The Bolzano–Weierstrass theorem says that there exists at least one accumulation point of S, where the concept of accumulation point is based on the usual topology induced by the conventional metric. When the metric does not satisfy the symmetric condition, Wu [18] has proposed two different concepts of open balls given by

$$B^{\triangleleft}(x; r) = \{y \in X : d(x, y) < r\}$$

and

$$B^{\triangleright}(x; r) = \{y \in X : d(y, x) < r\},$$

which can induces two respective topologies as follows

$$\tau^{\triangleleft} = \{O^{\triangleleft} \subseteq X : x \in O^{\triangleleft} \text{ if and only if there exist } r > 0 \text{ such that } x \in B^{\triangleleft}(x; r) \subseteq O^{\triangleleft}\}.$$

and

$$\tau^{\triangleright} = \{O^{\triangleright} \subseteq X : x \in O^{\triangleright} \text{ if and only if there exist } r > 0 \text{ such that } x \in B^{\triangleright}(x; r) \subseteq O^{\triangleright}\}.$$

In this case, we can similarly define the concepts of $\triangleleft$-accumulation point and $\triangleright$-accumulation point based on the open balls $B^{\triangleleft}(x; r)$ and $B^{\triangleright}(x; r)$, respectively. Therefore, we can similarly obtain the $\triangleleft$-type of Bolzano–Weierstrass theorem and $\triangleright$-type of Bolzano–Weierstrass theorem by considering the $\triangleleft$-accumulation point and $\triangleright$-accumulation point, respectively, which is used to present the completeness in $\mathbb{R}$.

Example 1. *We are going to claim that every $>$-Cauchy sequence in $\mathbb{R}_+$ is convergent in the sense of $x_n \xrightarrow{d^{\triangleright}} x$ with respect to a pre-metric $d : \mathbb{R}_+ \times \mathbb{R}_+ \to \mathbb{R}_+$ defined by*

$$d(x,y) = \begin{cases} x & \text{if } x > y \\ 0 & \text{if } x = y \\ 2y - x & \text{if } x < y, \end{cases}$$

where the symmetric condition is not satisfied and d satisfies the ⋈-triangle inequality.

Let $T = \{x_1, x_2, \cdots, x_n, \cdots\}$ be a >-Cauchy sequence in $\mathbb{R}_+$. We are going to show that T is ▷-bounded. Given $\epsilon = 1$, there is an integer N such that $d(x_n, x_N) < 1$ for each $n > N$. This means that $x_n \in B^▷(x_N; 1)$ for each $n \geq N$. We define

$$r = 1 + \max\{d(x_1, 0), \cdots, d(x_N, 0)\}.$$

- For $n \leq N$, we have

$$d(x_n, 0) \leq \max\{d(x_1, 0), \cdots, d(x_N, 0)\} < r.$$

- For $n > N$, using the ⋈-triangle inequality, we have

$$d(x_n, 0) \leq d(x_n, x_N) + d(x_N, 0) < 1 + d(x_N, 0) \leq r.$$

Then, we see that $T \subseteq B^▷(0; r)$, which says that T is ▷-bounded. Using the Bolzano–Weierstrass theorem, the sequence T has a ▷-accumulation point $x^* \in \mathbb{R}_+$. Next we are going to show that $x_n \xrightarrow{d^▷} x^*$. Given any $\epsilon > 0$, there exists an integer N such that $n > m \geq N$ implies $d(x_n, x_m) < \epsilon/2$. Since x^* is a ▷-accumulation point of the sequence T, it follows that the open ball $B^▷(x^*; \epsilon/2)$ contains a point x_m for $m \geq N$, i.e., $d(x_m, x^*) < \epsilon/2$. Therefore, for $n > N$, using the ⋈-triangle inequality, we have

$$d(x_n, x^*) \leq d(x_n, x_m) + d(x_m, x^*) < \frac{\epsilon}{2} + \frac{\epsilon}{2} = \epsilon,$$

which shows that $x_n \xrightarrow{d^▷} x^*$. In other words, the pre-metric space $(\mathbb{R}, d)$ is $(>, ▷)$-complete

Example 2. *Continued from Example 1, we are going to claim that every <-Cauchy sequence in $\mathbb{R}_+$ is convergent in the sense of $x_n \xrightarrow{d^◁} x$. Let $T = \{x_1, x_2, \cdots, x_n, \cdots\}$ be a <-Cauchy sequence in $\mathbb{R}_+$. We are going to show that T is ◁-bounded. Given $\epsilon = 1$, there is an integer N such that $d(x_N, x_n) < 1$ for each $n > N$. This means that $x_n \in B^◁(x_N; 1)$ for each $n \geq N$. We define*

$$r = 1 + \max\{d(0, x_1), \cdots, d(0, x_N)\}.$$

- For $n \leq N$, we have

$$d(0, x_n) \leq \max\{d(0, x_1), \cdots, d(0, x_N)\} < r.$$

- For $n > N$, using the ⋈-triangle inequality, we have

$$d(0, x_n) \leq d(0, x_N) + d(x_N, x_n) < 1 + d(0, x_N) \leq r.$$

Then, we see that $T \subseteq B^◁(0; r)$, which says that T is ◁-bounded. Using the Bolzano–Weierstrass theorem, the sequence T has a ◁-accumulation point $x° \in \mathbb{R}_+$. Next we are going to show that $x_n \xrightarrow{d^◁} x°$. Given any $\epsilon > 0$, there exists an integer N such that $n > m \geq N$ implies $d(x_m, x_n) < \epsilon/2$. Since $x°$ is a ◁-accumulation point of the sequence T, it follows that the open ball $B^◁(x°; \epsilon/2)$ contains a point x_m for $m \geq N$, i.e., $d(x°, x_m) < \epsilon/2$. Therefore, for $n > N$, using the ⋈-triangle inequality, we have

$$d(x^\circ, x_n) \leq d(x^\circ, x_m) + d(x_m, x_n) < \frac{\epsilon}{2} + \frac{\epsilon}{2} = \epsilon,$$

which shows that $x_n \xrightarrow{d^\triangleleft} x^\circ$. In other words, the pre-metric space $(\mathbb{R}, d)$ is $(<, \triangleleft)$-complete

Example 3. *Continued from Examples 1 and 2, we are going to claim that the pre-metric space $(\mathbb{R}, d)$ is simultaneously $\triangleright$-complete and $\triangleleft$-complete. Let $T = \{x_1, x_2, \cdots, x_n, \cdots\}$ be a Cauchy sequence in $\mathbb{R}_+$. It means that T is both a $>$-Cauchy sequence and $<$-Cauchy sequence in $\mathbb{R}_+$. Examples 1 and 2 say that there exist x^* and x° satisfying $x_n \xrightarrow{d^\triangleright} x^*$ and $x_n \xrightarrow{d^\triangleleft} x^\circ$. In other words, the pre-metric space $(\mathbb{R}, d)$ is simultaneously $\triangleright$-complete and $\triangleleft$-complete. We also remark that $x^* \neq x^\circ$ in general.*

4. Banach Contraction Principle for Pre-Metric Spaces

Let $T : X \to X$ be a function from a nonempty set X into itself. If $T(x) = x$, we say that $x \in X$ is a fixed point of T. The well-known Banach contraction principle says that any functions that are a contraction on X has a fixed point when X is taken to be a complete metric space. In this paper, we study the Banach contraction principle when X is taken to be a complete pre-metric space.

Definition 6. *Let (X, d) be a pre-metric space. A function $T : (X, d) \to (X, d)$ is called a contraction on X when there is a real number $0 < \alpha < 1$ satisfying*

$$d(T(x), T(y)) \leq \alpha d(x, y)$$

for any $x, y \in X$.

Given any initial element $x_0 \in X$, using the function T, we consider the iterative sequence $\{x_n\}_{n=1}^\infty$ as follows:

$$x_1 = T(x_0), x_2 = T(x_1) = T^2(x_0), \cdots, x_n = T^n(x_0), \cdots. \quad (5)$$

We are going to show that the sequence $\{x_n\}_{n=1}^\infty$ can converge to a fixed point of T under some suitable conditions.

Theorem 1 (Banach Contraction Principle Using the $\triangleleft$-Triangle Inequality). *Let (X, d) be a $(>, \triangleright)$-complete pre-metric space or $(<, \triangleright)$-complete pre-metric space such that the $\triangleleft$-triangle inequality is satisfied. Suppose that the function $T : (X, d) \to (X, d)$ is a contraction on X. Then T has a unique fixed point $x \in X$. Moreover, the fixed point x is obtained by the following limit*

$$d(x_n, x) \to 0 \text{ as } n \to \infty,$$

where the sequence $\{x_n\}_{n=1}^\infty$ is generated according to (5).

Proof. Given any initial element $x_0 \in X$, according to (5), we can generate the iterative sequence $\{x_n\}_{n=1}^\infty$. The purpose is to show that $\{x_n\}_{n=1}^\infty$ is both a $>$-Cauchy sequence and $<$-Cauchy sequence. Since T is a contraction on X, without having the symmetric condition, we have the two cases as follows:

$$d(x_{m+1}, x_m) = d(T(x_m), T(x_{m-1})) \leq \alpha d(x_m, x_{m-1})$$
$$= \alpha d(T(x_{m-1}), T(x_{m-2})) \leq \alpha^2 d(x_{m-1}, x_{m-2})$$
$$\leq \cdots \leq \alpha^m d(x_1, x_0)$$

and

$$d(x_m, x_{m+1}) \leq \alpha^m d(x_0, x_1).$$

For $m > n$, since the $\triangleleft$-triangle inequality is assumed to be satisfied, according to the third observation in Remark 1, we obtain

$$d(x_m, x_n) \leq d(x_{m-1}, x_m) + d(x_{m-1}, x_{m-2}) + \cdots + d(x_{n+1}, x_n)$$
$$\leq \alpha^{m-1} \cdot d(x_0, x_1) + \left(\alpha^{m-2} + \cdots + \alpha^n\right) \cdot d(x_1, x_0)$$
$$\leq \left(\alpha^{m-1} + \alpha^{m-2} + \cdots + \alpha^n\right) \cdot \max\{d(x_0, x_1), d(x_1, x_0)\}$$
$$= \alpha^n \cdot \frac{1 - \alpha^{m-n}}{1 - \alpha} \cdot \max\{d(x_0, x_1), d(x_1, x_0)\}$$

and

$$d(x_n, x_m) \leq \alpha^n \cdot \frac{1 - \alpha^{m-n}}{1 - \alpha} \cdot \max\{d(x_0, x_1), d(x_1, x_0)\}.$$

Since $0 < \alpha < 1$, we have $1 - \alpha^{m-n} < 1$ in the numerator. Therefore, we obtain

$$d(x_m, x_n) \leq \frac{\alpha^n}{1 - \alpha} \cdot \max\{d(x_0, x_1), d(x_1, x_0)\} \to 0 \text{ as } n \to \infty$$

and

$$d(x_n, x_m) \to 0 \text{ as } n \to \infty,$$

which shows that $\{x_n\}_{n=1}^{\infty}$ is both a $>$-Cauchy sequence and $<$-Cauchy sequence. Since X is $(>, \triangleright)$-complete or $(<, \triangleright)$-complete, there exists $x \in X$ satisfying $d(x_n, x) \to 0$ as $n \to \infty$.

Now, we are going to claim that x is indeed a fixed point. We have

$$d(x, T(x)) \leq d(x_m, x) + d(x_m, T(x)) \text{ (using the } \triangleleft\text{-triangle inequality)}$$
$$= d(x_m, x) + d(T(x_{m-1}), T(x))$$
$$\leq d(x_m, x) + \alpha d(x_{m-1}, x) \text{ (using the contraction)}$$

which implies $d(x, T(x)) = 0$ as $m \to \infty$. We conclude that $T(x) = x$ by the condition of pre-metric space. The uniqueness will also be obtained. Assume that there is another fixed point $\bar{x}$ of T, i.e., $\bar{x} = T(\bar{x})$. The contraction of function T says that

$$d(\bar{x}, x) = d(T(\bar{x}), T(x)) \leq \alpha d(\bar{x}, x).$$

Since $0 < \alpha < 1$, we conclude that $d(\bar{x}, x) = 0$, i.e., $\bar{x} = x$. This completes the proof. □

Theorem 2 (Banach Contraction Principle Using the $\triangleright$-Triangle Inequality). *Let (X, d) be a $(>, \triangleleft)$-complete pre-metric space or $(<, \triangleleft)$-complete pre-metric space such that the $\triangleright$-triangle inequality is satisfied. Suppose that the function $T : (X, d) \to (X, d)$ is a contraction on X. Then T has a unique fixed point $x \in X$. Moreover, the fixed point x is obtained by the following limit*

$$d(x, x_n) \to 0 \text{ as } n \to \infty,$$

where the sequence $\{x_n\}_{n=1}^{\infty}$ is generated according to (5).

Proof. Given any initial element $x_0 \in X$, according to (5), we can generate the iterative sequence $\{x_n\}_{n=1}^{\infty}$. The purpose is to show that $\{x_n\}_{n=1}^{\infty}$ is both a $>$-Cauchy sequence and $<$-Cauchy sequence. From the proof of Theorem 1, the contraction of function T says that

$$d(x_{m+1}, x_m) \leq \alpha^m d(x_1, x_0) \text{ and } d(x_m, x_{m+1}) \leq \alpha^m d(x_0, x_1).$$

For $m > n$, since the $\triangleright$-triangle inequality is satisfied, according to the second observation in Remark 1, we obtain

$$d(x_m, x_n) \leq d(x_m, x_{m-1}) + d(x_{m-2}, x_{m-1}) + \cdots + d(x_n, x_{n+1})$$
$$\leq \alpha^{m-1} \cdot d(x_1, x_0) + \left(\alpha^{m-2} + \cdots + \alpha^n\right) \cdot d(x_0, x_1)$$
$$\leq \left(\alpha^{m-1} + \alpha^{m-2} + \cdots + \alpha^n\right) \cdot \max\{d(x_0, x_1), d(x_1, x_0)\}$$
$$= \alpha^n \cdot \frac{1 - \alpha^{m-n}}{1 - \alpha} \cdot \max\{d(x_0, x_1), d(x_1, x_0)\}$$

and
$$d(x_n, x_m) \leq \alpha^n \cdot \frac{1 - \alpha^{m-n}}{1 - \alpha} \cdot \max\{d(x_0, x_1), d(x_1, x_0)\},$$

which also imply
$$d(x_m, x_n) \to 0 \text{ and } d(x_n, x_m) \to 0 \text{ as } n \to \infty.$$

Therefore $\{x_n\}_{n=1}^\infty$ is both a $>$-Cauchy sequence and $<$-Cauchy sequence. Since X is $(>,\triangleleft)$-complete or $(<,\triangleleft)$-complete, there exists $x \in X$ satisfying $d(x, x_n) \to 0$ as $n \to \infty$.

Regarding the uniqueness, we have
$$d(x, T(x)) \leq d(x, x_m) + d(T(x), x_m) \text{ (using the } \triangleright\text{-triangle inequality)}$$
$$= d(x, x_m) + d(T(x), T(x_{m-1}))$$
$$\leq d(x, x_m) + \alpha d(x, x_{m-1}) \text{ (using the contraction)}$$

which implies $d(x, T(x)) = 0$ as $m \to \infty$. We conclude that $T(x) = x$. The uniqueness can also be obtained from the argument in the proof of Theorem 1. This completes the proof. □

Theorem 3 (Banach Contraction Principle Using the $\bowtie$-Triangle Inequality). *Let (X, d) be a pre-metric space such that the $\bowtie$-triangle inequality is satisfied. We also assume that any one of the following conditions is satisfied:*

- (X, d) is simultaneously $(>, \triangleright)$-complete and $(>, \triangleleft)$-complete;
- (X, d) is simultaneously $(>, \triangleright)$-complete and $(<, \triangleleft)$-complete;
- (X, d) is simultaneously $(<, \triangleright)$-complete and $(>, \triangleleft)$-complete;
- (X, d) is simultaneously $(<, \triangleright)$-complete and $(<, \triangleleft)$-complete;
- (X, d) is simultaneously $\triangleright$-complete and $\triangleleft$-complete.

Suppose that the function $T : (X, d) \to (X, d)$ is a contraction on X. Then T has a unique fixed point $x \in X$. Moreover, the fixed point x is obtained by the following limits
$$d(x_n, x) \to 0 \text{ or } d(x, x_n) \to 0 \text{ as } n \to \infty,$$

where the sequence $\{x_n\}_{n=1}^\infty$ is generated according to (5).

Proof. Given any initial element $x_0 \in X$, according to (5), we can generate the iterative sequence $\{x_n\}_{n=1}^\infty$. The purpose is to show that $\{x_n\}_{n=1}^\infty$ is both a $>$-Cauchy sequence and $<$-Cauchy sequence. From the proof of Theorem 1, the contraction of function T says that
$$d(x_{m+1}, x_m) \leq \alpha^m d(x_1, x_0) \text{ and } d(x_m, x_{m+1}) \leq \alpha^m d(x_0, x_1).$$

For $m > n$, since the $\bowtie$-triangle inequality is satisfied, according to the first observation in Remark 1, we obtain
$$d(x_m, x_n) \leq d(x_m, x_{m-1}) + d(x_{m-1}, x_{m-2}) + \cdots + d(x_{n+1}, x_n)$$
$$\leq \left(\alpha^{m-1} + \alpha^{m-2} + \cdots + \alpha^n\right) \cdot d(x_1, x_0)$$
$$= \alpha^n \cdot \frac{1 - \alpha^{m-n}}{1 - \alpha} \cdot d(x_1, x_0)$$

and

$$d(x_n, x_m) \leq d(x_{m-1}, x_m) + d(x_{m-2}, x_{m-1}) + \cdots + d(x_n, x_{n+1})$$
$$\leq \left(\alpha^{m-1} + \alpha^{m-2} + \cdots + \alpha^n\right) \cdot d(x_0, x_1)$$
$$= \alpha^n \cdot \frac{1 - \alpha^{m-n}}{1 - \alpha} \cdot d(x_0, x_1),$$

which imply

$$d(x_m, x_n) \to 0 \text{ and } d(x_n, x_m) \to 0 \text{ as } n \to \infty.$$

This proves that $\{x_n\}_{n=1}^{\infty}$ is both a $>$-Cauchy sequence and $<$-Cauchy sequence. It follows that $\{x_n\}_{n=1}^{\infty}$ is a Cauchy sequence.

Assume that X is simultaneously $(>,\triangleright)$-complete and $(<,\triangleleft)$-complete. Then there exists $x^*, x^\circ \in X$ satisfying $d(x_n, x^*) \to 0$ and $d(x^\circ, x_n) \to 0$ as $n \to \infty$. Now, we have

$$d(x^\circ, T(x^*)) \leq d(x^\circ, x_m) + d(x_m, T(x^*)) \text{ (using the } \bowtie\text{-triangle inequality)}$$
$$= d(x^\circ, x_m) + d(T(x_{m-1}), T(x^*))$$
$$\leq d(x^\circ, x_m) + \alpha d(x_{m-1}, x^*) \text{ (using the contraction)},$$

which implies $d(x^\circ, T(x^*)) = 0$ as $m \to \infty$. Therefore, we obtain $T(x^*) = x^\circ$. Now, we have

$$d(T(x^\circ), x^*) \leq d(T(x^\circ), x_m) + d(x_m, x^*) \text{ (using the } \bowtie\text{-triangle inequality)}$$
$$= d(T(x^\circ), T(x_{m-1})) + d(x_m, x^*)$$
$$\leq \alpha d(x^\circ, x_{m-1}) + d(x_m, x^*) \text{ (using the contraction)},$$

which implies $d(T(x^\circ), x^*) = 0$ as $m \to \infty$. We also obtain $T(x^\circ) = x^*$. Now, we have

$$T^2(x^*) = T(T(x^*)) = T(x^\circ) = x^* \text{ and } T^2(x^\circ) = T(T(x^\circ)) = T(x^*) = x^\circ. \tag{6}$$

This shows that x^* and x° are fixed points of the composition mapping $T \circ T \equiv T^2$. The contraction of function T says that

$$d(x^*, x^\circ) = d(T^2(x^*), T^2(x^\circ)) \leq \alpha d(T(x^*), T(x^\circ)) \leq \alpha^2 d(x^*, x^\circ).$$

Since $0 < \alpha < 1$, i.e., $0 < \alpha^2 < 1$, we conclude that $d(x^*, x^\circ) = 0$, i.e., $x^* = x^\circ$. This also says that $x^* = x^\circ$ is a fixed point of T.

The uniqueness can be obtained using the argument in the proof of Theorem 1. For the other three conditions, we can similarly obtain the desired results. This completes the proof. □

Example 4. *Continued from Example 3, since the pre-metric space $(\mathbb{R}, d)$ is simultaneously $\triangleright$-complete and $\triangleleft$-complete, any function $T : (\mathbb{R}, d) \to (\mathbb{R}, d)$ that is a contraction on $\mathbb{R}$ has a unique fixed point. The concrete examples regarding functions that are contraction on $\mathbb{R}$ can be obtained from the literature.*

Theorem 4 (Banach Contraction Principle Using the $\diamond$-Triangle Inequality). *Let (X, d) be a pre-metric space such that the $\diamond$-triangle inequality is satisfied. We also assume that any one of the following conditions is satisfied:*

- *(X, d) is simultaneously $(>,\triangleright)$-complete and $(>,\triangleleft)$-complete;*
- *(X, d) is simultaneously $(>,\triangleright)$-complete and $(<,\triangleleft)$-complete;*
- *(X, d) is simultaneously $(<,\triangleright)$-complete and $(>,\triangleleft)$-complete;*
- *(X, d) is simultaneously $(<,\triangleright)$-complete and $(<,\triangleleft)$-complete.*

Suppose that the function $T : (X, d) \to (X, d)$ is a contraction on X. Then T has a unique fixed point $x \in X$. Moreover, the fixed point x is obtained by the following limits

$$d(x_n, x) \to 0 \text{ or } d(x, x_n) \to 0 \text{ as } n \to \infty,$$

where the sequence $\{x_n\}_{n=1}^{\infty}$ is generated according to (5).

Proof. Given any initial element $x_0 \in X$, according to (5), we can generate the iterative sequence $\{x_n\}_{n=1}^{\infty}$. The purpose is to show that $\{x_n\}_{n=1}^{\infty}$ is both a $>$-Cauchy sequence and $<$-Cauchy sequence. From the proof of Theorem 1, the contraction of function T says that

$$d(x_{m+1}, x_m) \leq \alpha^m d(x_1, x_0) \text{ and } d(x_m, x_{m+1}) \leq \alpha^m d(x_0, x_1).$$

For $m > n$, since the $\diamond$-triangle inequality is satisfied, according to the fourth observation in Remark 1 by assuming $m - n \equiv p$ is an even number, we obtain

$$d(x_m, x_n) \leq d(x_n, x_{n+1}) + d(x_{n+1}, x_{n+2}) + d(x_{n+3}, x_{n+2}) + d(x_{n+3}, x_{n+4}) + d(x_{n+5}, x_{n+4})$$
$$+ d(x_{n+5}, x_{n+6}) + \cdots + d(x_{m-1}, x_m)$$
$$\leq \alpha^n d(x_0, x_1) + \alpha^{n+1} d(x_0, x_1) + \alpha^{n+2} d(x_1, x_0) + \alpha^{n+3} d(x_0, x_1) + \alpha^{n+4} d(x_1, x_0)$$
$$+ \alpha^{n+5} d(x_0, x_1) + \cdots + \alpha^{m-1} d(x_0, x_1)$$
$$\leq \left(\alpha^{m-1} + \alpha^{m-2} + \cdots + \alpha^n \right) \cdot \max\{d(x_0, x_1), d(x_1, x_0)\}$$
$$= \alpha^n \cdot \frac{1 - \alpha^{m-n}}{1 - \alpha} \cdot \max\{d(x_0, x_1), d(x_1, x_0)\}$$

and

$$d(x_n, x_m) \leq d(x_{n+1}, x_n) + d(x_{n+2}, x_{n+1}) + d(x_{n+2}, x_{n+3}) + d(x_{n+4}, x_{n+3}) + d(x_{n+4}, x_{n+5})$$
$$+ d(x_{n+6}, x_{n+5}) + \cdots + d(x_m, x_{m-1})$$
$$\leq \alpha^n d(x_1, x_0) + \alpha^{n+1} d(x_1, x_0) + \alpha^{n+2} d(x_0, x_1) + \alpha^{n+3} d(x_1, x_0) + \alpha^{n+4} d(x_0, x_1)$$
$$+ \alpha^{n+5} d(x_1, x_0) + \cdots + \alpha^{m-1} d(x_1, x_0)$$
$$\leq \left(\alpha^{m-1} + \alpha^{m-2} + \cdots + \alpha^n \right) \cdot \max\{d(x_0, x_1), d(x_1, x_0)\}$$
$$= \alpha^n \cdot \frac{1 - \alpha^{m-n}}{1 - \alpha} \cdot \max\{d(x_0, x_1), d(x_1, x_0)\}$$

which imply

$$d(x_m, x_n) \to 0 \text{ and } d(x_n, x_m) \to 0 \text{ as } n \to \infty.$$

This proves that $\{x_n\}_{n=1}^{\infty}$ is both a $>$-Cauchy sequence and $<$-Cauchy sequence.

Assume that X is simultaneously $(>, \triangleright)$-complete and $(<, \triangleleft)$-complete. Then there exists $x^*, x^\circ \in X$ satisfying $d(x_n, x^*) \to 0$ and $d(x^\circ, x_n) \to 0$ as $n \to \infty$. Now, we have

$$d(x^*, T(x^\circ)) \leq d(x_m, x^*) + d(T(x^\circ), x_m) \text{ (using the $\diamond$-triangle inequality)}$$
$$= d(x_m, x^*) + d(T(x^\circ), T(x_{m-1}))$$
$$\leq d(x_m, x^*) + \alpha d(x^\circ, x_{m-1}) \text{ (using the contraction)},$$

which implies $d(x^*, T(x^\circ)) = 0$ as $m \to \infty$. We conclude that $T(x^\circ) = x^*$. We also have

$$d(T(x^*), x^\circ) \leq d(x_m, T(x^*)) + d(x^\circ, x_m) \text{ (using the $\diamond$-triangle inequality)}$$
$$= d(T(x_{m-1}), T(x^*)) + d(x^\circ, x_m)$$
$$\leq \alpha d(x_{m-1}, x^*) + d(x^\circ, x_m) \text{ (using the contraction)},$$

which implies $d(T(x^*), x^\circ) = 0$ as $m \to \infty$. We conclude that $T(x^*) = x^\circ$. The remaining proof follows from the same argument in the proof of Theorem 3. This completes the proof. □

5. Meir–Keeler Type of Fixed Point Theorems for Pre-Metric Spaces

In the sequel, we are going to establish the Meir–Keeler type of fixed point theorems for pre-metric spaces. First of all, we consider the different contraction.

Definition 7. *Let (X, d) be a pre-metric space. A function $T : (X, d) \to (X, d)$ is called a weakly strict contraction on X when the following conditions are satisfied:*

- $d(x, y) = 0$ implies $d(T(x), T(y)) = 0$;
- $d(x, y) \neq 0$ implies $d(T(x), T(y)) < d(x, y)$.

It is clear to see that if T is a contraction on X, then it is also a weakly strict contraction on X.

Theorem 5 (Fixed Points Using the ◁-Triangle Inequality). *Let (X, d) be a $(>, \triangleright)$-complete (resp. $(<, \triangleright)$-complete) pre-metric space such that the ◁-triangle inequality is satisfied. Suppose that the function $T : (X, d) \to (X, d)$ is a weakly strict contraction on X, and that $\{T^n(x_0)\}_{n=1}^{\infty}$ forms a $>$-Cauchy sequence (resp. $<$-Cauchy sequence) for some $x_0 \in X$. Then, the function T has a unique fixed point $x \in X$. Moreover, the fixed point x is obtained by the following limit*

$$d(T^n(x_0), x) \to 0 \text{ as } n \to \infty.$$

Proof. Since $\{T^n(x_0)\}_{n=1}^{\infty}$ is a $>$-Cauchy sequence, the $(>, \triangleright)$-completeness says that there exists $x \in X$ satisfying $d(T^n(x_0), x) \to 0$ as $n \to \infty$. In other words, given any $\epsilon > 0$, there exists an integer N satisfying $d(T^n(x_0), x) < \epsilon$ for $n \geq N$. Regarding $d(T^n(x_0), x)$, we consider two different cases as follows.

- Suppose that $d(T^n(x_0), x) = 0$. Then, the weakly strict contraction of T says that

$$d(T^{n+1}(x_0), T(x)) = 0 < \epsilon.$$

- Suppose that $d(T^n(x_0), x) \neq 0$. Then, the weakly strict contraction of T says that

$$d(T^{n+1}(x_0), T(x)) < d(T^n(x_0), x) < \epsilon \text{ for } n \geq N.$$

The above two cases conclude that $d(T^{n+1}(x_0), T(x)) \to 0$ as $n \to \infty$. Using the ◁-triangle inequality, we obtain

$$d(T(x), x) \leq d\left(T^{n+1}(x_0), T(x)\right) + d\left(T^{n+1}(x_0), x\right) \to 0 \text{ as } n \to \infty,$$

which shows that $d(T(x), x) = 0$, i.e., $T(x) = x$. In other words, x is a fixed point.

Regarding the uniqueness, suppose that $\bar{x}$ is another fixed point of T. i.e., $T(\bar{x}) = \bar{x}$. Since $x \neq \bar{x}$, the weakly strict contraction of T says that

$$d(x, \bar{x}) = d(T(x), T(\bar{x})) < d(x, \bar{x}).$$

This contradiction shows that $\bar{x}$ cannot be a fixed point of T.

When $\{T^n(x_0)\}_{n=1}^{\infty}$ is a $<$-Cauchy sequence, using the $(<, \triangleright)$-completeness, we can similarly obtain the desired results. This completes the proof. □

Theorem 6 (Fixed Points Using the ▷-Triangle Inequality). *Let (X, d) be a $(>, \triangleleft)$-complete (resp. $(<, \triangleleft)$-complete) pre-metric space such that the ▷-triangle inequality is satisfied. Suppose that the function $T : (X, d) \to (X, d)$ is a weakly strict contraction on X, and that $\{T^n(x_0)\}_{n=1}^{\infty}* forms a $>$-Cauchy sequence (resp. $<$-Cauchy sequence) for some $x_0 \in X$. Then, the function T has a unique fixed point $x \in X$. Moreover, the fixed point x is obtained by the following limit

$$d(x, T^n(x_0)) \to 0 \text{ as } n \to \infty.$$

Proof. Since $\{T^n(x_0)\}_{n=1}^\infty$ is a $>$-Cauchy sequence, the $(>, \triangleleft)$-completeness says that there exists $x \in X$ satisfying $d(x, T^n(x_0)) \to 0$ as $n \to \infty$. From the proof of Theorem 5, the weakly strict contraction of T can similarly show that $d(T(x), T^{n+1}(x_0)) \to 0$ as $n \to \infty$. Using the $\triangleright$-triangle inequality, we obtain

$$d(T(x), x) \leq d\Big(T(x), T^{n+1}(x_0)\Big) + d\Big(x, T^{n+1}(x_0)\Big) \to 0 \text{ as } n \to \infty,$$

which says that $d(T(x), x) = 0$, i.e., $T(x) = x$. This shows that x is a fixed point. The remaining proof follows from the proof of Theorem 5. This completes the proof. □

Theorem 7 (Fixed Points Using the $\bowtie$-Triangle Inequality). *Let (X, d) be a simultaneously $(>, \triangleright)$-complete and $(>, \triangleleft)$-complete (resp. $(<, \triangleright)$-complete and $(<, \triangleleft)$-complete) pre-metric space such that the $\bowtie$-triangle inequality is satisfied. Suppose that the function $T : (X, d) \to (X, d)$ is a weakly strict contraction on X, and that $\{T^n(x_0)\}_{n=1}^\infty$ forms a $>$-Cauchy sequence (resp. $<$-Cauchy sequence) for some $x_0 \in X$, then T has a unique fixed point $x \in X$. Moreover, the fixed point x is obtained by the following limits*

$$d(T^n(x_0), x) \to 0 \text{ or } d(x, T^n(x_0)) \to 0 \text{ as } n \to \infty.$$

Proof. Since $\{T^n(x_0)\}_{n=1}^\infty$ is a $>$-Cauchy sequence, the $(>, \triangleright)$-completeness says that there exists $x^* \in X$ satisfying $d(T^n(x_0), x^*) \to 0$ as $n \to \infty$. The $(>, \triangleleft)$-completeness also says that there exists another $x^\circ \in X$ satisfying $d(x^\circ, T^n(x_0)) \to 0$ as $n \to \infty$. From the proof of Theorem 5, the weakly strict contraction of T can similarly show that $d(T^{n+1}(x_0), T(x^*)) \to 0$ and $d(T(x^\circ), T^{n+1}(x_0)) \to 0$ as $n \to \infty$. Using the $\bowtie$-triangle inequality, we obtain

$$d(T(x^\circ), x^*) \leq d\Big(T(x^\circ), T^{n+1}(x_0)\Big) + d\Big(T^{n+1}(x_0), x^*\Big) \to 0 \text{ as } n \to \infty,$$

which says that $d(T(x^\circ), x^*) = 0$, i.e., $T(x^\circ) = x^*$. On the other hand, we also have

$$d(x^\circ, T(x^*)) \leq d\Big(x^\circ, T^{n+1}(x_0)\Big) + d\Big(T^{n+1}(x_0), T(x^*)\Big) \to 0 \text{ as } n \to \infty,$$

which says that $T(x^*) = x^\circ$. By referring to (6), it follows that x^* and x° are fixed points of the composition mapping $T \circ T \equiv T^2$. Suppose that $x^* \neq x^\circ$. We want to claim $T(x^*) \neq T(x^\circ)$. Assume that it is not true, i.e., $T(x^*) = T(x^\circ)$. Then, we shall have

$$x^\circ = T(x^*) = T(x^\circ) = x^*,$$

which contradicts $x^* \neq x^\circ$. The weakly strict contraction of T also says that

$$d(x^*, x^\circ) = d(T^2(x^*), T^2(x^\circ)) < d(T(x^*), T(x^\circ)) < d(x^*, x^\circ).$$

This contradiction shows that $x^* = x^\circ$, and says that $x^* = x^\circ$ is a fixed point of T. The uniqueness can be obtained from the proof of Theorem 5

When $\{T^n(x_0)\}_{n=1}^\infty$ is a $<$-Cauchy sequence, using the $(<, \triangleright)$-completeness and $(<, \triangleleft)$-completeness, we can similarly obtain the desired results. This completes the proof. □

Theorem 8 (Fixed Points Using the $\diamond$-Triangle Inequality). *Let (X, d) be a simultaneously $(>, \triangleright)$-complete and $(>, \triangleleft)$-complete (resp. $(<, \triangleright)$-complete and $(<, \triangleleft)$-complete) pre-metric space such that the $\diamond$-triangle inequality is satisfied. Suppose that the function $T : (X, d) \to$*

(X, d) is a weakly strict contraction on X, and that $\{T^n(x_0)\}_{n=1}^{\infty}$ forms a $>$-Cauchy sequence (resp. $<$-Cauchy sequence) for some $x_0 \in X$. Then T has a unique fixed point $x \in X$. Moreover, the fixed point x is obtained by the following limits

$$d(T^n(x_0), x) \to 0 \text{ or } d(x, T^n(x_0)) \to 0 \text{ as } n \to \infty.$$

Proof. From the proof of Theorem 7, there exist $x^*, x^\circ \in X$ satisfying $d(T^n(x_0), x^*) \to 0$, $d(x^\circ, T^n(x_0)) \to 0$, $d(T^{n+1}(x_0), T(x^*)) \to 0$ and $d(T(x^\circ), T^{n+1}(x_0)) \to 0$ as $n \to \infty$. Using the $\diamond$-triangle inequality, we can obtain

$$d(x^*, T(x^\circ)) \leq d\left(T^{n+1}(x_0), x^*\right) + d\left(T(x^\circ), T^{n+1}(x_0)\right) \to 0 \text{ as } n \to \infty,$$

which says that $d(x^*, T(x^\circ)) = 0$, i.e., $T(x^\circ) = x^*$. We also have

$$d(T(x^*), x^\circ) \leq d\left(T^{n+1}(x_0), T(x^*)\right) + d\left(x^\circ, T^{n+1}(x_0)\right) \to 0 \text{ as } n \to \infty,$$

which says that $T(x^*) = x^\circ$. The remaining proof follows from the similar argument in the proof of Theorem 7. This completes the proof. □

Next, we consider the different fixed point theorems based on the weakly uniformly strict contraction that was proposed by Meir and Keeler [12].

Definition 8. *Let (X, d) be a pre-metric space. A function $T : (X, d) \to (X, d)$ is called a weakly uniformly strict contraction on X when the following conditions are satisfied:*
- $d(x, y) = 0$ implies $d(T(x), T(y)) = 0$;
- *given any $\epsilon > 0$, there exists $\delta > 0$ such that $\epsilon \leq d(x, y) < \epsilon + \delta$ implies $d(T(x), T(y)) < \epsilon$ for any $x, y \in X$ with $d(x, y) \neq 0$.*

Remark 2. *We observe that if T is a weakly uniformly strict contraction on X, then T is also a weakly strict contraction on X.*

Lemma 1. *Let (X, d) be a pre-metric space, and let $T : (X, d) \to (X, d)$ be a weakly uniformly strict contraction on X. Then, the sequences $\{d(T^n(x), T^{n+1}(x))\}_{n=1}^{\infty}$ and $\{d(T^{n+1}(x), T^n(x))\}_{n=1}^{\infty}$ are decreasing to zero for any $x \in X$.*

Proof. For convenience, we simply write $T^n(x) = x_n$ for all n. Let $c_n = d(x_n, x_{n+1})$. Regarding $d(x_{n-1}, x_n) \geq 0$, we consider two different cases as follows.
- Suppose that $d(x_{n-1}, x_n) \neq 0$. By Remark 2, we have

$$c_n = d(x_n, x_{n+1}) = d(T^n(x), T^{n+1}(x)) < d(T^{n-1}(x), T^n(x)) = d(x_{n-1}, x_n) = c_{n-1}.$$

- Suppose that $d(x_{n-1}, x_n) = 0$. Then, by the first condition of Definition 8, we have

$$c_n = d(T^n(x), T^{n+1}(x)) = d(T(x_{n-1}), T(x_n)) = 0 \leq c_{n-1}.$$

The above two cases conclude that the sequence $\{c_n\}_{n=1}^{\infty}$ is decreasing. We also consider the following two cases.
- Let m be the first index in the sequence $\{x_n\}_{n=1}^{\infty}$ satisfying $d(x_{m-1}, x_m) = 0$. Then, we want to claim

$$c_{m-1} = c_m = c_{m+1} = \cdots = 0.$$

Using the first condition of Definition 8, we have

$$0 = d(T(x_{m-1}), T(x_m)) = d(T^m(x), T^{m+1}(x)) = d(x_m, x_{m+1}) = c_m.$$

Since $d(x_m, x_{m+1}) = 0$, using the similar argument, we can also obtain $c_{m+1} = 0$ and $d(x_{m+1}, x_{m+2}) = 0$. This shows that the sequence $\{c_n\}_{n=1}^{\infty}$ is indeed decreasing to zero.

- Suppose that $d(x_m, x_{m+1}) \neq 0$ for all $m \geq 1$. Since the sequence $\{c_n\}_{n=1}^{\infty}$ is decreasing, we assume that $c_n \downarrow \epsilon > 0$, i.e., $c_n \geq \epsilon > 0$ for all n. Therefore, there exists $\delta > 0$ satisfying $\epsilon \leq c_m < \epsilon + \delta$ for some m, i.e., $\epsilon \leq d(x_m, x_{m+1}) < \epsilon + \delta$. Using the second condition of Definition 8, it follows that

$$c_{m+1} = d(x_{m+1}, x_{m+2}) = d(T^{m+1}(x), T^{m+2}(x)) = d(T(x_m), T(x_{m+1})) < \epsilon,$$

which contradicts $c_{m+1} \geq \epsilon$.

Therefore, we conclude that the sequence $\{d(T^n(x), T^{n+1}(x))\}_{n=1}^{\infty}$ is indeed decreasing to zero for any $x \in X$. We can similarly show that the sequence $\{d(T^{n+1}(x), T^n(x))\}_{n=1}^{\infty}$ is decreasing to zero for any $x \in X$. This completes the proof. □

Theorem 9 (Meir–Keeler Type of Fixed Points Using the ◁-Triangle Inequality). *Let (X, d) be a $(>, \triangleright)$-complete pre-metric space such that the ◁-triangle inequality is satisfied, and let $T : (X, d) \to (X, d)$ be a weakly uniformly strict contraction on X. Then T has a unique fixed point. Moreover, the fixed point x is obtained by the following limit*

$$d(T^n(x_0), x) \to 0 \text{ as } n \to \infty \text{ for some } x_0.$$

Proof. According to Theorem 5 and Remark 2, we just need to claim that if T is a weakly uniformly strict contraction, then $\{T^n(x_0)\}_{n=1}^{\infty} = \{x_n\}_{n=1}^{\infty}$ is a $>$-Cauchy sequence for $x_0 \in X$. Suppose that $\{x_n\}_{n=1}^{\infty}$ is not a $>$-Cauchy sequence. Then, there exists $2\epsilon > 0$ such that, given any integer N, there exist $n > m \geq N$ satisfying $d(x_n, x_m) > 2\epsilon$. We are going to lead to a contradiction. The weakly uniformly strict contraction of T says that there exists $\delta > 0$ satisfying

$$\epsilon \leq d(x, y) < \epsilon + \delta \text{ implies } d(T(x), T(y)) < \epsilon \text{ for any } x, y \in X \text{ with } d(x, y) \neq 0.$$

Let $\delta' = \min\{\delta, \epsilon\}$. We want to show

$$\epsilon \leq d(x, y) < \epsilon + \delta' \text{ implies } d(T(x), T(y)) < \epsilon \text{ for any } x, y \in X \text{ with } d(x, y) \neq 0. \quad (7)$$

Indeed, when $\delta' = \epsilon$, i.e., $\epsilon < \delta$, we have $\epsilon + \delta' = \epsilon + \epsilon < \epsilon + \delta$.

Let $c_n = d(x_n, x_{n+1})$ and $\bar{c}_n = d(x_{n+1}, x_n)$. Since the sequences $\{c_n\}_{n=1}^{\infty}$ and $\{\bar{c}_n\}_{n=1}^{\infty}$ are decreasing to zero by Lemma 1, we can find a common integer N satisfying

$$c_N < \delta'/3 \text{ and } \bar{c}_N < \delta'/3. \quad (8)$$

For $n > m \geq N$, we have

$$d(x_n, x_m) > 2\epsilon \geq \epsilon + \delta', \quad (9)$$

which implicitly says that $d(x_n, x_m) \neq 0$. Since the sequence $\{\bar{c}_n\}_{n=1}^{\infty}$ is decreasing by Lemma 1 again, we can obtain

$$d(x_{m+1}, x_m) = \bar{c}_m \leq \bar{c}_N < \frac{\delta'}{3} \leq \frac{\epsilon}{3} < \epsilon. \quad (10)$$

For j with $m < j \leq n$, using the ◁-triangle inequality, it follows that

$$d(x_{j+1}, x_m) \leq d(x_j, x_{j+1}) + d(x_j, x_m). \quad (11)$$

We want to cliam that there exists an integer j with $m < j \leq n$ satisfying $d(x_j, x_m) \neq 0$ and
$$\epsilon + \frac{2\delta'}{3} < d(x_j, x_m) < \epsilon + \delta'. \tag{12}$$

Let $\gamma_j = d(x_j, x_m)$ for $j = m+1, \cdots, n$. Using (9) and (10), we have
$$\gamma_{m+1} < \epsilon \text{ and } \gamma_n > \epsilon + \delta'. \tag{13}$$

Let j_0 be an index satisfying
$$j_0 = \max\left\{ j \in [m+1, n] : \gamma_j \leq \epsilon + \frac{2\delta'}{3} \right\}. \tag{14}$$

Then, from (13), we see that $m + 1 \leq j_0 < n$, which also says that j_0 is well-defined. By the definition of j_0, it follows that $j_0 + 1 \leq n$ and $\gamma_{j_0+1} > \epsilon + \frac{2\delta'}{3}$, which also says that $d(x_{j_0+1}, x_m) \neq 0$. Therefore, the expression (12) will be sound if we can show
$$\epsilon + \frac{2\delta'}{3} < \gamma_{j_0+1} < \epsilon + \delta'.$$

Suppose that this is not true, i.e., $\gamma_{j_0+1} \geq \epsilon + \delta'$. From (11), we have
$$\frac{\delta'}{3} > c_N \geq c_{j_0} = d(x_{j_0}, x_{j_0+1}) \geq \gamma_{j_0+1} - \gamma_{j_0} \geq \epsilon + \delta' - \epsilon - \frac{2\delta'}{3} = \frac{\delta'}{3}.$$

This contradiction says that the expression (12) is sound. Since $d(x_j, x_m) \neq 0$, using (7), it follows that (12) implies
$$d(x_{j+1}, x_{m+1}) = d(T(x_j), T(x_m)) < \epsilon. \tag{15}$$

Using the ◁-triangle inequality and referring to (2), we can obtain
$$\begin{aligned}
d(x_j, x_m) &\leq d(x_{j+1}, x_j) + d(x_{j+1}, x_{m+1}) + d(x_{m+1}, x_m) \\
&= \bar{c}_j + d(x_{j+1}, x_{m+1}) + \bar{c}_m < \bar{c}_j + \epsilon + \bar{c}_m \text{ (by (15))} \\
&\leq \bar{c}_N + \epsilon + \bar{c}_N \text{ (since } \{\bar{c}_n\}_{n=1}^\infty \text{ is decreasing)} \\
&< \frac{\delta'}{3} + \epsilon + \frac{\delta'}{3} \text{ (by (8))} \\
&= \epsilon + \frac{2\delta'}{3},
\end{aligned}$$

which contradicts (12). This contradiction shows that every sequence $\{T^n(x)\}_{n=1}^\infty = \{x_n\}_{n=1}^\infty$ is a >-Cauchy sequence. Using Theorem 5, the proof is complete. □

Theorem 10 (Meir–Keeler Type of Fixed Points Using the ▷-Triangle Inequality). *Let (X, d) be a $(>, \triangleleft)$-complete pre-metric space such that the ▷-triangle inequality is satisfied, and let $T : (X, d) \to (X, d)$ be a weakly uniformly strict contraction on X. Then T has a unique fixed point. Moreover, the fixed point x is obtained by the following limit*
$$d(x, T^n(x_0)) \to 0 \text{ as } n \to \infty \text{ for some } x_0.$$

Proof. According to Theorem 6 and Remark 2, we just need to claim that if T is a weakly uniformly strict contraction, then $\{T^n(x_0)\}_{n=1}^\infty = \{x_n\}_{n=1}^\infty$ is a <-Cauchy sequence for $x_0 \in X$. Suppose that $\{x_n\}_{n=1}^\infty$ is not a <-Cauchy sequence. Then, there exists $2\epsilon > 0$

such that, given any integer N, there exist $n > m \geq N$ satisfying $d(x_m, x_n) > 2\epsilon$. Let $\delta' = \min\{\delta, \epsilon\}$. For $n > m \geq N$, we have

$$d(x_m, x_n) > 2\epsilon \geq \epsilon + \delta', \tag{16}$$

which implicitly says that $d(x_m, x_n) \neq 0$. Let $c_n = d(x_n, x_{n+1})$ and $\bar{c}_n = d(x_{n+1}, x_n)$. Since the sequence $\{c_n\}_{n=1}^\infty$ is decreasing by Lemma 1, we obtain

$$d(x_m, x_{m+1}) = c_m \leq c_N < \frac{\delta'}{3} \leq \frac{\epsilon}{3} < \epsilon. \tag{17}$$

For j with $m < j \leq n$, using the $\triangleright$-triangle inequality, we also have

$$d(x_m, x_{j+1}) \leq d(x_m, x_j) + d(x_{j+1}, x_j). \tag{18}$$

We want to cliam that there exists an integer j with $m < j \leq n$ satisfying $d(x_m, x_j) \neq 0$ and

$$\epsilon + \frac{2\delta'}{3} < d(x_m, x_j) < \epsilon + \delta'. \tag{19}$$

Let $\gamma_j = d(x_m, x_j)$ for $j = m+1, \cdots, n$. From (16) and (17), we can also obtain (13). Let j_0 be defined in (14). Then, the expression (19) will be sound if we can show that

$$\epsilon + \frac{2\delta'}{3} < \gamma_{j_0+1} < \epsilon + \delta'.$$

Suppose that this is not true, i.e., $\gamma_{j_0+1} \geq \epsilon + \delta'$. From (18) and (8), it follows that

$$\frac{\delta'}{3} > \bar{c}_N \geq \bar{c}_{j_0} = d(x_{j_0+1}, x_{j_0}) \geq \gamma_{j_0+1} - \gamma_{j_0} \geq \epsilon + \delta' - \epsilon - \frac{2\delta'}{3} = \frac{\delta'}{3}.$$

This contradiction says that (19) is sound. Since $d(x_m, x_j) \neq 0$, using (7), it follows that (19) implies

$$d(x_{m+1}, x_{j+1}) = d(T(x_m), T(x_j)) < \epsilon. \tag{20}$$

Using the $\triangleright$-triangle inequality and referring to (1), we can obtain

$$d(x_m, x_j) \leq d(x_j, x_{j+1}) + d(x_{m+1}, x_{j+1}) + d(x_m, x_{m+1})$$
$$= c_j + d(x_{m+1}, x_{j+1}) + c_m < c_j + \epsilon + c_m \text{ (using Equation (20))}$$
$$\leq c_N + \epsilon + c_N \text{ (since } \{c_n\}_{n=1}^\infty \text{ is decreasing)}$$
$$< \frac{\delta'}{3} + \epsilon + \frac{\delta'}{3} \text{ (using Equation (8))}$$
$$= \epsilon + \frac{2\delta'}{3},$$

which contradicts (19). This contradiction shows that every sequence $\{T^n(x)\}_{n=1}^\infty = \{x_n\}_{n=1}^\infty$ is a $<$-Cauchy sequence. Using Theorem 6, the proof is complete. □

Theorem 11 (Meir–Keeler Type of Fixed Points Using the $\bowtie$-Triangle Inequality). *Let (X, d) be a pre-metric space such that the $\bowtie$-triangle inequality is satisfied. We also assume that any one of the following conditions is satisfied:*

- *(X, d) is simultaneously $(>, \triangleright)$-complete and $(>, \triangleleft)$-complete;*
- *(X, d) is simultaneously $(<, \triangleright)$-complete and $(<, \triangleleft)$-complete;*
- *(X, d) is simultaneously $\triangleright$-complete and $\triangleleft$-complete.*

Suppose that $T: (X, d) \to (X, d)$ is a weakly uniformly strict contraction on X. Then T has a unique fixed point. Moreover, the fixed point x is obtained by the following limits

$$d(T^n(x_0), x) \to 0 \text{ or } d(x, T^n(x_0)) \to 0 \text{ as } n \to \infty.$$

Proof. According to Theorem 7 and Remark 2, we just need to claim that if T is a weakly uniformly strict contraction, then $\{T^n(x_0)\}_{n=1}^{\infty} = \{x_n\}_{n=1}^{\infty}$ is both a <-Cauchy sequence and >-Cauchy sequence for $x_0 \in X$. Suppose that $\{x_n\}_{n=1}^{\infty}$ is not a <-Cauchy sequence. Then, there exists $2\epsilon > 0$ such that, given any integer N, there exist $n > m \geq N$ satisfying $d(x_m, x_n) > 2\epsilon$. We are going to follow the similar proof of Theorem 10.

Let $\delta' = \min\{\delta, \epsilon\}$, and let $\gamma_j = d(x_m, x_j)$ for $j = m+1, \cdots, n$. For j with $m < j \leq n$, using the ⋈-triangle inequality, we have

$$d(x_m, x_{j+1}) \leq d(x_m, x_j) + d(x_j, x_{j+1}),$$

which implies

$$\frac{\delta'}{3} > c_N \geq c_{j_0} = d(x_{j_0}, x_{j_0+1}) \geq \gamma_{j_0+1} - \gamma_{j_0} \geq \epsilon + \delta' - \epsilon - \frac{2\delta'}{3} = \frac{\delta'}{3}.$$

This contradiction shows that there exists an integer j with $m < j \leq n$ satisfying $d(x_m, x_j) \neq 0$ and

$$\epsilon + \frac{2\delta'}{3} < d(x_m, x_j) < \epsilon + \delta', \tag{21}$$

which implies

$$d(x_{m+1}, x_{j+1}) = d(T(x_m), T(x_j)) < \epsilon. \tag{22}$$

Using the ⋈-triangle inequality, we can obtain

$$\begin{aligned}
d(x_m, x_j) &\leq d(x_m, x_{m+1}) + d(x_{m+1}, x_{j+1}) + d(x_{j+1}, x_j) \\
&= c_m + d(x_{m+1}, x_{j+1}) + \bar{c}_j < c_m + \epsilon + \bar{c}_j \text{ (by (22))} \\
&\leq c_N + \epsilon + \bar{c}_N \text{ (since } \{c_n\}_{n=1}^{\infty} \text{ and } \{\bar{c}_n\}_{n=1}^{\infty} \text{ are decreasing)} \\
&< \frac{\delta'}{3} + \epsilon + \frac{\delta'}{3} \text{ (by (8))} \\
&= \epsilon + \frac{2\delta'}{3},
\end{aligned}$$

which contradicts (21). This contradiction shows that every sequence $\{T^n(x)\}_{n=1}^{\infty} = \{x_n\}_{n=1}^{\infty}$ is a <-Cauchy sequence.

Suppose that $\{x_n\}_{n=1}^{\infty}$ is not a >-Cauchy sequence. Then, there exists $2\epsilon > 0$ such that, given any integer N, there exist $n > m \geq N$ satisfying $d(x_n, x_m) > 2\epsilon$. Let $\delta' = \min\{\delta, \epsilon\}$, and let $\gamma_j = d(x_j, x_m)$ for $j = m+1, \cdots, n$. For j with $m < j \leq n$, using the ⋈-triangle inequality, we have

$$d(x_{j+1}, x_m) \leq d(x_{j+1}, x_j) + d(x_j, x_m),$$

which implies

$$\frac{\delta'}{3} > \bar{c}_N \geq \bar{c}_{j_0} = d(x_{j_0+1}, x_{j_0}) \geq \gamma_{j_0+1} - \gamma_{j_0} \geq \epsilon + \delta' - \epsilon - \frac{2\delta'}{3} = \frac{\delta'}{3}.$$

This contradiction shows that there exists an integer j with $m < j \leq n$ satisfying $d(x_j, x_m) \neq 0$ and

$$\epsilon + \frac{2\delta'}{3} < d(x_j, x_m) < \epsilon + \delta', \tag{23}$$

which implies
$$d(x_{j+1}, x_{m+1}) = d(T(x_j), T(x_m)) < \epsilon. \tag{24}$$

Using the ⋈-triangle inequality, we can obtain

$$\begin{aligned}
d(x_j, x_m) &\leq d(x_j, x_{j+1}) + d(x_{j+1}, x_{m+1}) + d(x_{m+1}, x_m) \\
&= c_j + d(x_{j+1}, x_{m+1}) + \bar{c}_m < c_j + \epsilon + \bar{c}_m \text{ (by (24))} \\
&\leq c_N + \epsilon + \bar{c}_N \text{ (since } \{c_n\}_{n=1}^\infty \text{ and } \{\bar{c}_n\}_{n=1}^\infty \text{ are decreasing)} \\
&< \frac{\delta'}{3} + \epsilon + \frac{\delta'}{3} \text{ (by (8))} \\
&= \epsilon + \frac{2\delta'}{3},
\end{aligned}$$

which contradicts (23). This contradiction shows that every sequence $\{T^n(x)\}_{n=1}^\infty = \{x_n\}_{n=1}^\infty$ is a $>$-Cauchy sequence. Using Theorem 7, the proof is complete. □

Example 5. *Continued from Example 3, since the pre-metric space $(\mathbb{R}, d)$ is simultaneously ▷-complete and ◁-complete, any function $T : (\mathbb{R}, d) \to (\mathbb{R}, d)$ that is a weakly uniformly strict contraction on $\mathbb{R}$ has a unique fixed point. The concrete examples regarding functions that are weakly uniformly strict contraction on $\mathbb{R}$ can be obtained from the literature.*

We finally remark that the Meir–Keeler type of fixed point theorem based on the ◇-triangle inequality cannot be obtained by using an argument similar to Theorem 11. In other words, we need to design a different argument to obtain the Meir–Keeler type of fixed point theorem based on the ◇-triangle inequality. It is also possible that we cannot establish the Meir–Keeler type of fixed point theorem based on the ◇-triangle inequality. Therefore, this problem remains open and could be the subject future research.

Funding: This research received no external funding.

Institutional Review Board Statement: Not applicable.

Informed Consent Statement: Not applicable.

Data Availability Statement: Not applicable.

Conflicts of Interest: The author declares no conflict of interest.

References

1. Wilson, W.A. On Semi-metric spaces. *Am. J. Math.* **1931**, *53*, 361–373. [CrossRef]
2. Alegre, C.; Marín, J. Modified w-distances on quasi-metric spaces and a fixed point theorem on complete quasi-metric spaces. *Topol. Its Appl.* **2016**, *203*, 32–41. [CrossRef]
3. Ali-Akbari, M.; Pourmahdian, M. Completeness of hyperspaces of compact subsets of quasi-metric spaces. *Acta Math. Hung.* **2010**, *127*, 260–272. [CrossRef]
4. Cao, J.; Rodríguez-López, J. On hyperspace topologies via distance functionals in quasi-metric spaces. *Acta Math. Hung.* **2006**, *112*, 249–268. [CrossRef]
5. Cobzas, S. Completeness in quasi-metric spaces and Ekeland Variational Principle. *Topol. Its Appl.* **2011**, *158*, 1073–1084. [CrossRef]
6. Collins Agyingi, A.; Haihambo, P.; Künzi, H.-P.A. Endpoints in T_0-quasi-metric spaces. *Topol. Its Appl.* **2014**, *168*, 82–93. [CrossRef]
7. Doitchinov, D. On completeness in quasi-metric spaces. *Topol. Its Appl.* **1988**, *30*, 127–148. [CrossRef]
8. Künzi, H.-P.A. A construction of the B-completion of a T_0-quasi-metric space. *Topol. Its Appl.* **2014**, *170*, 25–39. [CrossRef]
9. Künzi, H.-P.A.; Kivuvu, C.M. The B-completion of a T_0-quasi-metric space. *Topol. Its Appl.* **2009**, *156*, 2070–2081. [CrossRef]
10. Künzi, H.-P.A.; Yildiz, F. Convexity structures in T_0-quasi-metric spaces. *Topol. Its Appl.* **2016**, *200*, 2–18. [CrossRef]
11. Matthews, S.G. Partial metric topology. *Ann. N. Y. Acad. Sci.* **1994**, *728*, 183–197. [CrossRef]
12. Meir, A.; Keeler, E. A Theorem on Contraction Mappings. *J. Math. Anal. Appl.* **1969**, *28*, 326–329. [CrossRef]
13. Romaguera, S.; Antonino, J.A. On convergence complete strong quasi-metrics. *Acta Math. Hung.* **1994**, *64*, 65–73. [CrossRef]
14. Triebel, H. A new approach to function spaces on quasi-metric spaces. *Rev. Mater. Complut.* **2005**, *18*, 7–48. [CrossRef]
15. Ume, J.S. A minimization theorem in quasi-metric spaces and its applications. *Int. J. Math. Math. Sci.* **2002**, *31*, 443–447. [CrossRef]
16. Vitolo, P. A representation theorem for quasi-metric spaces. *Topol. Its Appl.* **1995**, *65*, 101–104. [CrossRef]
17. Wilson, W.A. On Quasi-metric spaces. *Am. J. Math.* **1931**, *53*, 675–684. [CrossRef]

18. Wu, H.-C. Pre-Metric Spaces Along with Different Types of Triangle Inequalities. *Axiom* **2018**, *7*, 34. [CrossRef]
19. Pavlović, M.V.; Radenović, S. A note on the Meir-Keeler theorem in the context of b-metric spaces. *Vojnoteh. Glas./Mil. Tech. Cour.* **2019**, *67*, 1–12. [CrossRef]

Article

Fixed Points Results for Various Types of Tricyclic Contractions

Mustapha Sabiri, Abdelhafid Bassou *, Jamal Mouline and Taoufik Sabar

Laboratory of Algebra, Analysis and Applications (L3A), Departement of Mathematics and Computer Science, Faculty of Sciences Ben M'sik, Hassan II University of Casablanca, Casablanca 20000, Morocco; sabiri10mustapha@gmail.com (M.S.); mouline61@gmail.com (J.M.); sabarsaw@gmail.com (T.S.)
* Correspondence: hbassou@gmail.com

Abstract: In this paper, we introduce four new types of contractions called in this order Kanan-S-type tricyclic contraction, Chattergea-S-type tricyclic contraction, Riech-S-type tricyclic contraction, Cirić-S-type tricyclic contraction, and we prove the existence and uniqueness for a fixed point for each situation.

Keywords: fixed points; S-type tricyclic contraction; metric spaces

1. Introduction

It is well known that the Banach contraction principle was published in 1922 by S. Banach as follows:

Theorem 1. *Let (X,d) be a complete metric space and a self mapping $T : X \longrightarrow X$. If there exists $k \in [0,1)$ such that, for all $x, y \in X$, $d(Tx, Ty) \leq kd(x, y)$, then T has a unique fixed point in X.*

The Banach contraction principle has been extensively studied and different generalizations were obtained.

In 1968 [1], Kannan established his famous extension of this contraction.

Theorem 2. *Ref. [1] Let (X, d) be a complete metric space and a self mapping $T : X \longrightarrow X$. If T satisfies the following condition:*

$$d(Tx, Ty) \leq k[d(x, Tx) + d(y, Ty)] \quad \text{for all} \quad x, y \in X \quad \text{where} \quad 0 < k < \frac{1}{2},$$

then T has a fixed point in X.

A similar contractive condition has been introduced by Chattergea in 1972 [2] as follows:

Theorem 3. *Ref. [2] Let $T : X \longrightarrow X$, where (X, d) is a complete metric space. If there exists $0 < k < \frac{1}{2}$ such that*

$$d(Tx, Ty) \leq k[d(y, Tx) + d(Ty, x)] \quad \text{for all} \quad x, y \in X,$$

then T has a fixed point in X.

We can also find another extension of the Banach contraction principle obtained by S. Reich, Kannan in 1971 [3].

Theorem 4. *Ref. [3] Let $T : X \longrightarrow X$, where (X, d) is a complete metric space. If there exists $0 < k < \frac{1}{3}$ such that*

$$d(Tx, Ty) \leq k[d(x, y) + d(x, Tx) + d(y, Ty)] \quad \text{for all} \quad x, y \in X,$$

then T has a fixed point in X.

In addition, in the same year, Ciric gave the following extension [4].

Theorem 5. *Ref. [4] Let $T : X \longrightarrow X$, where (X,d) a complete metric space. If there exists $k \in [0,1)$ such that*

$$d(Tx, Ty) \leq k Max[d(x,y), d(x,Tx), d(y,Ty), d(y,Tx), d(Ty,x)] \quad \text{for all} \quad x, y \in X,$$

then T has a fixed point in X.

Many authors have investigated these situations and many results were proved (see [5–13]).

In this article, we prove the uniqueness and existence of the fixed points in different types contractions for a self mapping T defined on the union of tree closed subsets of a complete metric space with k in different intervals.

2. Preliminaries

In best approximation theory, the concept of tricyclic mappings extends that of ordinary cyclic mappings. Moreover, in the case where two of the sets, say A and C, coincide, we find a cyclic mapping which is also a self-map, and, hence, a best proximity point result for a tricyclic mappings means also a fixed point and a best proximity point result for a self-map and a cyclic mapping.

Definition 1. *Let A, B be nonempty subsets of a metric space (X,d). A mapping $T : A \cup B \longrightarrow A \cup B$ is said to be cyclic if :*

$$T(A) \subseteq B, T(B) \subseteq A.$$

In 2003, Kirk et al. [14] proved that, if $T : A \cup B \longrightarrow A \cup B$ is cyclic and, for some $k \in (0,1), d(Tx, Ty) \leq kd(x,y)$ for all $x \in A, y \in B$, then $A \cap B \neq \emptyset$, and T has a unique fixed point in $A \cap B$.

In 2017, Sabar et al. [15] proved a similar result for tricyclic mappings and introduced the concept of tricyclic contractions.

Theorem 6. *Ref. [15] Let A, B and C be nonempty closed subsets of a complete metric space (X,d), and let a mapping $T : A \cup B \cup C \longrightarrow A \cup B \cup C$. If $T(A) \subseteq B, T(B) \subseteq C$ and $T(C) \subseteq A$ and there exists $k \in (0,1)$ such that $D(Tx, Ty, Tz) \leq kD(x,y,z)$ for all $(x,y,z) \in A \times B \times C$, then $A \cap B \cap C$ is nonempty and T has a unique fixed point in $A \cap B \cap C$,*

where $D(x,y,z) = d(x,y) + d(x,z) + d(y,z)$.

Definition 2. *Ref. [15] Let A, B and C be nonempty subsets of a metric space (X,d). A mapping $T : A \cup B \cup C \longrightarrow A \cup B \cup C$ is said to be tricyclic contracton if there exists $0 < k < 1$ such that:*
1. $T(A) \subseteq B, T(B) \subseteq C$ and $T(C) \subseteq A$.
2. $D(Tx, Ty, Tz) \leq kD(x,y,z) + (1-k)\delta(A,B,C)$ for all $(x,y,z) \in A \times B \times C$.

where $\delta(A, B, C) = \inf\{D(x,y,z) : x \in A, y \in B, z \in C\}$

Very Recently, Sabiri et al. introduced an extension of the aforementioned mappings and called them p-cyclic contractions [16].

3. Main Results

Definition 3. *Let A, B and C be nonempty subsets of a metric space (X,d). A mapping $T : A \cup B \cup C \longrightarrow A \cup B \cup C$ is said to be a Kannan-S-type tricyclic contraction, if there exists $k \in \left(0, \frac{1}{3}\right)$ such that*

1. $T(A) \subseteq B, T(B) \subseteq C, T(C) \subseteq A$.
2. $D(Tx, Ty, Tz) \leq k[d(x, Tx) + d(y, Ty) + d(z, Tz)]$ for all $(x, y, z) \in A \times B \times C$.

We give an example to show that a map can be a tricyclic contraction but not a Kannan-S-type tricyclic contraction.

Example 1. Let X be $\mathbb{R}^2$ normed by the norm $\| (x,y) \| = |x| + |y|$, and $A = [1,2] \times \{0\}$, $B = \{0\} \times [-2,-1]$, $C = [-2,-1] \times \{0\}$, then

$$\delta(A, B, C) = D((1,0), (0,-1), (-1,0)) = 6.$$

Put $T : A \cup B \cup C \longrightarrow A \cup B \cup C$ such that

$$T(x,0) = \left(0, -\frac{x+2}{3}\right) \quad \text{if } (x,0) \in A,$$

$$T(0,y) = \left(\frac{y-2}{3}, 0\right) \quad \text{if } (0,y) \in B,$$

$$T(z,0) = \left(-\frac{z-2}{3}, 0\right) \quad \text{if } (z,0) \in C,$$

We have $T(A) \subseteq B, T(B) \subseteq C$ and $T(C) \subseteq A$, and

$$\begin{aligned}
D(T(x,0), T(0,y), T(z,0)) &= D\left((0, -\frac{x+2}{3}), (\frac{y-2}{3}, 0), (-\frac{z-2}{3}, 0)\right) \\
&= \frac{2}{3}(x - y - z) + 4 \\
&= \frac{1}{3}D((x,0), (0,y), (z,0)) + 4 \\
&= \frac{1}{3}D((x,0), (0,y), (z,0)) + (1 - \frac{1}{3})\delta(A, B, C)
\end{aligned}$$

for all $(x,0) \in A, (0,y) \in B, (z,0) \in C$.
On the other hand,

$$D(T(2,0), T(0,-2), T(-2,0)) = D\left((0, -\frac{4}{3}), (\frac{-4}{3}, 0), (\frac{4}{3}, 0)\right) = 8$$

and

$$d((2,0), T(2,0)) + d((0,-2), T(0,-2)) + d((-2,0), T(-2,0)) = 10,$$

which implies that
$D(T(2,0), T(0,-2), T(-2,0))$

$$> \frac{1}{3}[d((2,0), T(2,0)) + d((0,-2), T(0,-2)) + d((-2,0), T(-2,0))]$$

Then, T is tricyclic contraction but not a Kannan-S-type tricyclic contraction.

Now, we give an example for which T is a Kannan-S-type tricyclic contraction but not a tricyclic contraction.

Example 2. Let $X = \mathbb{R}$ with the usual metric. Let $A = B = C = [0,1]$, then $\delta(A, B, C) = 0$.
Put $T : A \cup B \cup C \longrightarrow A \cup B \cup C$ such that

$$Tx = \frac{1}{6} \text{ if } 0 \leq x < 1, \quad Tx = \frac{1}{4} \text{ if } x = 1$$

For $x = 1, y = 1$ and $z = \frac{23}{24}$, we have

$$D(T(1), T(1), T(\frac{23}{24})) = D(\frac{1}{4}, \frac{1}{4}, \frac{1}{6}) = 2d(\frac{1}{4}, \frac{1}{6}) = \frac{1}{6}.$$

and

$$D(1, 1, \frac{23}{24}) = 2d(1, \frac{23}{24}) = \frac{1}{12}.$$

Then, T is not tricyclic contraction.
However T is a Kannan-S-type tricyclic contraction. Indeed:

- If $x = y = z = 1$, we have

$$D(T(1), T(1), T(1)) = 0 \leq \frac{9}{4}k$$

for all $k \geq 0$, then for $0 \leq k < \frac{1}{3}$.

- If $x \in [0, 1), y \in [0, 1)$ and $z \in [0, 1)$, we have

$$D(Tx, Ty, Tz) = 0 \leq k(d(x, \frac{1}{6}) + d(y, \frac{1}{6}) + d(z, \frac{1}{6}))$$

for all $k \geq 0$, then for $0 \leq k < \frac{1}{3}$.

- If $x = 1, y \in [0, 1)$ and $z \in [0, 1)$, we have

$$D(T_1, Ty, Tz) = D(\frac{1}{4}, \frac{1}{6}, \frac{1}{6}) = \frac{1}{6}$$

and

$$d(1, T(1)) + d(y, Ty) + d(z, Tz) = \frac{3}{4} + d(y, \frac{1}{6}) + d(z, \frac{1}{6}),$$

then, for $k = \frac{2}{9}$, we have

$$D(T(1), T(y, Tz) \leq k(d(1, T(1)) + d(y, Ty) + d(z, Tz)).$$

- If $x = 1, y = 1$ and $z \in [0, 1)$, we have

$$D(T(1), T(1), Tz) = D(\frac{1}{4}, \frac{1}{4}, \frac{1}{6}) = \frac{1}{6}$$

and

$$d(1, T(1)) + d(1, T(1)) + d(z, Tz) = \frac{3}{2} + d(z, \frac{1}{6}).$$

Then, for $k = \frac{2}{9}$, we have

$$D(T(1), T(1), Tz) \leq k(d(1, T(1)) + d(1, T(1)) + d(z, Tz)).$$

Consequently, for $k = \frac{2}{9}$, we have :

$$D(Tx, Ty, Tz) \leq k(d(x, Tx) + d(y, Ty) + d(z, Tz)) \text{ for all } (x, y, z) \in A \times B \times C.$$

Theorem 7. *Let A, B and C be nonempty closed subsets of a complete metric space (X, d), and let $T : A \cup B \cup C \longrightarrow A \cup B \cup C$ be a Kannan-S-type tricyclic contraction. Then, T has a unique fixed point in $A \cap B \cap C$.*

Proof. Fix $x \in A$. We have

$$d\left(T^3x, T^2x\right) \leq D\left(T^3x, T^2x, Tx\right) \leq k\left[d\left(T^2x, T^3x\right) + d\left(Tx, T^2x\right) + d(x, Tx)\right].$$

Then,
$$d(T^3x, T^2x) \leq k\left[d(T^2x, T^3x) + d(Tx, T^2x) + d(x, Tx)\right],$$
which implies
$$d(T^3x, T^2x) \leq \frac{k}{(1-k)}\left[d(Tx, T^2x) + d(x, Tx)\right].$$

Similarly, we have
$$d(T^2x, Tx) \leq \frac{k}{(1-k)}\left[d(T^3x, T^2x) + d(x, Tx)\right]$$
$$d(T^2x, Tx) \leq \frac{k}{(1-k)}\left[\frac{k}{(1-k)}\left[d(Tx, T^2x) + d(x, Tx)\right] + d(x, Tx)\right]$$
$$\Longrightarrow d(T^2x, Tx) \leq \frac{k}{1-2k}(d(x, Tx)).$$

Then,
$$d(T^2x, Tx) \leq t d(x, Tx) \text{ where } t = \frac{k}{1-2k} \text{ and } t \in (0,1),$$
which implies
$$d(T^{n+1}x, T^n x) \leq t^n d(x, Tx), \text{ for all } n \geq 1$$

Consequently,
$$\sum_{n=1}^{+\infty} d(T^{n+1}x, T^n x) \leq (\sum_{n=1}^{+\infty} t^n) d(x, Tx) < +\infty$$

implies that $\{T^n x\}$ is a Cauchy sequence in (X, d). Hence, there exists $z \in A \cup B \cup C$ such that $T^n x \longrightarrow z$. Notice that $\{T^{3n}x\}$ is a sequence in A, $\{T^{3n-1}x\}$ is a sequence in C and $\{T^{3n-2}x\}$ is a sequence in B and that both sequences tend to the same limit z. Regarding the fact that A, B and C are closed, we conclude $z \in A \cap B \cap C$, hence $A \cap B \cap C \neq \emptyset$.
To show that z is a fixed point, we must show that $Tz = z$. Observe that

$$\begin{aligned} d(Tz, z) &= \lim d(Tz, T^{3n}x) \leq \lim D(T^{3n}x, T^{3n-1}x, Tz) \\ &\leq \lim k[d(T^{3n-1}x, T^{3n}x) + d(T^{3n-2}x, T^{3n-1}x) + d(z, Tz)] \\ &\leq k d(Tz, z), \end{aligned}$$

which is equivalent to
$$(1-k)d(Tz, z) = 0.$$

Since $k \in \left(0, \frac{1}{3}\right)$, then $d(Tz, z) = 0$, which implies $Tz = z$.

To prove the uniqueness of z,, assume that there exists $w \in A \cup B \cup C$ such that $w \neq z$ and $Tw = w$. Taking into account that T is tricyclic, we get $w \in A \cap B \cap C$. We have

$$d(z, w) = d(Tz, Tw) \leq D(Tz, Tw, Tw) \leq k[d(z, Tz) + d(w, Tw) + d(w, Tw)] = 0$$

which implies $d(z, w) = 0$. We get that $z = w$ and hence z is the unique fixed point of T. □

Example 3. Let X be $\mathbb{R}^2$ normed by the norm $\| (x, y) \| = |x| + |y|$, let $A = \{0\} \times [0, +1]$, $B = [0, +1] \times \{0\}$, $C = \{0\} \times [-1, 0]$ and let $T : A \cup B \cup C \longrightarrow A \cup B \cup C$ be defined by

$$T(0, x) = \left(\frac{x}{6}, 0\right) \quad \text{if } (0, x) \in A,$$

$$T(y, 0) = \left(0, \frac{-y}{6}\right) \quad \text{if } (y, 0) \in B,$$

$$T(0,z) = \left(0, \frac{-z}{6}\right) \quad \text{if } (0,z) \in C.$$

We have
$$T(A) \subseteq B, T(B) \subseteq C \text{ and } T(C) \subseteq A$$

In addition, for all $(0,x) \in A, (y,0) \in B, (0,z) \in C$, we have

$$D(T(0,x), T(y,0), T(0,z)) = D\left(\left(\frac{x}{6},0\right), \left(0, \frac{-y}{6}\right), \left(0, \frac{-z}{6}\right)\right) = \frac{1}{3}(x+y-z)$$

In addition, we have

$$d((0,x), T(0,x)) + d((y,0), T(y,0)) + d((0,z), T(0,z)) = \frac{7}{6}(x+y-z)$$

This implies

$$D(T(0,x), T(y,0), T(0,z)) = \frac{2}{7}[d((0,x), T(0,x)) + d((y,0), T(y,0)) + d((0,z), T(0,z))].$$

Then, T is a Kannan-S-type tricyclic contraction, and T has a unique fixed point $(0,0)$ in $A \cap B \cap C$.

Corollary 1. *Let (X,d) be a complete metric space and a self mapping $T: X \longrightarrow X$. If there exists $k \in \left(0, \frac{1}{3}\right)$ such that*

$$D(Tx, Ty, Tz) \leq k[d(x, Tx) + d(y, Ty) + d(z, Tz)]$$

for all $(x,y,z) \in X^3$, then T has a unique fixed point.

Now, we shall define another type of a tricyclic contraction.

Definition 4. *Let A, B and C be nonempty subsets of a metric space (X,d). A mapping $T: A \cup B \cup C \longrightarrow A \cup B \cup C$ is said to be a Chattergea-S-type tricyclic contraction if $T(A) \subseteq B, T(B) \subseteq C, T(C) \subseteq A$, and there exist $k \in \left(0, \frac{1}{3}\right)$ such that $D(Tx, Ty, Tz) \leq k[d(y, Tx) + d(z, Ty) + d(x, Tz)]$ for all $(x,y,z) \in A \times B \times C$.*

Theorem 8. *Let A, B and C be nonempty closed subsets of a complete metric space (X,d), and let $T: A \cup B \cup C \longrightarrow A \cup B \cup C$ be a Chattergea-S-type tricyclic contraction. Then, T has a unique fixed point in $A \cap B \cap C$.*

Proof. Fix $x \in A$. We have

$$D\left(Tx, T^2x, T^3x\right) \leq k\left[d(Tx, Tx) + d\left(T^2x, T^2x\right) + d\left(T^3x, x\right)\right]$$

which implies

$$D\left(T^3x, T^2x, Tx\right) \leq kd\left(T^3x, x\right)$$

so

$$d\left(T^3x, T^2x\right) \leq k\left[d\left(T^3x, T^2x\right) + d\left(T^2x, Tx\right) + d(Tx, x)\right] \text{ (by the triangular inequality)}$$

$$\implies d\left(T^3x, T^2x\right) \leq \frac{k}{(1-k)}\left[d\left(Tx, T^2x\right) + d(x, Tx)\right]$$

and

$$d\left(T^2x, Tx\right) \leq D\left(T^3x, T^2x, Tx\right) \leq \frac{k}{(1-k)}\left[d\left(T^3x, T^2x\right) + d(x, Tx)\right]$$

$$\implies d\left(T^2x, Tx\right) \leq \frac{k}{(1-k)}\left[\frac{k}{(1-k)}\left[d\left(Tx, T^2x\right) + d(x, Tx)\right] + d(x, Tx)\right]$$

$$\implies d\left(T^2x, Tx\right) \leq \frac{k}{1-2k}(d(x, Tx))$$

Then,
$$d\left(T^2x, Tx\right) \leq td(x, Tx) \text{ where } t = \frac{k}{1-2k} \text{ and } t \in (0,1),$$

which implies
$$d\left(T^{n+1}x, T^nx\right) \leq t^n d(x, Tx)$$

for all $n \geq 1$. Consequently,

$$\sum_{n=1}^{+\infty} d\left(T^{n+1}x, T^nx\right) \leq (\sum_{n=1}^{+\infty} t^n) d(x, Tx) < +\infty$$

implies that $\{T^nx\}$ is a Cauchy sequence in (X,d). Hence, there exists $z \in A \cup B \cup C$ such that $T^nx \longrightarrow z$. Notice that $\{T^{3n}x\}$ is a sequence in A, $\{T^{3n-1}x\}$ is a sequence in C, and $\{T^{3n-2}x\}$ is a sequence in B and that both sequences tend to the same limit z. Regarding that A, B and C are closed, we conclude $z \in A \cap B \cap C$, hence $A \cap B \cap C \neq \emptyset$.

To show that z is a fixed point, we must show that $Tz = z$. Observe that

$$\begin{aligned}d(Tz, z) &= \lim d\left(Tz, T^{3n}x\right) \leq \lim D\left(Tz, T^{3n}x, T^{3n-1}x\right) \\ &\leq \lim k[d\left(T^{3n-1}x, Tz\right) + \left(T^{3n-2}x, T^{3n}x\right) + d(z, T^{3n-1}x)] \leq kd(Tz, z),\end{aligned}$$

which is equivalent to $(1-k)d(Tz, z) = 0$. Since $k \in \left(1, \frac{1}{3}\right)$, then $d(Tz, z) = 0$, which implies $Tz = z$.

To prove the uniqueness of z, assume that there exists $w \in A \cup B \cup C$ such that $w \neq z$ and $Tw = w$. Taking into account that T is tricyclic, we get $w \in A \cap B \cap C$.

We have
$$\begin{aligned}d(z, w) &= d(Tz, Tw) \leq D(Tz, Tw, Tw) \\ &\leq k[d(Tz, w) + d(Tw, w) + d(Tw, z)] \\ &\leq 2kd(z, w).\end{aligned}$$

Then, $d(z, w) = 0$. We conclude that $z = w$ and hence z is the unique fixed point of T. □

Corollary 2. *Let (X, d) be a complete metric space and a self mapping $T : X \longrightarrow X$. If there exists $k \in \left(0, \frac{1}{3}\right)$ such that*

$$D(Tx, Ty, Tz) \leq k[d(y, Tx) + d(z, Ty) + d(x, Tz)]$$

for all $(x, y, z) \in X^3$, then T has a unique fixed point.

In this step, we define a Reich-S-type tricyclic contraction.

Definition 5. *Let A, B and C be nonempty subsets of a metric space (X, d). A mapping $T : A \cup B \cup C \longrightarrow A \cup B \cup C$ is said to be a Reich-S-type tricyclic contraction if there exists $k \in \left(0, \frac{1}{7}\right)$ such that:*

1. $T(A) \subseteq B, T(B) \subseteq C, T(C) \subseteq A$.

2. $D(Tx, Ty, Tz) \leq k[D(x,y,z) + d(x,Tx) + d(y,Ty) + d(z,Tz)]$ for all $(x,y,z) \in A \times B \times C$.

Theorem 9. *Let A, B and C be nonempty closed subsets of a complete metric space (X,d), and let $T: A \cup B \cup C \longrightarrow A \cup B \cup C$ be a Reich-S-type tricyclic contraction. Then, T has a unique fixed point in $A \cap B \cap C$.*

Proof. Fix $x \in A$. We have

$$\begin{aligned}
d(T^2x, T^3x) &\leq D(Tx, T^2x, T^3x) \\
&\leq k\Big[D(x, Tx, T^2x) + d(T^2x, T^3x) + d(Tx, T^2x) + d(x, Tx)\Big] \\
\Longrightarrow d(T^2x, T^3x)(1-k) &\leq k[2d(T^2x, Tx) + 2d(x, Tx) + d(T^2x, x)]
\end{aligned}$$

$$\begin{aligned}
\Longrightarrow d(T^2x, T^3x) &\leq \frac{k}{1-k}\Big[2d(T^2x, Tx) + 2d(x, Tx) + d(T^2x, x)\Big] \\
&\leq \frac{k}{1-k}\Big[2d(T^2x, Tx) + 2d(x, Tx) + d(T^2x, Tx) + d(Tx, x)\Big] \\
&\leq \frac{k}{1-k}\Big[3d(T^2x, Tx) + 3d(x, Tx)\Big]
\end{aligned}$$

$$\Longrightarrow d(T^2x, T^3x) \leq \frac{3k}{1-k}[d(T^2x, Tx) + d(x, Tx)]$$

and

$$d(T^2x, Tx) \leq D(Tx, T^2x, T^3x) \leq k\Big[D(x, Tx, T^2x) + d(T^2x, T^3x) + d(Tx, T^2x) + d(x, Tx)\Big]$$

$$\Longrightarrow d(T^2x, Tx) \leq k\Big[3d(T^2x, Tx) + 3d(x, Tx) + d(T^2x, T^3x)\Big]$$

$$\Longrightarrow d(T^2x, Tx)(1 - 3k) \leq k[d(T^2x, T^3x) + 3d(x, Tx)]$$

$$\Longrightarrow d(T^2x, Tx) \leq \frac{k}{1-3k}d(T^2x, T^3x) + \frac{3k}{1-3k}d(x, Tx)$$

$$\Longrightarrow d(T^2x, Tx) \leq \frac{k}{1-3k}\frac{3k}{1-k}[d(T^2x, Tx) + d(x, Tx)] + \frac{3k}{1-3k}d(x, Tx)$$

$$\Longrightarrow d(T^2x, Tx) \leq \frac{3k^2}{(1-3k)(1-k)}d(T^2x, Tx) + \Big(\frac{3k^2}{(1-3k)(1-k)} + \frac{3k}{(1-3k)}\Big)d(x, Tx)$$

$$\Longrightarrow d(T^2x, Tx)\Big(1 - \frac{3k^2}{(1-3k)(1-k)}\Big) \leq \frac{3k^2 + 3k(1-k)}{(1-3k)(1-k)}d(x, Tx)$$

$$\Longrightarrow d(T^2x, Tx)\Big((1-3k)(1-k) - 3k^2\Big) \leq (3k^2 + 3k(1-k))d(x, Tx)$$

$$\Longrightarrow d(T^2x, Tx)(1 - 4k) \leq 3kd(x, Tx)$$

$$\Longrightarrow d(T^2x, Tx) \leq \frac{3k}{(1-4k)}d(x, Tx).$$

Then,

$$d(T^2x, Tx) \leq td(x, Tx) \text{ where } t = \frac{3k}{(1-4k)} \text{ and } t \in (0,1),$$

which implies

$$d(T^{n+1}x, T^nx) \leq t^n d(x, Tx),$$

consequently
$$\sum_{n=1}^{+\infty} d\left(T^{n+1}x, T^n x\right) \leq (\sum_{n=1}^{+\infty} t^n) d(x, Tx) < +\infty$$

This implies that $\{T^n x\}$ is a Cauchy sequence in (X, d). Hence, there exists $z \in A \cup B \cup C$ such that $T^n x \longrightarrow z$. Notice that $\{T^{3n} x\}$ is a sequence in A, $\{T^{3n-1} x\}$ is a sequence in C and $\{T^{3n-2} x\}$ is a sequence in B and that both sequences tend to the same limit z. Regarding the fact that A, B and C are closed, we conclude that $z \in A \cap B \cap C$, hence $A \cap B \cap C \neq \emptyset$.

To show that z is a fixed point, we must show that $Tz = z$. Observe that

$$\begin{aligned}
d(Tz, z) &= \lim d\left(Tz, T^{3n} x\right) \\
&\leq \lim D\left(T^{3n} x, T^{3n-1} x, Tz\right) \\
&\leq \lim k[d\left(T^{3n-1} x, T^{3n-2} x\right) + d\left(T^{3n-1} x, z\right) + d(T^{3n-2} x, z) \\
&+ d\left(T^{3n-1} x, T^{3n} x\right) + d\left(T^{3n-2} x, T^{3n-1} x\right) + d(z, Tz)] \\
&\leq k d(Tz, z),
\end{aligned}$$

which is equivalent to $(1 - k) d(Tz, z) = 0$.

Since $k \in \left(0, \frac{1}{7}\right)$, then $d(Tz, z) = 0$, which implies $Tz = z$.

To prove the uniqueness of z, assume that there exists $w \in A \cup B \cup C$ such that $w \neq z$ and $Tw = w$. Taking into account that T is tricyclic, we get $w \in A \cap B \cap C$.

$$\begin{aligned}
d(z, w) &= d(Tz, Tw) \\
&\leq D(Tz, Tw, Tw) \\
&\leq k[2d(z, w) + d(w, w) + d(z, Tz) + d(Tw, w) + d(Tw, w)] \\
&\leq 2k d(z, w)
\end{aligned}$$

implies $d(z, w) = 0$. We conclude that $z = w$ and hence z is the unique fixed point of T. □

Example 4. *We take the same example 3.*
Let X be $\mathbb{R}^2$ normed by the norm $\| (x, y) \| = |x| + |y|$,

$$A = \{0\} \times [0, +1], B = [0, +1] \times \{0\}, C = \{0\} \times [-1, 0]$$

and let $T : A \cup B \cup C \longrightarrow A \cup B \cup C$ be defined by

$$T(0, x) = \left(\frac{x}{6}, 0\right) \quad \text{if } (0, x) \in A,$$

$$T(y, 0) = \left(0, \frac{-y}{6}\right) \quad \text{if } (y, 0) \in B,$$

$$T(0, z) = \left(0, \frac{-z}{6}\right) \quad \text{if } (0, z) \in C,$$

We have T is tricyclic and for all $(0, x) \in A, (y, 0) \in B, (0, z) \in C$,

$$\begin{aligned}
D(T(0, x), T(y, 0), T(0, z)) &= D\left(\left(\frac{x}{6}, 0\right), \left(0, \frac{-y}{6}\right), \left(0, \frac{-z}{6}\right)\right) \\
&= \frac{1}{3}(x + y - z).
\end{aligned}$$

In addition, we have

$$D((0,x),(y,0),(0,z)) + d((0,x), T(0,x)) + d((y,0), T(y,0)) + d((0,z), T(0,z))$$
$$= 2(x+y-z) + \frac{7}{6}(x+y-z) = \frac{19}{6}(x+y-z).$$

Then,

$$\begin{aligned}
D(T(0,x), T(y,0), T(0,z)) &= \frac{2}{19}(D((0,x),(y,0),(0,z)) + d((0,x), T(0,x)) \\
&\quad + d((y,0), T(y,0)) + d((0,z), T(0,z))) \\
&\leq \frac{1}{7}(D((0,x),(y,0),(0,z)) + d((0,x), T(0,x)) \\
&\quad + d((y,0), T(y,0)) + d((0,z), T(0,z)))
\end{aligned}$$

This implies that T is a Reich-S-type tricyclic contraction, and T has a unique fixed point $(0,0)$ in $A \cap B \cap C$.

Corollary 3. *Let (X, d) a complete metric space and a self mapping $T : X \longrightarrow X$. If there exists $k \in \left(0, \frac{1}{7}\right)$ such that*

$$D(Tx, Ty, Tz) \leq k[D(x,y,z) + d(x, Tx) + d(y, Ty) + d(z, Tz)]$$

for all $(x,y,z) \in X^3$, then T has a unique fixed point in X.

The next tricyclic contraction considered in this section is the Cirić-S-type tricyclic contraction defined below.

Definition 6. *Let A, B and C be nonempty subsets of a metric space (X, d), $T : A \cup B \cup C \longrightarrow A \cup B \cup C$ be a Cirié-S-type tricyclic contraction, if there exists $k \in (0,1)$ such that*
1. $T(A) \subseteq B, T(B) \subseteq C, T(C) \subseteq A$
2. $D(Tx, Ty, Tz) \leq kM(x,y,z)$ *for all $(x,y,z) \in A \times B \times C$.*

where $M(x,y,z) = \max\{D(x,y,z), d(x,Tx), d(y,Ty), d(z,Tz)\}$

The fixed point theorem of the Cirić-S-type tricyclic contraction reads as follows.

Theorem 10. *Let A, B and C be nonempty closed subsets of a complete metric space (X, d), and let $T : A \cup B \cup C \longrightarrow A \cup B \cup C$ be a Cirić-S- type tricyclic contraction, then T has a unique fixed point in $A \cap B \cap C$.*

Proof. Taking $x \in A$, we have $D(Tx, Ty, Tz) \leq kM(x,y,z)$ for all $(x,y,z) \in A \times B \times C$. If $M(x,y,z) = D(x,y,z)$, Theorem 7 implies the desired result.
Consider the case $M(x,y,z) = d(x, Tx)$. We have:

$$D\left(Tx, T^2x, T^3x\right) \leq kd(x, Tx) \implies d\left(Tx, T^2x\right) \leq kd(x, Tx)$$
$$\implies d\left(T^n x, T^{n+1} x\right) \leq k^n d(x, Tx)$$

Consequently,

$$\sum_{n=1}^{+\infty} d\left(T^{n+1}x, T^n x\right) \leq (\sum_{n=1}^{+\infty} k^n) d(x, Tx) < +\infty$$

which implies that $\{T^n x\}$ is a Cauchy sequence in (X, d). Hence, there exists $z \in A \cup B \cup C$ such that $T^n x \longrightarrow z$. Notice that $\{T^{3n}x\}$ is a sequence in A, $\{T^{3n-1}x\}$ is a sequence in C, and $\{T^{3n-2}x\}$ is a sequence in B and that both sequences tend to the same limit z; regarding the fact that A, B and C are closed, we conclude $z \in A \cap B \cap C$, hence $A \cap B \cap C \neq \emptyset$.

To show that z is a fixed point, we must show that $Tz = z$. Observe that

$$d(Tz, z) = \lim d\left(Tz, T^{3n}x\right) \leq \lim D\left(T^{3n}x, T^{3n-1}x, Tz\right) \leq kd(Tz, z),$$

which is equivalent to $(1-k)d(Tz, z) = 0$. Since $k \in (0, 1)$, then $d(Tz, z) = 0$, which implies $Tz = z$.

To prove the uniqueness of z, assume that there exists $w \in A \cup B \cup C$ such that $w \neq z$ and $Tw = w$.

Taking into account that T is tricyclic, we get $w \in A \cap B \cap C$.

$d(z, w) = d(Tz, Tw) \leq D(Tz, Tw, Tw) \leq kd(z, Tz) = 0$ implies $d(z, w) = 0$. We conclude that $z = w$ and hence z is the unique fixed point of T.

Consider the case $M(x, y, z) = d(y, Ty)$. We have :

$$D\left(Tx, T^2x, T^3x\right) \leq kd\left(Tx, T^2x\right) \implies d\left(Tx, T^2x\right) \leq kd\left(Tx, T^2x\right) < d\left(Tx, T^2x\right),$$

which is impossible since $k \in (0, 1)$

Consider the case $M(x, y, z) = d(z, Tz)$. We have:

$$D\left(Tx, T^2x, T^3x\right) \leq kd\left(T^2x, T^3x\right) \implies d\left(T^2x, T^3x\right) \leq kd\left(T^2x, T^3x\right) < d\left(T^2x, T^3x\right),$$

which is impossible since $k \in (0, 1)$. □

Corollary 4. *Let A, B and C be a nonempty subset of a complete metric space (X, d) and let a mapping $T : A \cup B \cup C \longrightarrow A \cup B \cup C$. If there exists $k \in (0, 1)$ such that*
1. $T(A) \subseteq B, T(B) \subseteq C, T(C) \subseteq A$.
2. $D(Tx, Ty, Tz) \leq k \max\{D(x, y, z), d(x, Tx)\} \; \forall (x, y, z) \in A \times B \times C$.

Then, T has a unique fixed point in $A \cap B \cap C$.

Author Contributions: Conceptualization, M.S. and A.B.; validation, M.S., J.M. and A.B.; writing—original draft preparation, M.S. and J.M.; writing—review and editing, A.B. and T.S.; supervision, J.M. and A.B.; project administration, M.S. and J.M. All authors have read and agreed to the published version of the manuscript.

Funding: This research received no external funding.

Institutional Review Board Statement: Not applicable.

Informed Consent Statement: Not applicable.

Data Availability Statement: Not applicable.

Conflicts of Interest: The authors declare no conflict of interest.

References

1. Kannan, R. Some results on fixed points. *Bull. Calcutta. Math. Soc.* **1968**, *60*, 71–76.
2. Chatterjea, S.K. Fixed point theorems. *C. R. Acad. Bulgare Sci.* **1972**, *25*, 727–730. [CrossRef]
3. Reich, S. Kannan's. Fixed point theorem. *Boll. Dell'Unione Mat. Ital.* **1971**, *4*, 459–465.
4. Cirić, L.B. A generalization of Bannach's contraction priciple. *Proc. Am. Math. Soc.* **1974**, *45*, 267–273. [CrossRef]
5. Reich, S. Some remarks concerning contraction mappings, canad. *Math. Bull.* **1971**, *14*, 121–124. [CrossRef]
6. Eldred, A.A.; Veeramani, P. Existance and converence of best proximity points. *J. Math. Anal. Appl.* **2006**, *323*, 1001–1006. [CrossRef]
7. Krapinar, E.; Erhan, I.M. Best on different Type Contractions. *Appl. Math. Inf. Sci.* **2011**, *5*, 558–569.
8. Petric, M.A. Best proximity point theorems for weak cyclic Kannan contractions. *Filmat* **2011**, *25*, 145–154. [CrossRef]
9. Wong, C. On Kannan maps. *Proc. Am. Math. Soc.* **1975**, *47*, 105–111. [CrossRef]
10. Todorcević, V. *Harmonic Quasiconformal Mappings and Hyper-Bolic Type Metrics*; Springer: Cham, Switzerland, 2019.
11. Radenović, S. Some remarks on mappings satisfying cyclical con- tractive conditions. *Afr. Mat.* **2016**, *27*, 291–295. [CrossRef]

12. Zaslavski, A.J. Two fixed point results for a class of mappings of contractive type. *J. Nonlinear Var. Anal.* **2018**, *2*, 113–119.
13. Reich, S.; Zaslavski, A.J. Monotone contractive mappings. *J. Nonlinear Var. Anal.* **2017**, *1*, 391–401.
14. Kirk, W.A.; Srinivasan, P.S.; Veeramani, P. Fixed point fo mappings satisfyaing cyclical contractive conditions. *Fixed Point Theory* **2003**, *4*, 79–89.
15. Sabar, T.; Bassou, M.A. Best proximity point of tricyclic contraction. *Adv. Fixed Point Ttheory* **2017**, *7*, 512–523.
16. Sabiri, M.; Mouline, J.; Bassou, A.; Sabar, T. A New Best Approximation Result in (S) Convex Metric spaces. *Int. J. Math. Math. Sci.* **2020**, *2020*, 4367482. [CrossRef]

Article

Common Fixed Point Results for Almost $\mathcal{R}_g$-Geraghty Type Contraction Mappings in b_2-Metric Spaces with an Application to Integral Equations

Samera M. Saleh [1], Salvatore Sessa [2,*], Waleed M. Alfaqih [3] and Fawzia Shaddad [4]

1. Department of Mathematics, Faculty of Science, Taiz University, Taiz 6803, Yemen; samirasaleh2007@yahoo.com
2. Dipartimento di Architettura, Università degli Studi di Napoli Federico II, Via Toledo 402, 80134 Napoli, Italy
3. Department of Mathematics, Hajjah University, Hajjah 1729, Yemen; waleedmohd2016@gmail.com
4. Department of Mathematics, Sana'a University, Sana'a 1247, Yemen; fzsh99@gmail.com
* Correspondence: sessa@unina.it

Abstract: In this paper, we define almost $\mathcal{R}_g$-Geraghty type contractions and utilize the same to establish some coincidence and common fixed point results in the setting of b_2-metric spaces endowed with binary relations. As consequences of our newly proved results, we deduce some coincidence and common fixed point results for almost g-α-η Geraghty type contraction mappings in b_2-metric spaces. In addition, we derive some coincidence and common fixed point results in partially ordered b_2-metric spaces. Moreover, to show the utility of our main results, we provide an example and an application to non-linear integral equations.

Keywords: b_2-metric space; fixed point; binary relation; almost $\mathcal{R}_g$-Geraghty type contraction

MSC: 47H10; 54H25

1. Introduction

The extension of fixed point theory to generalized structures, such as cone metric spaces, partial metric spaces, b-metric spaces and 2-metric spaces has received much attention. 2-metric space is a generalized metric space which was introduced by Gähler in [1]. Unlike the ordinary metric, the 2-metric is not a continuous function. The topology induced by 2-metric space is called 2-metric topology which is generated by the set of all open spheres with two centers. It is easy to observe that 2-metric space is not topologically equivalent to an ordinary metric. Hence, there is not any relationship between the results obtained in 2-metric spaces and the correspondence results in metric spaces. For fixed point results in the setting of 2-metric spaces, the readers may refer to [2–5] and references therein.

The concept of b-metric spaces was introduced by Czerwik [6,7] which is a generalization of the usual metric spaces and 2-metric spaces as well. Several papers have dealt with fixed point theory for single-valued and multi-valued operators in b-metric spaces have been obtained (see, e.g., [8–10]).

In 2014, Mustafa et al. [11] introduced the notion of b_2-metric spaces, as a generalization of both 2-metric and b-metric spaces.

On the other hand, the branch of related metric (metric space endowed with a binary relation) fixed point theory is a relatively new area was initiated by Turinici [12]. Recently, this direction of research is undertaken by several researchers such as: Bhaskar and Lakshmikantham [13], Samet and Turinici [14], Ben-El-Mechaiekh [15], Imdad et al. [16,17] and some others.

The aims of this paper are as follows:

- to define almost $\mathcal{R}_g$-Geraghty type contractions;

- to establish some coincidence and common fixed point results in the setting of b_2-metric spaces endowed with binary relations;
- to deduce some fixed point and common fixed point results in partially ordered b_2-metric spaces;
- to provide an example which shows the utility of our main results;
- to apply our newly proven results to non-linear integral equations.

2. Preliminaries

Definition 1 ([11]). *Let X be a non-empty set, $s \geq 1$ a given real number and $d : X^3 \to \mathbb{R}$ be a map satisfying the following conditions:*

(i) *for every pair of distinct points $x, y \in X$, there exists a point $z \in X$ such that $d(x, y, z) \neq 0$;*
(ii) *if at least two of three points x, y, z are the same, then $d(x, y, z) = 0$;*
(iii) $d(x, y, z) = d(x, z, y) = d(y, x, z) = d(y, z, x) = d(z, x, y) = d(z, y, x)$, *for all $x, y, z \in X$;*
(iv) $d(x, y, z) \leq s[d(x, y, w) + d(y, z, w) + d(z, x, w)]$, *for all $x, y, z, w \in X$.*

Then d is called a b_2-metric on X and (X, d) is called a b_2-metric space with parameter s.

Obviously, for $s = 1$, b_2-metric reduces to 2-metric.

Example 1. *Let (X, d) be a 2-metric space and $\rho(x, y, w) = (d(x, y, w))^p$, where $p \geq 1$ is a real number. We see that ρ is a b_2-metric with $s = 3^{p-1}$. In view of the convexity of $f(x) = x^p$, on $[0, \infty)$ for $p \geq 1$ and Jensen inequality, we have*

$$(a + b + c)^p \leq 3^{p-1}(a^p + b^p + c^p).$$

Therefore, condition (iv) of Definition 1 is satisfied and ρ is a b_2-metric on X.

Definition 2 ([11]). *Let $\{x_n\}$ be a sequence in a b_2-metric space (X, d). Then*

(i) $\{x_n\}$ *is said to be b_2-convergent and converges to $x \in X$, written $\lim_{n \to \infty} x_n = x$, if for all $a \in X$, $\lim_{n \to \infty} d(x_n, x, a) = 0$.*
(ii) $\{x_n\}$ *is said to be b_2-Cauchy in X if for all $a \in X$, $\lim_{m,n \to \infty} d(x_m, x_n, a) = 0$.*
(iii) (X, d) *is said to be b_2-complete if every b_2-Cauchy sequence is a b_2-convergent sequence.*

Definition 3 ([11]). *Let (X, d) and $(\bar{X}, \bar{d})$ be two b_2-metric spaces and let $f : X \to \bar{X}$ be a mapping. Then f is said to be b_2-continuous at a point $z \in X$ if for a given $\varepsilon > 0$, there exists $\delta > 0$ such that $x \in X$ and $d(z, x, a) < \delta$ for all $a \in X$ imply that $\bar{d}(fz, fx, a) < \varepsilon$. The mapping f is b_2-continuous on X if it is b_2-continuous at all $z \in X$.*

Proposition 1 ([11]). *Let (X, d) and $(\bar{X}, \bar{d})$ be two b_2-metric spaces. Then a mapping $f : X \to \bar{X}$ is b_2-continuous at a point $x \in X$ if it is b_2-sequentially continuous at x, that is, whenever $\{x_n\}$ is b_2-convergent to x, $\{f(x_n)\}$ is b_2-convergent to $f(x)$.*

Lemma 1 ([11]). *Let (X, d) be a b_2-metric space. Suppose that $\{x_n\}$ and $\{y_n\}$ are b_2-converge to x and y, respectively. Then, we have*

$$\frac{1}{s^2}d(x, y, a) \leq \liminf_{n \to \infty} d(x_n, y_n, a) \leq \limsup_{n \to \infty} d(x_n, y_n, a) \leq s^2 d(x, y, a) \quad \text{for all } a \in X.$$

In particular, if $y_n = y$, is constant, then

$$\frac{1}{s}d(x, y, a) \leq \liminf_{n \to \infty} d(x_n, y, a) \leq \limsup_{n \to \infty} d(x_n, y, a) \leq s d(x, y, a) \quad \text{for all } a \in X.$$

Definition 4. *Let f and g be two self mappings on a non-empty set X. If $w = fx = gx$ for some $x \in X$, then x is called a coincidence point of f and g and w is called a point of coincidence of f and g.*

Definition 5 ([18]). *Two self mappings f and g are said to be weakly compatible if they commute at their coincidence points, that is, $fx = gx$ implies that $fgx = gfx$.*

Lemma 2 ([19]). *Let f and g be weakly compatible self mappings of a non-empty set X. If f and g have a unique point of coincidence $w = fx = gx$, then w is the unique common fixed point of f and g.*

A non-empty subset $\mathcal{R}$ of $X \times X$ is said to be a binary relation on X. Trivially, $X \times X$ is a binary relation on X known as the universal relation. For simplicity, we will write $x\mathcal{R}y$ whenever $(x, y) \in \mathcal{R}$ and write $x\mathcal{R}^\# y$ whenever $x\mathcal{R}y$ and $x \neq y$. Observe that $\mathcal{R}^\#$ is also a binary relation on X and $\mathcal{R}^\# \subseteq \mathcal{R}$. The elements x and y of X are said to be $\mathcal{R}$-comparable if $x\mathcal{R}y$ or $y\mathcal{R}x$, this is denoted by $[x, y] \in \mathcal{R}$.

Definition 6. *A binary relation $\mathcal{R}$ on X is said to be:*
(i) *reflexive if $x\mathcal{R}x$ for all $x \in X$;*
(ii) *transitive if, for any $x, y, z \in X$, $x\mathcal{R}y$ and $y\mathcal{R}z$ imply $x\mathcal{R}z$; antisymmetric if, for any $x, y \in X$, $x\mathcal{R}y$ and $y\mathcal{R}x$ imply $x = y$;*
(iii) *preorder if it is reflexive and transitive;*
(iv) *partial order if it is reflexive, transitive and antisymmetric.*

Let X be a nonempty set, $\mathcal{R}$ a binary relation on X and $Y \subseteq X$. Then the restriction of $\mathcal{R}$ to Y is denoted by $\mathcal{R}|_Y$ and is defined by $\mathcal{R} \cap Y^2$. The inverse of $\mathcal{R}$ is denoted by $\mathcal{R}^{-1}$ and is defined by $\mathcal{R}^{-1} = \{(x, y) \in X \times X : (y, x) \in \mathcal{R}\}$ and $\mathcal{R}^s = \mathcal{R} \cup \mathcal{R}^{-1}$.

Definition 7 ([20]). *Let X be a non-empty set and $\mathcal{R}$ a binary relation on X. A sequence $\{x_n\} \subseteq X$ is said to be an $\mathcal{R}$-preserving sequence if $x_n \mathcal{R} x_{n+1}$ for all $n \in \mathbb{N}_0$.*

Definition 8 ([20]). *Let X be a non-empty set and $f : X \to X$. A binary relation $\mathcal{R}$ on X is said to be f-closed if for all $x, y \in X$, $x\mathcal{R}y$ implies $fx\mathcal{R}fy$.*

Definition 9 ([20]). *Let X be a non-empty set and $f, g : X \to X$. A binary relation $\mathcal{R}$ on X is said to be (f, g)-closed if for all $x, y \in X$, $gx\mathcal{R}gy$ implies $fx\mathcal{R}fy$.*

Definition 10 ([20]). *Let (X, d) be a metric space and $\mathcal{R}$ a binary relation on X. We say that X is $\mathcal{R}$-complete if every $\mathcal{R}$-preserving Cauchy sequence in X converges to a limit in X.*

Remark 1. *Every complete metric space is $\mathcal{R}$-complete, whatever the binary relation $\mathcal{R}$. Particularly, under the universal relation, the notion of $\mathcal{R}$-completeness coincides with the usual completeness.*

Definition 11 ([21]). *Let (X, d) be a metric space, $\mathcal{R}$ a binary relation on X, $f : X \to X$ and $x \in X$. We say that f is $\mathcal{R}$-continuous at x if, for any $\mathcal{R}$-preserving sequence $\{x_n\} \subseteq X$ such that $x_n \to x$, we have $fx_n \to fx$. Moreover, f is called $\mathcal{R}$-continuous if it is $\mathcal{R}$-continuous at each point of X.*

Remark 2. *Every continuous mapping is $\mathcal{R}$-continuous, whatever the binary relation $\mathcal{R}$. Particularly, under the universal relation, the notion of $\mathcal{R}$-continuity coincides with the usual continuity.*

Definition 12 ([21]). *Let (X, d) be a metric space, $\mathcal{R}$ a binary relation on X, $f, g : X \to X$ and $x \in X$. We say that f is $(g, \mathcal{R})$-continuous at x if, for any sequence $\{x_n\} \subseteq M$ such that $\{gx_n\}$ is $\mathcal{R}$-preserving and $gx_n \to gx$, we have $fx_n \to fx$. Moreover, f is called $(g, \mathcal{R})$-continuous if it is $(g, \mathcal{R})$-continuous at each point of X.*

Observe that on setting $g = I$, Definition 12 reduces to Definition 11.

Remark 3. *Every g-continuous mapping is $(g, \mathcal{R})$-continuous, whatever the binary relation $\mathcal{R}$. Particularly, under the universal relation, the notion of $(g, \mathcal{R})$-continuity coincides with the usual g-continuity.*

Definition 13 ([21]). Let (X, d) be a metric space, $\mathcal{R}$ be a binary relation on X and $f, g : X \to X$. We say that the pair (f, g) is $\mathcal{R}$-compatible if for any sequence $\{x_n\} \subseteq X$ such that $\{fx_n\}$ and $\{gx_n\}$ are $\mathcal{R}$-preserving and $\lim_{n\to\infty} gx_n = \lim_{n\to\infty} fx_n = x \in X$, we have $\lim_{n\to\infty} d(gfx_n, fgx_n) = 0$.

Remark 4. *Every compatible pair is $\mathcal{R}$-compatible, whatever the binary relation $\mathcal{R}$. Particularly, under the universal relation, the notion of $\mathcal{R}$-compatibility coincides with the usual compatibility.*

Definition 14 ([20]). Let (X, d) be a metric space. A binary relation $\mathcal{R}$ on X is said to be d-self-closed if for any $\mathcal{R}$-preserving sequence $\{x_n\} \subseteq X$ such that $x_n \to x$, there exists a subsequence $\{x_{n_k}\}$ of $\{x_n\}$ such that $[x_{n_k}, x] \in \mathcal{R}$ for all $k \in \mathbb{N}_0$.

3. Common Fixed Point Results for Almost $\mathcal{R}_g$-Geraghty Type Contraction Mappings

Lemma 3. *Let (X, d) be a b_2-metric space endowed with a binary relation $\mathcal{R}$ and $f, g : X \to X$ such that $f(X) \subseteq g(X)$, with $\mathcal{R}$ is (f, g)-closed and $\mathcal{R}|_{g(X)}$ is transitive. Assume that there exists $x_0 \in X$ such that $gx_0 \mathcal{R} fx_0$. Define a sequence $\{x_n\}$ in X by $fx_n = gx_{n+1}$ for $n \geq 0$. Then*

$$gx_m \mathcal{R} gx_n \text{ and } fx_m \mathcal{R} fx_n \quad \text{for all } m, n \in \mathbb{N}_0 \text{ with } m < n.$$

Proof. Since there exists $x_0 \in X$ such that $gx_0 \mathcal{R} fx_0$, $fx_n = gx_{n+1}$, and $\mathcal{R}$ is (f, g)-closed, we deduce that $gx_0 \mathcal{R} gx_1$, then $gx_1 = fx_0 \mathcal{R} fx_1 = gx_2$. By continuing this process, we get $gx_n \mathcal{R} gx_{n+1}$ for all $n \in \mathbb{N}$. Suppose that $m < n$, so $gx_m \mathcal{R} gx_{m+1}$ and $gx_{m+1} \mathcal{R} gx_{m+2}$, by $\mathcal{R}$ is g-transitive we have $gx_m \mathcal{R} gx_{m+2}$. Again, since $gx_m \mathcal{R} gx_{m+2}$ and $gx_{m+2} \mathcal{R} gx_{m+3}$, we get that $gx_m \mathcal{R} gx_{m+3}$. By continuing this process, we obtain $gx_m \mathcal{R} gx_n$. for all $m, n \in \mathbb{N}$ with $m < n$. In similar way and since $f(X) \subseteq g(X)$, we conclude $fx_m \mathcal{R} fx_n$ for all $m, n \in \mathbb{N}$ with $m < n$. □

In 1973, Geraghty [22] introduced the class F of all functions $\beta : [0, \infty) \to [0, 1)$ which satisfy that $\lim_{n\to\infty} \beta(t_n) = 1$ implies $\lim_{n\to\infty} t_n = 0$. In addition, the author proved a fixed point result, generalizing the Banach contraction principle. Afterwards, there are many results about fixed point theorems by using such functions in this class. Đukić et al. [23] obtained fixed point results of this kind in b-metric and from [23] we denote Ω to the family of all functions $\beta_s : [0, \infty) \to [0, \frac{1}{s})$ for a real number $s \geq 1$, which satisfy the condition

$$\lim_{n\to\infty} \beta_s(t_n) = \frac{1}{s} \quad \text{implies} \quad \lim_{n\to\infty} t_n = 0.$$

Definition 15. Let (X, d) be a b_2-metric space and $f, g : X \to X$. Suppose for all $x, y, a \in X$,

$$M(x, y, a) = \max\left\{d(gx, gy, a), d(gx, fx, a), d(gy, fy, a), \frac{d(gx, fy, a) + d(gy, fx, a)}{2s}\right\},$$

and

$$N(x, y, a) = \min\{d(gx, fx, a), d(gy, fy, a), d(gx, fy, a), d(gy, fx, a)\}.$$

We say that f is almost $\mathcal{R}_g$-Geraghty type contraction mapping if there exist $L \geq 0$ and $\beta_s \in \Omega$ such that

$$d(fx, fy, a) \leq \beta_s(M(x, y, a)) M(x, y, a) + LN(x, y, a), \tag{1}$$

for all $x, y, a \in X$, with $gx \mathcal{R} gy$, $fx \mathcal{R}^* fy$.

Definition 16. Let (X, d) be a b_2-metric space and $f : X \to X$. Suppose for all $x, y, a \in X$,

$$M(x, y, a) = \max\left\{d(x, y, a), d(x, fx, a), d(y, fy, a), \frac{d(x, fy, a) + d(y, fx, a)}{2s}\right\},$$

and

$$N(x, y, a) = \min\{d(x, fx, a), d(y, fy, a), d(x, fy, a), d(y, fx, a)\}.$$

We say that f is almost $\mathcal{R}$-Geraghty type contraction mapping if there exist $L \geq 0$ and $\beta_s \in \Omega$ such that

$$d(fx, fy, a) \leq \beta_s(M(x, y, a))M(x, y, a) + LN(x, y, a), \tag{2}$$

for all $x, y, a \in X$, with $x\mathcal{R}y$, $fx\mathcal{R}^n fy$.

Now, we present our main result as follows:

Theorem 1. Let (X, d) be a b_2-metric space endowed with a binary relation $\mathcal{R}$ and $f, g : X \to X$ such that $f(X) \subseteq g(X)$, $g(X)$ is a b_2-complete subspace of X. Assume that f is almost $\mathcal{R}_g$-Geraghty type contraction mapping and the following conditions hold:
(i) there exists x_0 in X such that $gx_0 \mathcal{R} fx_0$;
(ii) $\mathcal{R}$ is (f, g)-closed and $\mathcal{R}|_{g(X)}$ is transitive;
(iii) $\mathcal{R}|_{g(X)}$ is d-self closed provided (1) holds for all $x, y, a \in X$ with $gx\mathcal{R}gy$ and $fx\mathcal{R}^n fy$.
Then f and g have a coincidence point in X.

Proof. Let $x_0 \in X$ such that $gx_0 \mathcal{R} fx_0$. The proof is finished if $gx_0 = fx_0$ and x_0 is a coincidence point of f and g. Let us take $gx_0 \neq fx_0$, then since $f(X) \subseteq g(X)$ we can choose $x_1 \in X$ such that $fx_0 = gx_1$. Continuing this process, we can define a sequence $\{gx_n\}$ in X by $fx_n = gx_{n+1}$, for all $n \in \mathbf{N}_0$.

We divide the proof into three steps as follows.

Step 1: We claim that $\lim_{n \to \infty} d(gx_n, gx_{n+1}, a) = 0$. From Lemma 3, we have $\{gx_n\}$ is $\mathcal{R}$-preserving sequence that is $gx_n\mathcal{R}gx_{n+1}$ and $fx_n\mathcal{R}fx_{n+1}$, for all $n \in \mathbf{N}_0$. If $fx_{n_0} = fx_{n_0+1}$, for some $n_0 \in \mathbf{N}_0$, then x_{n_0+1} is a coincidence point of f and g. Suppose that $fx_n \neq fx_{n+1}$, for all $n \in \mathbf{N}_0$. Therefore, from (1), we obtain

$$
\begin{aligned}
d(gx_{n+1}, gx_{n+2}, gx_n) &= d(fx_n, fx_{n+1}, gx_n) \\
&\leq \beta_s(M(x_n, x_{n+1}, gx_n))M(x_n, x_{n+1}, gx_n) + LN(x_n, x_{n+1}, gx_n) \to (*)
\end{aligned}
$$

where

$$
\begin{aligned}
M(x_n, x_{n+1}, gx_n) &= \max\Big\{d(gx_n, gx_{n+1}, gx_n), d(gx_n, fx_n, gx_n), d(gx_{n+1}, fx_{n+1}, gx_n), \\
&\qquad \frac{d(gx_n, fx_{n+1}, gx_n) + d(gx_{n+1}, fx_n, gx_n)}{2s}\Big\} \\
&= \max\Big\{d(gx_n, gx_{n+1}, gx_n), d(gx_n, gx_{n+1}, gx_n), d(gx_{n+1}, gx_{n+2}, gx_n), \\
&\qquad \frac{d(gx_n, gx_{n+2}, gx_n) + d(gx_{n+1}, gx_{n+1}, gx_n)}{2s}\Big\} \\
&= d(gx_{n+1}, gx_{n+2}, gx_n),
\end{aligned}
$$

and

$$
\begin{aligned}
N(x_n, x_{n+1}, gx_n) &= \min\{d(gx_n, fx_n, gx_n), d(gx_{n+1}, fx_{n+1}, gx_n), d(gx_n, fx_{n+1}, gx_n), \\
&\qquad d(gx_{n+1}, fx_n, gx_n)\} = 0.
\end{aligned}
$$

If $d(gx_{n+1}, gx_{n+2}, gx_n) \neq 0$ for some $n \in \mathbf{N}_0$, then we have (due to (*))

$$d(gx_{n+1}, gx_{n+2}, gx_n) \leq \beta_s(d(gx_{n+1}, gx_{n+2}, gx_n))d(gx_{n+1}, gx_{n+2}, gx_n),$$

yielding that

$$d(gx_{n+1}, gx_{n+2}, gx_n) - \beta_s(d(gx_{n+1}, gx_{n+2}, gx_n))d(gx_{n+1}, gx_{n+2}, gx_n) \leq 0,$$

or

$$d(gx_{n+1}, gx_{n+2}, gx_n)[1 - \beta_s(d(gx_{n+1}, gx_{n+2}, gx_n))] \leq 0 \to (**).$$

Divide both sides in (**) by $d(gx_{n+1}, gx_{n+2}, gx_n) \neq 0$, we obtain

$$1 - \beta_s(d(gx_{n+1}, gx_{n+2}, gx_n)) \leq 0,$$

or

$$\beta_s(d(gx_{n+1}, gx_{n+2}, gx_n)) \geq 1,$$

a contradiction [as $\beta_s : [0, \infty) \to [0, \frac{1}{s})$ and $s \geq 1$ so $\beta_s(c) < \frac{1}{s} \leq 1$, that is $\beta_s(c) < 1$ for all $c \in [0, \infty)$]. Therefore, we must have

$$d(gx_{n+1}, gx_{n+2}, gx_n) = 0, \quad \text{for all } n \in \mathbf{N}_0. \tag{3}$$

Thus, by the rectangle inequality and (3) we get

$$d(gx_n, gx_{n+2}, a) \leq s[d(gx_n, gx_{n+1}, a) + d(gx_{n+1}, gx_{n+2}, a)], \tag{4}$$

for all $n \in \mathbf{N}_0, a \in X$. Using (4), Lemma 3 and (1) we have

$$\begin{aligned} d(gx_{n+1}, gx_{n+2}, a) &= d(fx_n, fx_{n+1}, a) \\ &\leq \beta_s(M(x_n, x_{n+1}, a))M(x_n, x_{n+1}, a) + LN(x_n, x_{n+1}, a). \end{aligned} \tag{5}$$

Observe that

$$M(x_n, x_{n+1}, a) = \max\{d(gx_n, gx_{n+1}, a), d(gx_{n+1}, gx_{n+2}, a)\},$$

and

$$\begin{aligned} N(x_n, x_{n+1}, a) &= \min\{d(gx_n, fx_n, a), d(gx_{n+1}, fx_{n+1}, a), d(gx_n, fx_{n+1}, a), d(gx_{n+1}, fx_n, a)\} \\ &= \min\{d(gx_n, gx_{n+1}, a), d(gx_{n+1}, gx_{n+2}, a), d(gx_n, gx_{n+2}, a), d(gx_{n+1}, gx_{n+1}, a)\} \\ &= 0. \end{aligned}$$

Now, if $M(x_n, x_{n+1}, a) = d(gx_{n+1}, gx_{n+2}, a)$, then from (5) we have

$$d(gx_{n+1}, gx_{n+2}, a) \leq \beta_s(d(gx_{n+1}, gx_{n+2}, a))d(gx_{n+1}, gx_{n+2}, a) < d(gx_{n+1}, gx_{n+2}, a),$$

a contradiction. Hence, $M(x_n, x_{n+1}, a) = d(gx_n, gx_{n+1}, a)$, and

$$d(gx_{n+1}, gx_{n+2}, a) \leq \beta_s(d(gx_n, gx_{n+1}, a))d(gx_n, gx_{n+1}, a) < d(gx_n, gx_{n+1}, a), \tag{6}$$

for all $n \in \mathbf{N}_0$ and $a \in X$, which implies that the sequence $\{d(gx_n, gx_{n+1}, a)\}$ is strictly decreasing of positive numbers. Hence, there exists $\delta \geq 0$ such that $\lim_{n \to \infty} d(gx_n, gx_{n+1}, a) = \delta$. Suppose that $\delta > 0$. So, taking the limit as $n \to \infty$, from (6) we obtain

$$\frac{1}{s}\delta \leq \delta \leq \lim_{n \to \infty} \beta_s(d(gx_n, gx_{n+1}, a))\delta \leq \frac{1}{s}\delta.$$

Hence,
$$\lim_{n\to\infty} \beta_s(d(gx_n, gx_{n+1}, a)) = \frac{1}{s}.$$
From the property of β_s, we conclude that $\lim_{n\to\infty} d(gx_n, gx_{n+1}, a) = 0$ a contradiction, hence, $\delta = 0$ and
$$\lim_{n\to\infty} d(gx_n, gx_{n+1}, a) = 0. \tag{7}$$

Step 2: We claim that $d(gx_i, gx_j, gx_k) = 0$ for all $i, j, k \in \mathbf{N}_0$. Since $\{d(gx_n, gx_{n+1}, a)\}$ is strictly decreasing and $d(gx_0, gx_1, gx_0) = 0$, we conclude that $d(gx_n, gx_{n+1}, gx_0) = 0$, for all $n \in \mathbf{N}_0$.

Since $d(gx_{m-1}, gx_m, gx_m) = 0$ for all $m \in \mathbf{N}_0$ and $\{d(gx_n, gx_{n+1}, a)\}$ is strictly decreasing we obtain that

$$d(gx_n, gx_{n+1}, gx_m) = 0, \quad \text{for all } n \geq m - 1. \tag{8}$$

For $0 \leq n < m - 1$, we have $m - 1 \geq n + 1$, so from (8) we have

$$d(gx_{m-1}, gx_m, gx_{n+1}) = d(gx_{m-1}, gx_m, gx_n) = 0. \tag{9}$$

Thus, by the rectangle inequality, $d(gx_n, gx_{n+1}, gx_{n+1}) = 0$, and using (9) we obtain

$$\begin{aligned}
d(gx_n, gx_{n+1}, gx_m) &\leq s[d(gx_n, gx_{n+1}, gx_{m-1}) + d(gx_{n+1}, gx_m, gx_{m-1}) + d(gx_m, gx_n, gx_{m-1})] \\
&= sd(gx_n, gx_{n+1}, gx_{m-1}) \\
&\leq sd(gx_n, gx_{n+1}, gx_{n+1}) = 0.
\end{aligned}$$

Therefore, we get

$$d(gx_n, gx_{n+1}, gx_m) = 0, \quad \text{for all } 0 \leq n < m - 1. \tag{10}$$

Hence, from (8) and (10) we have

$$d(gx_n, gx_{n+1}, gx_m) = 0, \quad \text{for all } n, m \in \mathbf{N}_0.$$

Now, for all $i, j, k \in \mathbf{N}_0$, $i < j$ and $d(gx_i, gx_j, gx_{j-1}) = d(gx_k, gx_j, gx_{j-1}) = 0$, applying the rectangle inequality we get

$$\begin{aligned}
d(gx_i, gx_j, gx_k) &\leq s[d(gx_i, gx_j, gx_{j-1}) + d(gx_j, gx_k, gx_{j-1}) + d(gx_k, gx_i, gx_{j-1})] \\
&= sd(gx_k, gx_i, gx_{j-1}) \\
&\leq s^2 d(gx_k, gx_i, gx_{j-2}) \leq \ldots \leq s^{j-i} d(gx_k, gx_i, gx_i) = 0.
\end{aligned}$$

Therefore, for all $i, j, k \in \mathbf{N}_0$, we have

$$d(gx_i, gx_j, gx_k) = 0. \tag{11}$$

Step 3: We show that $\{gx_n\}$ is a b_2-Cauchy sequence. Suppose to the contrary that $\{gx_n\}$ is not a b_2-Cauchy sequence. Then there is $\varepsilon > 0$ such that for an integer k there exist integers $n(k), m(k)$ with $n(k) > m(k) > k$ such that

$$d(gx_{m(k)}, gx_{n(k)}, a) \geq \varepsilon, \tag{12}$$

for every integer k, let $n(k)$ be the least positive integer with $n(k) > m(k)$, satisfying (12) and such that

$$d(gx_{m(k)}, gx_{n(k)-1}, a) < \varepsilon. \tag{13}$$

Using the rectangle inequality, (11) and (12) we have

$$\varepsilon \leq d(gx_{m(k)}, gx_{n(k)}, a) \leq s[d(gx_{m(k)}, gx_{n(k)-1}, a) + d(gx_{n(k)}, gx_{n(k)-1}, a)].$$

Again, using the rectangle inequality and (11) in the above inequality, it follows that

$$\varepsilon \leq s^2[d(gx_{m(k)}, gx_{m(k)-1}, a) + d(gx_{m(k)-1}, gx_{n(k)-1}, a)] + sd(gx_{n(k)}, gx_{n(k)-1}, a)].$$

In addition,

$$d(gx_{m(k)-1}, gx_{n(k)-1}, a) \leq s[d(gx_{m(k)-1}, gx_{m(k)}, a) + d(gx_{n(k)-1}, gx_{m(k)}, a)].$$

Taking the upper limit as $k \to \infty$, in the above three inequalities and from (7) and (13) it follows that

$$\varepsilon \leq \limsup_{k \to \infty} d(gx_{m(k)}, gx_{n(k)}, a) < s\varepsilon, \tag{14}$$

$$\frac{\varepsilon}{s^2} \leq \limsup_{k \to \infty} d(gx_{m(k)-1}, gx_{n(k)-1}, a) < s\varepsilon, \tag{15}$$

$$\frac{\varepsilon}{s^3} \leq \limsup_{k \to \infty} d(gx_{m(k)}, gx_{n(k)-1}, a) < \varepsilon. \tag{16}$$

Again, using the rectangle inequality, (11) and (12) we get

$$d(gx_{m(k)-1}, gx_{n(k)}, a) \leq s[d(gx_{m(k)-1}, gx_{n(k)-1}, a) + d(gx_{n(k)}, gx_{n(k)-1}, a)],$$

$$\varepsilon \leq d(gx_{m(k)}, gx_{n(k)}, a) \leq s[d(gx_{m(k)}, gx_{m(k)-1}, a) + d(gx_{n(k)}, gx_{m(k)-1}, a)].$$

Taking the upper limit as $k \to \infty$, in the above two inequalities and from (7) and (15), we get

$$\frac{\varepsilon}{s} \leq \limsup_{k \to \infty} d(gx_{m(k)-1}, gx_{n(k)}, a) < s^2\varepsilon. \tag{17}$$

Now, from Lemma 3 we have $fx_{m(k)-1} \mathcal{R}^n fx_{n(k)-1}$ for all $m(k), n(k) \in \mathbb{N}_0$ with $m(k) < n(k)$. Hence, from (1) we conclude that

$$d(gx_{m(k)}, gx_{n(k)}, a) = d(fx_{m(k)-1}, fx_{n(k)-1}, a)$$
$$\leq \beta_s(M(x_{m(k)-1}, x_{n(k)-1}, a))M(x_{m(k)-1}, x_{n(k)-1}, a) + LN(x_{m(k)-1}, x_{n(k)-1}, a), \tag{18}$$

where

$$M(x_{m(k)-1}, x_{n(k)-1}, a) = \max\{d(gx_{m(k)-1}, gx_{n(k)-1}, a), d(gx_{m(k)-1}, fx_{m(k)-1}, a),$$
$$d(gx_{n(k)-1}, fx_{n(k)-1}, a), \frac{d(gx_{m(k)-1}, fx_{n(k)-1}, a) + d(gx_{n(k)-1}, fx_{m(k)-1}, a)}{2s}\},$$
$$= \max\Big\{ d(gx_{m(k)-1}, gx_{n(k)-1}, a), d(gx_{m(k)-1}, gx_{m(k)}, a), d(gx_{n(k)-1}, gx_{n(k)}, a),$$
$$\frac{d(gx_{m(k)-1}, gx_{n(k)}, a) + d(gx_{n(k)-1}, gx_{m(k)}, a)}{2s} \Big\}, \tag{19}$$

and

$$N(x_{m(k)-1}, x_{n(k)-1}, a) = \min\{d(gx_{m(k)-1}, gx_{m(k)}, a), d(gx_{n(k)-1}, gx_{n(k)}, a), d(gx_{m(k)-1}, gx_{n(k)}, a),$$
$$d(gx_{n(k)-1}, gx_{m(k)}, a)\}. \tag{20}$$

Taking the upper limit as $k \to \infty$, in (19), (20) and using (7), (15)–(17) it follows that

$$\frac{\varepsilon}{s^2} \leq \limsup_{k \to \infty} M(x_{m(k)-1}, x_{n(k)-1}, a) < s\varepsilon, \tag{21}$$

and

$$\limsup_{k \to \infty} N(x_{m(k)-1}, x_{n(k)-1}, a) = 0. \tag{22}$$

Now, taking the upper limit as $k \to \infty$ in (18) and using (14), (21) and (22), we conclude that

$$\frac{1}{s} = \frac{\varepsilon}{s\varepsilon} \leq \frac{\limsup_{k\to\infty} d(gx_{m(k)}, gx_{n(k)}, a)}{\limsup_{k\to\infty} M(x_{m(k)-1}, x_{n(k)-1}, a)} \leq \limsup_{k\to\infty} \beta_s(M(x_{m(k)-1}, x_{n(k)-1}, a)) \leq \frac{1}{s}.$$

Thus, $\limsup_{k\to\infty} \beta_s(M(x_{m(k)-1}, x_{n(k)-1}, a)) = \frac{1}{s}$. Hence, $\limsup_{k\to\infty} M(x_{m(k)-1}, x_{n(k)-1}, a) = 0$, which is a contradiction. Therefore, $\{gx_n\}$ is a b_2-Cauchy sequence. As $g(X)$ is b_2-complete subspace of X, then there exist $z \in X$ such that

$$\lim_{n\to\infty} gx_n = \lim_{n\to\infty} fx_n = gz. \tag{23}$$

Now, we show that z is a point of coincidence of f and g. From condition (iii), we have $\mathcal{R}|_{g(X)}$ is d-self closed and (1) holds for all $x, y, a \in X$ with $gx\mathcal{R}gy$ and $fx\mathcal{R}''fy$. As $\{gx_n\} \subseteq g(X)$, $\{gx_n\}$ is $\mathcal{R}|_{g(X)}$-preserving and $gx_n \to gz$ so there exists a subsequence $\{gx_{n(k)}\} \subseteq \{gx_n\}$ such that $gx_{n(k)}\mathcal{R}|_{g(X)}gz$ for all $k \in \mathbf{N}_0$ and since $\mathcal{R}$ is (f,g)-closed then $fx_{n(k)}\mathcal{R}|_{g(X)}fz$ for all $k \in \mathbf{N}_0$.

If $fx_{n(k)} = fz$ for all $k > k_0$, and $k_0, k \in \mathbf{N}_0$, then $\lim_{k\to\infty} fx_{n(k)} = fz$, and since $\lim_{n\to\infty} fx_n = gz$, we have $fz = gz$, that is z is a coincidence point of f and g.

On other hand, if $fx_{n(k)} \neq fz$ for all $k > k_0$, and $k_0, k \in \mathbf{N}_0$, then $fx_{n(k)}\mathcal{R}|_{g(X)}fz$ and $fx_{n(k)} \neq fz$ for all $k > k_0$, and $k_0, k \in \mathbf{N}_0$. Thus, $gx_{n(k)}\mathcal{R}|_{g(X)}gz$ and $fx_{n(k)}\mathcal{R}''|_{g(X)}fz$, and from (1), we have

$$d(gx_{n(k)+1}, fz, a) = d(fx_{n(k)}, fz, a) \leq \beta_s(M(x_{n(k)}, z, a))M(x_{n(k)}, z, a) + LN(x_{n(k)}, z, a), \tag{24}$$

where

$$M(x_{n(k)}, z, a) = \max\Big\{ d(gx_{n(k)}, gz, a), d(gx_{n(k)}, gx_{n(k)+1}, a), d(gz, fz, a),$$
$$\frac{d(gx_{n(k)}, fz, a) + d(gz, gx_{n(k)+1}, a)}{2s} \Big\}, \tag{25}$$

and

$$N(x_{n(k)}, z, a) = \min\{d(gx_{n(k)}, gx_{n(k)+1}, a), d(gz, fz, a), d(gx_{n(k)}, fz, a), d(gz, gx_{n(k)+1}, a)\}. \tag{26}$$

Letting $k \to \infty$ in (25), (26), we get

$$\limsup_{k\to\infty} M(x_{n(k)}, z, a) = \max\Big\{ d(gz, fz, a), \frac{\limsup_{k\to\infty} d(gx_{n(k)}, fz, a)}{2s} \Big\},$$

and

$$\limsup_{k\to\infty} N(x_{n(k)}, z, a) = 0. \tag{27}$$

From Lemma 1, we have

$$\frac{d(gz, fz, a)}{s} \leq \limsup_{k\to\infty} d(gx_{n(k)}, fz, a) \leq sd(gz, fz, a). \tag{28}$$

Thus,

$$\max\{d(gz, fz, a), \frac{d(gz, fz, a)}{2s^2}\} \leq \limsup_{k\to\infty} M(x_{n(k)}, z, a) \leq \max\{d(gz, fz, a), \frac{d(gz, fz, a)}{2}\},$$

yields,
$$\limsup_{k \to \infty} M(x_{n(k)}, z, a) = d(gz, fz, a), \tag{29}$$

Again, taking the upper limit as $k \to \infty$, in (24) and using Lemma 1, (27) and (29), we get

$$\begin{aligned}
\frac{d(gz, fz, a)}{s} &\leq \limsup_{k \to \infty} d(gx_{n(k)+1}, fz, a) \\
&\leq \limsup_{k \to \infty} \beta_s(M(x_{n(k)}, z, a)) \limsup_{k \to \infty} M(x_{n(k)}, z, a) \\
&\leq \limsup_{k \to \infty} \beta_s(M(x_{n(k)}, z, a)) d(gz, fz, a) \\
&\leq \frac{1}{s} d(gz, fz, a).
\end{aligned}$$

Hence, $\limsup_{k \to \infty} \beta_s(M(x_{n(k)}, z, a)) = \frac{1}{s}$, so from the property of β_s we conclude that $\limsup_{k \to \infty} M(x_{n(k)}, z, a) = 0$ implies $d(gz, fz, a) = 0$ for all $a \in X$. That is, $gz = fz$. This shows that f and g have a coincidence point. □

The next theorem shows that under some additional hypotheses we can deduce the existence and uniqueness of a common fixed point.

Theorem 2. *In addition to the hypotheses of Theorem 1, suppose that f and g are weakly compatible and for all coincidence points u, v of f and g, there exists $w \in X$ such that $gu\mathcal{R}gw$ and $gv\mathcal{R}gw$. Then f and g have a unique common fixed point.*

Proof. The set of coincidence points of f and g is not empty due to Theorem 1. Suppose that u and v are two coincidence points of f and g, that is, $fu = gu$ and $fv = gv$. We will show that $gu = gv$. By our assumption, there exists $w \in X$ such that

$$gu\mathcal{R}gw \quad \text{and} \quad gv\mathcal{R}gw. \tag{30}$$

Now, proceeding similarly to the proof of Theorem 1, we can define a sequence $\{w_n\}$ in X as $fw_n = gw_{n+1}$ for all $n \in \mathbf{N}_0$ and $w_0 = w$, with $\lim_{n \to \infty} d(gw_n, gw_{n+1}, a) = 0$. Since $gu\mathcal{R}gw_0$ ($gv\mathcal{R}gw_0$) and $\mathcal{R}$ is (f, g)-closed, we conclude that $fu\mathcal{R}fw_0 (fv\mathcal{R}fw_0)$. Hence, $gu\mathcal{R}gw_1(gv\mathcal{R}gw_1)$. By induction, we have

$$gu\mathcal{R}gw_n \quad \text{and} \quad gv\mathcal{R}gw_n, \quad \forall n \in \mathbf{N}_0. \tag{31}$$

From (1) and using (31), we obtain

$$d(gu, gw_{n+1}, a) = d(fu, fw_n, a) \leq \beta_s(M(u, w_n, a))M(u, w_n, a) + LN(u, w_n, a), \tag{32}$$

where

$$\begin{aligned}
M(u, w_n, a) &= \max\left\{d(gu, gw_n, a), d(gu, fu, a), d(gw_n, fw_n, a), \frac{d(gu, fw_n, a) + d(gw_n, fu, a)}{2s}\right\}, \\
&= \max\left\{d(gu, gw_n, a), d(gw_n, gw_{n+1}, a), \frac{d(gu, gw_{n+1}, a) + d(gw_n, gu, a)}{2s}\right\},
\end{aligned}$$

and

$$\begin{aligned}
N(u, w_n, a) &= \min\{d(gu, fu, a), d(gw_n, fw_n, a), d(gu, fw_n, a), d(gw_n, fu, a)\} \\
&= \min\{d(gu, gu, a), d(gw_n, gw_{n+1}, a), d(gu, gw_{n+1}, a), d(gw_n, gu, a)\} = 0.
\end{aligned}$$

Hence,

$$d(gu, gw_{n+1}, a) \leq \beta_s(M(u, w_n, a))M(u, w_n, a)$$
$$< \frac{1}{s}M(u, w_n, a) \leq M(u, w_n, a).$$

Since,

$$d(gu, gw_{n+1}, a) < M(u, w_n, a)$$
$$= \max\left\{d(gu, gw_n, a), d(gw_n, gw_{n+1}, a), \frac{d(gu, gw_{n+1}, a) + d(gw_n, gu, a)}{2s}\right\}$$
$$= \max\left\{d(gu, gw_n, a), d(gw_n, gw_{n+1}, a)\right\}.$$

Thus,

$$M(u, w_n, a) = \max\left\{d(gu, gw_n, a), d(gw_n, gw_{n+1}, a)\right\}.$$

(Case1): if $M(u, w_n, a) = d(gu, gw_n, a)$, then

$$d(gu, gw_{n+1}, a) \leq \beta_s(d(gu, gw_n, a))d(gu, gw_n, a) < \frac{1}{s}d(gu, gw_n, a) \leq d(gu, gw_n, a), \quad (33)$$

it follows that $d(gu, gw_{n+1}, a) < d(gu, gw_n, a)$. Thus, $\{d(gu, gw_n, a)\}$ is strictly decreasing. Hence, there exists $\gamma \geq 0$ such that $\lim_{n \to \infty} d(gu, gw_n, a) = \gamma$. Letting $n \to \infty$ in (33), we obtain

$$\frac{\gamma}{s} \leq \gamma = \lim_{n \to \infty} d(gu, gw_{n+1}, a) \leq \lim_{n \to \infty} \beta_s(d(gu, gw_n, a)) \lim_{n \to \infty} d(gu, gw_n, a)$$
$$\leq \lim_{n \to \infty} \beta_s(d(gu, gw_n, a))\gamma$$
$$\leq \frac{\gamma}{s},$$

this implies

$$\frac{1}{s} \leq \lim_{n \to \infty} \beta_s(d(gu, gw_n, a)) < \frac{1}{s}.$$

Thus,

$$\lim_{n \to \infty} \beta_s(d(gu, gw_n, a)) = \frac{1}{s}.$$

From the property of β_s, we conclude that $\lim_{n \to \infty} d(gu, gw_n, a) = 0$.

(Case2): If $M(u, w_n, a) = d(gw_n, gw_{n+1}, a)$, then

$$d(gu, gw_{n+1}, a) \leq \beta_s(d(gw_n, gw_{n+1}, a))d(gw_n, gw_{n+1}, a).$$

Therefore,

$$\lim_{n \to \infty} d(gu, gw_{n+1}, a) \leq \lim_{n \to \infty} \beta_s(d(gw_n, gw_{n+1}, a)) \lim_{n \to \infty} d(gw_n, gw_{n+1}, a) = 0.$$

This yields $\lim_{n \to \infty} d(gu, gw_{n+1}, a) = 0$. Therefore, from all cases we conclude that

$$\lim_{n \to \infty} d(gu, gw_n, a) = 0. \quad (34)$$

Similarly, we can show that

$$\lim_{n \to \infty} d(gv, gw_n, a) = 0. \quad (35)$$

Hence, from (34) and (35), we obtain $gu = gv$. That is, f and g have a unique point of coincidence. From Lemma 2 f and g have a unique common fixed point. □

Now, we give an example to justify the hypotheses of Theorem 1.

Example 2. *Let $X = \{p, q, r, t\}$ be a set with b_2-metric $d : X^3 \to \mathcal{R}$ defined by*

$$d(p,q,r) = 0, \quad d(p,q,t) = 4, \quad d(p,r,t) = 1, \quad d(q,r,t) = 6,$$

with symmetry in all variables and if at least two of the arguments are equal then $d(x,y,a) = 0$. Then (X, d) is a complete b_2-metric space with $s = \frac{6}{5}$. Define a binary relation $\mathcal{R}$ on X by

$$\mathcal{R} = \{(p,p), (q,q), (r,r), (p,q), (q,r), (p,r), (r,p), (r,q)\}.$$

Define $f, g : X \to X$ and $\beta : (0, \infty) \to [0, 1)$ as follows:

$$f = \begin{pmatrix} p & q & r & t \\ p & p & r & t \end{pmatrix}, \quad g = \begin{pmatrix} p & q & r & t \\ p & r & q & t \end{pmatrix}, \quad \beta_s(t) = \frac{1+t}{s(1+2t)}$$

We show that all the hypotheses of Theorem 1 are satisfied. Clearly, (X, d) is a complete b_2-metric space and $f(X) \subseteq g(X)$, $g(X)$ is a b_2-complete subspace of X. $\mathcal{R} = \mathcal{R}|_{g(X)}$ is transitive. There is $r \in X$ such that $q = gr\mathcal{R}fr = r$. Since $\mathcal{R}|_{g(X)}$ is finite, then it is d-self closed. We show that $\mathcal{R}$ is (f, g)-closed, we study the nontrivial cases:

- $gp\mathcal{R}gr = p\mathcal{R}q \Rightarrow fp\mathcal{R}fr = p\mathcal{R}r \in \mathcal{R}, gr\mathcal{R}gq = q\mathcal{R}r \Rightarrow fr\mathcal{R}fq = r\mathcal{R}p,$
- $gp\mathcal{R}gq = p\mathcal{R}r \Rightarrow fp\mathcal{R}fq = p\mathcal{R}p, gq\mathcal{R}gp = r\mathcal{R}p \Rightarrow fq\mathcal{R}fp = p\mathcal{R}p,$
- $gq\mathcal{R}gr = r\mathcal{R}q \Rightarrow fq\mathcal{R}fr = p\mathcal{R}r.$

Now, we check the contractive condition 2. The nontrivial cases are when $a = t$, $(gp\mathcal{R}gr$ and $fp\mathcal{R}fr)$, $(gr\mathcal{R}gq$ and $fr\mathcal{R}fq)$ and $(gq\mathcal{R}gr$ and $fq\mathcal{R}fr)$.

In all three cases, we get $M(p,r,t) = M(r,q,t) = M(q,r,t) = 6, N(p,r,a) = N(r,q,t) = N(q,r,t) = 0$, and then

$$1 = d(fp, fr, t) = d(p, r, t) \leq \frac{35}{13} = \beta_s(6)6 = \beta_s(M(p,r,a))M(p,r,a) + LN(p,r,a),$$

$$1 = d(fr, fq, t) = d(r, p, t) \leq \frac{35}{13} = \beta_s(6)6 = \beta_s(M(r,q,a))M(r,q,a) + LN(r,q,a),$$

$$1 = d(fq, fr, t) = d(p, r, t) \leq \frac{35}{13} = \beta_s(6)6 = \beta_s(M(q,r,a))M(q,r,a) + LN(q,r,a),$$

Therefore, all the hypotheses of Theorem 1 are satisfied. Then f and g have two coincidence fixed points p and t. Noting that p, t are not $\mathcal{R}$-comparable so the uniqueness of coincidence point is not fulfilled.

By taking $g = I$ in Theorems 1 and 2 we deduce the following result.

Corollary 1. *Let (X, d) be a complete b_2-metric space endowed with a transitive binary relation $\mathcal{R} : X \to X$ and $f : X \to X$. Assume that f is almost $\mathcal{R}$-Geraghty type contraction mapping and the following conditions hold:*

(i) *there exists x_0 in X such that $x_0 \mathcal{R} f x_0$;*
(ii) *$\mathcal{R}$ is f-closed;*
(iii) *$\mathcal{R}$ is d-self closed provided (2) holds for all $x, y, a \in X$ with $fx\mathcal{R}^n fy$.*

Then f has a fixed point. Moreover, if for $u, v \in Fix(f)$, there exists $w \in X$ such that $u\mathcal{R}w$ and $v\mathcal{R}w$, then f has a unique fixed point.

4. Results for Almost g-α-η Geraghty Type Contraction Mappings in b_2-Metric Spaces

Fathollahi et al. [4] introduced the concepts of triangular 2-α-η-admissible mappings as follows.

Definition 17 ([4]). *Let (X,d) be a 2-metric space, $f : X \to X$ and $\alpha, \eta : X^3 \to [0, \infty)$. We say that f is a triangular 2-α-η-admissible mapping if for all $a \in X$,*

(i) $\alpha(x,y,a) \geq \eta(x,y,a)$ implies $\alpha(fx, fy, a) \geq \eta(fx, fy, a)$, $x, y \in X$,

(ii) $\begin{cases} \alpha(x,y,a) \geq \eta(x,y,a), \\ \alpha(y,z,a) \geq \eta(y,z,a), \end{cases}$ implies $\alpha(x,z,a) \geq \eta(x,z,a)$.

If we take $\eta(x,y,a) = 1$, then we say that f is a triangular 2-α-admissible mapping. In addition, if we take $\alpha(x,y,a) = 1$, then we say that f is a triangular 2-η-subadmissible mapping.

Motivated by Fathollahi [4], we define the following concepts.

Definition 18. *Let (X,d) be a b_2-metric space, $f, g : X \to X$ and $\alpha, \eta : X^3 \to [0, \infty)$. We say that f is a triangular g-b_2-α-η-admissible mapping if for all $a \in X$,*

(i) $\alpha(gx, gy, a) \geq \eta(gx, gy, a)$ implies $\alpha(fx, fy, a) \geq \eta(fx, fy, a)$, $x, y \in X$,

(ii) $\begin{cases} \alpha(gx, gy, a) \geq \eta(gx, gy, a), \\ \alpha(gy, gz, a) \geq \eta(gy, gz, a), \end{cases}$ implies $\alpha(gx, gz, a) \geq \eta(gx, gz, a)$, $x, y, z \in X$.

When $\eta(gx, gy, a) = 1$, we say that f is a triangular g-b_2-α-admissible mapping. In addition, when $\alpha(gx, gy, a) = 1$, we say that f is a triangular g-b_2-η-subadmissible mapping.

Definition 19. *Let (X,d) be a b_2-metric space with $s \geq 1$ and $f, g : X \to X$, $\alpha, \eta : X^3 \to [0, \infty)$. Suppose for all $x, y, a \in X$,*

$$M(x,y,a) = \max\left\{d(gx,gy,a), d(gx,fx,a), d(gy,fy,a), \frac{d(gx,fy,a) + d(gy,fx,a)}{2s}\right\},$$

and

$$N(x,y,a) = \min\{d(gx,fx,a), d(gy,fy,a), d(gx,fy,a), d(gy,fx,a)\}.$$

We say that f is almost g-α-η Geraghty type contraction mapping if there exist $L \geq 0$ and $\beta_s \in \Omega$ such that

$$\forall x, y \in X, \ \alpha(gx, gy, a) \geq \eta(gx, gy, a)$$
$$\Rightarrow d(fx, fy, a) \leq \beta_s(M(x,y,a))M(x,y,a) + LN(x,y,a), \tag{36}$$

for all $a \in X$.

Now, we state the following corollaries

Corollary 2. *Let (X,d) be a complete b_2-metric space and $f, g : X \to X$, such that $f(X) \subseteq g(X)$, $g(X)$ is a b_2-complete subspace of X. Assume that f is almost g-α-η Geraghty type contraction mapping and the following conditions hold:*

(i) *there exists x_0 in X such that $\alpha(gx_0, fx_0, a) \geq \eta(gx_0, fx_0, a)$ for all $a \in X$;*

(ii) *f is a triangular g-b_2-α-η-admissible mapping;*

(iii) *if $\{gx_n\}$ is a sequence in X such that $\alpha(gx_n, gx_{n+1}, a) \geq \eta(gx_n, gx_{n+1}, a)$ for all $a \in X$, $n \in \mathbf{N}_0$ and $gx_n \to gz$ as $n \to \infty$, then there exists a subsequence $\{gx_{n(k)}\}$ of $\{gx_n\}$ such that $\alpha(gx_{n(k)}, gz, a) \geq \eta(gx_{n(k)}, gz, a)$ for all $k \in \mathbf{N}_0$ and all $a \in X$.*

Then f and g have a coincidence point in X. Moreover, suppose that for all coincidence points u, v of f and g, there exists $w \in X$ such that $\alpha(gu, gw, a) \geq \eta(gu, gw, a)$ and $\alpha(gv, gw, a) \geq \eta(gv, gw, a)$ for all $a \in X$ and f, g are weakly compatible. Then f and g have a unique common fixed point.

Proof. Define $\mathcal{R}$ on X as

$$x\mathcal{R}y \iff \alpha(x,y,a) \geq \eta(x,y,a).$$

We note the following:

- since there exists $x_0 \in X$ such that $\alpha(gx_0, fx_0, a) \geq \eta(gx_0, fx_0, a)$ for all $a \in X$ then $gx_0\mathcal{R}fx_0$;
- if $gx\mathcal{R}gy$ then $\alpha(gx, gy, a) \geq \eta(gx, gy, a)$. As f is a triangular g-b_2-α-η-admissible mapping, $\alpha(fx, fy, a) \geq \eta(fx, fy, a)$ and so $fx\mathcal{R}fy$. Thus, $\mathcal{R}$ is (f, g)-closed;
- if $gx\mathcal{R}gy$ and $gy\mathcal{R}gz$, then $\alpha(gx, gy, a) \geq \eta(gx, gy, a)$ and $\alpha(gy, gz, a) \geq \eta(gy, gz, a)$. As f is a triangular g-b_2-α-η-admissible mapping, $\alpha(gx, gz, a) \geq \eta(gx, gz, a)$, that is, $gx\mathcal{R}gz$. Therefore, $\mathcal{R}|_{g(X)}$ is transitive;
- if $gx\mathcal{R}gy$, $fx\mathcal{R}^nfy$, then $\alpha(gx, gy, a) \geq \eta(gx, gy, a)$, $\alpha(fx, fy, a) \geq \eta(fx, fy, a)$. Since f is almost g-α-η Geraghty type contraction, so (1) holds;
- from (iii), we have $gx_n\mathcal{R}gx_{n+1}$ for all $n \in \mathbf{N}_0$ and $gx_n \to gz$ as $n \to \infty$, then there exists a subsequence $\{gx_{n(k)}\}$ of $\{gx_n\}$ such that $gx_{n(k)}\mathcal{R}gz$ for all $k \in \mathbf{N}_0$. Hence, all conditions of Theorem 1 are satisfied. Thus, f and g have a point of coincidence in X.

Finally, if for all coincidence points u, v of f and g, there exists $w \in X$ such that $\alpha(gu, gw, a) \geq \eta(gu, gw, a)$ and $\alpha(gv, gw, a) \geq \eta(gv, gw, a)$, then $gu\mathcal{R}gw$ and $gv\mathcal{R}gw$. That is, all hypotheses of Theorem 1 are satisfied. Therefore, f and g have a unique common fixed point. □

By taking $g = I$ in Definitions 18 and 19, we say that f is a triangular b_2-α-η-admissible mapping and f is almost α-η Geraghty type contraction mapping.

Now, we have the following corollary.

Corollary 3. *Let (X, d) be a complete b_2-metric space and $f : X \to X$. Assume that f is almost α-η Geraghty type contraction mapping and the following conditions hold:*

(i) *there exists x_0 in X such that $\alpha(x_0, fx_0, a) \geq \eta(x_0, fx_0, a)$ for all $a \in X$;*
(ii) *f is a triangular b_2-α-η-admissible mapping;*
(iii) *if $\{x_n\}$ is a sequence in X such that $\alpha(x_n, x_{n+1}, a) \geq \eta(x_n, x_{n+1}, a)$ for all $a \in X$, $n \in \mathbf{N}_0$ and $x_n \to z$ as $n \to \infty$, then there exists a subsequence $\{x_{n(k)}\}$ of $\{x_n\}$ such that $\alpha(x_{n(k)}, z, a) \geq \eta(x_{n(k)}, z, a)$ for all $k \in \mathbf{N}_0$ and all $a \in X$.*

Then f has a fixed point in X. Moreover, if for $u, v \in \text{Fix}(f)$ there exists $w \in X$ such that $\alpha(u, w, a) \geq \eta(u, w, a)$ and $\alpha(v, w, a) \geq \eta(v, w, a)$ for all $a \in X$, then f has a unique fixed point.

5. Fixed Point Results in Partially Ordered b_2-Metric Spaces

Fixed point theorems for monotone operators in ordered metric spaces are widely investigated and have found various applications in differential and integral equations. This trend was started by Turinici [12] in 1986. Ran and Reurings in [24] extended the Banach contraction principle in partially ordered sets with some applications to matrix equations. The obtained result in [24] was extended and refined by many authors (see, e.g., [25–27] and references therein). The aim of this section is to deduce our results in the context of partially ordered b_2-metric spaces. At first, we need to recall some concepts. Let X be a nonempty set. Then $(X, \preceq, d)$ is called a partially ordered b_2-metric space with $s \geq 1$ if (X, d) is a b_2-metric space and $(X, \preceq)$ is a partially ordered set.

Definition 20. *Let $(X, \preceq)$ be a partially ordered set and $x, y \in X$. Then x and y are called comparable if $x \preceq y$ or $y \preceq x$ holds.*

Definition 21. *Let $(X, \preceq)$ be a partially ordered set. A mapping f on X is said to be monotone non-decreasing if for all $x, y \in X$, $x \preceq y$ implies $fx \preceq fy$.*

Definition 22. Let $(X, \preceq)$ be a partially ordered set and $f, g : X \to X$. One says f is g-non-decreasing if for $x, y \in X$,

$$g(x) \preceq g(y) \quad \text{implies} \quad f(x) \preceq f(y).$$

By putting $\mathcal{R} = \preceq$ in Theorems 1 and 2, we get the following results.

Corollary 4. Let $(X, d, \preceq)$ be a complete partially ordered b_2-metric space. Assume that $f, g : X \to X$, are two mappings such that $f(X) \subseteq g(X)$, $g(X)$ is a b_2-complete subspace of X and f is a g-non-decreasing mapping. Suppose that there exists a function $\beta_s \in \Omega$ and $L \geq 0$ such that

$$d(fx, fy, a) \leq \beta_s(M(x, y, a))M(x, y, a) + LN(x, y, a), \tag{37}$$

where

$$M(x, y, a) = \max\left\{d(gx, gy, a), d(gx, fx, a), d(gy, fy, a), \frac{d(gx, fy, a) + d(gy, fx, a)}{2s}\right\},$$

and

$$N(x, y, a) = \min\{d(gx, fx, a), d(gy, fy, a), d(gx, fy, a), d(gy, fx, a)\},$$

for all $x, y, a \in X$ with $gx \preceq gy$. In addition, suppose that the following conditions hold:
(i) there exists x_0 in X such that $gx_0 \preceq fx_0$;
(ii) if $\{gx_n\}$ is a non-decreasing sequence in X with $gx_n \to gz$ as $n \to \infty$, then $gx_n \preceq gz$ for all $n \in \mathbf{N}_0$.

Then f and g have a coincidence point in X. Moreover, suppose that for all coincidence points u, v of f and g, there exists $w \in X$ such that $gu \preceq gw$ or $gv \preceq gw$ and f, g are weakly compatible. Then f and g have a unique common fixed point.

By taking $g = I$ in Corollary 4, we obtain the following corollary.

Corollary 5. Let $(X, d, \preceq)$ be a complete partially ordered b_2-metric space. Assume that $f : X \to X$ is a mapping satisfying the following conditions
(i) f is non-decreasing mapping;
(ii) there exist a function $\beta_s \in \Omega$ and $L \geq 0$ such that

$$d(fx, fy, a) \leq \beta_s(M(x, y, a))M(x, y, a) + LN(x, y, a), \tag{38}$$

where

$$M(x, y, a) = \max\left\{d(x, y, a), d(x, fx, a), d(y, fy, a), \frac{d(x, fy, a) + d(y, fx, a)}{2s}\right\},$$

and

$$N(x, y, a) = \min\{d(x, fx, a), d(y, fy, a), d(x, fy, a), d(y, fx, a)\},$$

for all $x, y, a \in X$ with $x \preceq y$;
(iii) there exists x_0 in X such that $x_0 \preceq fx_0$;
(iv) if $\{x_n\}$ is a non-decreasing sequence in X with $x_n \to z$ as $n \to \infty$, then $x_n \preceq z$ for all $n \in \mathbf{N}_0$.

Then f has a fixed point. Moreover, if $u, v \in Fix(f)$ such that there exists $w \in X$ with $u \preceq w$ and $v \preceq w$, then f has a unique fixed point. Then f has a fixed point. Moreover, if for every pair (u, v) of fixed points of f such that there exists $w \in X$ with $u \preceq w$ and $v \preceq w$, then f has a unique fixed point.

6. Application to Integral Equations

In this section, we study the existence of a solution for an integral equation using the results proved in Section 3. Let $X = (C[a,b], R)$ be the space of all real continuous functions on $[a,b]$ and $\rho : X \times X \to R^+$ defined by

$$\rho(x,y) = \max_{t \in [a,b]} |x(t) - y(t)|, \quad \forall x, y \in X.$$

Equip X with the 2-metric given by $\sigma : X^3 \to R^+$ which is defined by

$$\sigma(x,y,a) = \min\{\rho(x,y), \rho(y,a), \rho(a,x)\}, \quad \forall x, y, a \in X.$$

As (X, ρ) is a complete metric space, (X, σ) is a complete 2-metric space, according to Example 1, we define a b_2-metric on X by

$$d(x,y,a) = (\sigma(x,y,a))^2, \quad \forall x, y, a \in X.$$

It follows that (X, d) is a complete b_2-metric space with $s = 3$. Define a binary relation $\mathcal{R}$ on X by

$$\mathcal{R} = \{(x,y) \in X^2 : x(t) \le y(t) \text{ for all } t \in [a, \infty)\}. \tag{39}$$

Now, consider the integral equation:

$$x(t) = q(t) + \int_a^b h(t,s) A(s, x(s)) ds, \tag{40}$$

where $t \in [a,b] \subseteq R^+$. A solution of the Equation (40) is a function $x \in X = C[a,b]$. Assume that

(i) $h : [a,b] \times [a,b] \to [0, \infty)$, $q : [a,b] \to R$ and $A : [a,b] \times R \to R$ are continuous functions on $[a,b]$;

(ii) $\int_a^b h(t,s) dt \le r \le 1$;

(iii) there exists $x_0 \in X$ such that

$$x_0(t) \le q(t) + \int_a^b h(t,s) A(s, x_0(s)) ds.$$

(iv) A is nondecreasing in the second variable and for all $x, y, a \in X$, $s \in [a,b]$ there exists $0 < k < \frac{1}{\sqrt{3}}$ such that

$$\begin{aligned}
&\min \{|A(s,x(s)) - A(s,y(s))|, |A(s,x(s)) - a(s)|, |A(s,y(s)) - a(s)|\} \\
&\le |A(s,x(s)) - A(s,y(s))| \\
&\le k e^{-M(x,y,a)} \min\{|x(s) - y(s)|, |x(s) - a(s)|, |y(s) - a(s)|\},
\end{aligned}$$

where

$$M(x,y,a) = \max\left\{d(x,y,a), d(x,fx,a), d(y,fy,a), \frac{d(x,fy,a) + d(y,fx,a)}{2s}\right\}.$$

Now, we are equipped to state and prove our main result in this section.

Theorem 3. *Under the assumptions (i)–(iv), the integral Equation (40) has a solution in X.*

Proof. Define $f : X \to X$ by

$$fx(t) = q(t) + \int_a^b h(t,s)A(s,x(s))ds.$$

Observe that x is a solution for (40) if and only if x is a fixed point of f. Let $x,y,a \in X$ such that $x\mathcal{R}y$ for all $t \in [a,b]$. Since A is nondecreasing in the second variable, we have

$$\begin{aligned} fx(t) &= q(t) + \int_a^b h(t,s)A(s,x(s))ds \\ &\leq q(t) + \int_a^b h(t,s)A(s,y(s))ds \\ &= fy(t) \end{aligned}$$

Hence, $fx\mathcal{R}fy$ and $\mathcal{R}$ is f-closed. From Condition (iii), we conclude that $x_0 \leq fx_0$ for all $t \in [a,b]$, then $x_0 \mathcal{R} fx_0$. Now, for any $x,y,a \in X$ such that $fx\mathcal{R}^s fy$ we get

$$\begin{aligned} |fx(t) - fy(t)| &= |\int_a^b h(t,s)(A(s,x(s)) - A(s,y(s)))ds| \\ &\leq \int_a^b |h(t,s)||A(s,x(s)) - A(s,y(s))|ds \\ &\leq ke^{-M(x,y,a)} \int_a^b |h(t,s)|\min\{|x(s)-y(s)|, |x(s)-a(s)|, |y(s)-a(s)|\}ds \\ &\leq ke^{-M(x,y,a)} \int_a^b |h(t,s)|\min\{\max_{s\in[a,b]}|x(s)-y(s)|, \max_{s\in[a,b]}|x(s)-a(s)|, \\ & \qquad \max_{s\in[a,b]}|y(s)-a(s)|\}ds \\ &\leq ke^{-M(x,y,a)} \int_a^b |h(t,s)|\min\{\rho(x(s),y(s)), \rho(x(s),a(s)), \rho(y(s),a(s))\}ds \\ &\leq ke^{-M(x,y,a)} \int_a^b |h(t,s)|\sigma(x,y,a)ds \leq rke^{-M(x,y,a)}\sigma(x,y,a). \end{aligned}$$

Therefore,

$$\sigma(fx,fy,a) \leq \max_{t\in[a,b]} |fx(t) - fy(t)| \leq rke^{-M(x,y,a)}\sigma(x,y,a).$$

It follows that

$$d(fx,fy,a) \leq r^2k^2 e^{-2M(x,y,a)} d(x,y,a) \leq r^2k^2 e^{-2M(x,y,a)} M(x,y,a) \leq \frac{e^{-2M(x,y,a)}}{3} M(x,y,a).$$

Thus,

$$d(fx,fy,a) \leq \frac{e^{-2M(x,y,a)}}{3} M(x,y,a) + LN(x,y,a),$$

where

$$M(x,y,a) = \max\left\{d(x,y,a), d(x,fx,a), d(y,fy,a), \frac{d(x,fy,a) + d(y,fx,a)}{2s}\right\},$$

and

$$N(x,y,a) = \min\{d(x,fx,a), d(y,fy,a), d(x,fy,a), d(y,fx,a)\},$$

with $\beta_s(t) = \frac{e^{-2t}}{3}$ and $L \geq 0$. Then f is almost a $\mathcal{R}$-Geraghty type contraction. In addition, if $\{x_n\} \in X$ is an $\mathcal{R}$-preserving sequence such that $\lim_{n\to\infty} x_n = x \in X$, then $x_n \leq x$ for all n. Hence, $x_n \mathcal{R} x$, for all n. Therefore, all the hypotheses of Corollary 1 are satisfied. Hence, f has a fixed point which is a solution for the integral Equation (40) in $X = C([a,b], R)$. □

Author Contributions: All the authors contributed equally and significantly in writing this article. All authors have read and agreed to the published version of the manuscript.

Funding: This research received no external funding.

Institutional Review Board Statement: Not applicable.

Informed Consent Statement: Not applicable.

Data Availability Statement: Not applicable.

Acknowledgments: All the authors are grateful to the anonymous referees for their excellent suggestions, which greatly improved the presentation of the paper.

Conflicts of Interest: The authors declare no conflict of interest.

References

1. Gähler, S. 2-metrische Räume und ihre topologische Struktur. *Math. Nachrichten* **1963**, *26*, 115–148. [CrossRef]
2. Deshpande, B.; Chouhan, S. Common fixed point theorems for hybrid pairs of mappings with some weaker conditions in 2-metric spaces. *Fasc. Math.* **2011**, *46*, 37–55.
3. Dung, N.V.; Hang, V.T.L. Fixed point theorems for weak C-contractions in partially ordered 2-metric spaces. *Fixed Point Theory Appl.* **2013**, *2013*, 161. [CrossRef]
4. Fathollahi, S.; Hussain, N.; Khan, L.A. Fixed point results for modified weak and rational α-ψ-contractions in ordered 2-metric spaces. *Fixed Point Theory Appl.* **2014**, *2014*, 6. [CrossRef]
5. Naidu, S.V.R.; Prasad, J.R. Fixed point theorems in 2-metric spaces. *Indian J. Pure Appl. Math.* **1986**, *17*, 974–993.
6. Czerwik, S. Contraction mappings in b-metric spaces. *Acta Math. Inform. Univ. Ostrav.* **1993**, *1*, 5–11.
7. Czerwik, S. Nonlinear set-valued contraction mappings in b-metric spaces. *Atti Semin. Mat. Fis. Univ. Modena* **1998**, *46*, 263–276.
8. Aydi, H.; Bota, M.; Karapinar, E.; Moradi, S. A common fixed point for weak φ-contractions on b-metric spaces. *Fixed Point Theory* **2012**, *13*, 337–346.
9. Hussain, N.; Shah, M.-H. KKM mappings in cone b-metric spaces. *Comput. Math. Appl.* **2011**, *62*, 1677–1684. [CrossRef]
10. Roshan, J.R.; Parvaneh, V.; Sedghi, S.; Shobkolaei, N.; Shatanawi, W. Common fixed points of almost generalized $(\psi, \phi)_s$-contractive mappings in ordered b-metric spaces. *Fixed Point Theory Appl.* **2013**, *2013*, 159. [CrossRef]
11. Mustafa, Z.; Parvaneh, V.; Roshan, J.R.; Kadelburg, Z. b_2-Metric spaces and some fixed point theorems. *Fixed Point Theory Appl.* **2014**, *2014*, 144. [CrossRef]
12. Turinici, M. Abstract comparison principles and multivariable Gronwall-Bellman inequalities. *J. Math. Anal. Appl.* **1986**, *117*, 100–127. [CrossRef]
13. Bhaskar, T.G.; Lakshmikantham, V. Fixed point theorems in partially ordered metric spaces and applications. *Nonlinear Anal. Theory Methods Appl.* **2006**, *65*, 1379–1393. [CrossRef]
14. Samet, B.; Vetro, C.; Vetro, P. Fixed point theorems for α-ψ-contractive type mappings. *Nonlinear Anal.* **2012**, *75*, 2154–2165. [CrossRef]
15. Ben-El-Mechaiekh, H. The Ran–Reurings fixed point theorem without partial order: A simple proof. *J. Fixed Point Theory Appl.* **2014**, *16*, 373–383. [CrossRef]
16. Imdad, M.; Khan, Q.; Alfaqih, W.M.; Gubran, R. A relation theoretic $(F, \mathcal{R})$-contraction principle with applications to matrix equations. *Bull. Math. Anal. Appl.* **2018**, *10*, 1–12.
17. Gubran, R.; Imdad, M.; Khan, I.A.; Alfaqih, W.M. Order-theoretic common fixed point results for F-contractions. *Bull. Math. Anal. Appl.* **2018**, *10*, 80–88.
18. Jungck, G.; Rhoades, B.E. Fixed Points for set valued functions without continuity. *Indian J. Pure Appl. Math.* **1998**, *29*, 227–238.

19. Abbas, M.; Jungck, G. Common fixed point results for noncommuting mappings withoutcontinuity in cone metric spaces. *J. Math. Anal. Appl.* **2008**, *341*, 416–420. [CrossRef]
20. Alam, A.; Imdad, M. Relation-theoretic contraction principle. *J. Fixed Point Theory Appl.* **2015**, *17*, 693–702. [CrossRef]
21. Alam, A.; Imdad, M. Relation-theoretic metrical coincidence theorems, Filomat. *arXiv* **2017**, arXiv:1603.09159.
22. Geraghty, M. On contractive mappings. *Proc. Am. Math. Soc.* **1973**, *40*, 604–608. [CrossRef]
23. Dukić, D.; Kadelburg, Z.; Radenovixcx, S. Fixed points of Geraghty-type mappings in various generalized metric spaces. *Abstr. Appl. Anal.* **2011**, *2011*, 561245. [CrossRef]
24. Ran, A.C.M.; Reurings, M.C.B. A fixed point theorem in partially ordered sets and some applications to matrix equations. *Proc. Am. Math. Soc.* **2004**, *132*, 1435–1443. [CrossRef]
25. Agarwal, R.P.; El-Gebeily, M.A.; ÒRegan, D. Generalized contractions in partially ordered metric spaces. *Appl. Anal.* **2008**, *87*, 109–116. [CrossRef]
26. Harjani, J.; Lopez, B.; Sadarangani, K. A fixed point theorem for mappings satisfying a contractive condition of rational type on a partially ordered metric space. *Abstr. Appl. Anal.* **2010**, *2010*, 1–8. [CrossRef]
27. Petruşel, A.; Rus, I.A. Fixed point theorems in ordered L-spaces. *Proc. Am. Math. Soc.* **2006**, *134*, 411–418. [CrossRef]

MDPI
St. Alban-Anlage 66
4052 Basel
Switzerland
Tel. +41 61 683 77 34
Fax +41 61 302 89 18
www.mdpi.com

Axioms Editorial Office
E-mail: axioms@mdpi.com
www.mdpi.com/journal/axioms

www.ingramcontent.com/pod-product-compliance
Lightning Source LLC
LaVergne TN
LVHW070403100526
838202LV00014B/1377